10/03

SOUL
of the
SWORD

AN ILLUSTRATED HISTORY OF WEAPONRY

AND WARFARE FROM PREHISTORY TO THE PRESENT

ROBERT L. O'CONNELL

Illustrations by John Batchelor

THE FREE PRESS

New York London Toronto Sydney Singapore

ƒP

THE FREE PRESS
A Division of Simon & Schuster, Inc.
1230 Avenue of the Americas
New York, NY 10020

Copyright © 2002 by Robert L. O'Connell

All rights reserved, including the right of reproduction
in whole or in part in any form.

THE FREE PRESS and colophon are trademarks
of Simon & Schuster, Inc.

For information regarding special discounts for bulk purchases,
please contact Simon & Schuster Special Sales at
1-800-456-6798 or business@simonandschuster.com

Designed by Kim Llewellyn

All illustrations not otherwise credited below are by John Batchelor
Art Resource: 32, 69
Biblical Heritage International Publishers, Tel Aviv, Israel: 98
The British Museum: 57, 77, 79
The Cairo Museum: 22, 23
Corbis: 81
The Hague: 131
Historical Office, U.S. Army Aviation and Missile Command: 349
James Mellaart: 21
MEPL: 91
M. H. de Young Memorial Museum: 134
Needham Research Institute, Cambridge: 113
Peter Connolly: 39, 42, 104
Proceedings of the Prehistoric Society 43 (1977), p. 243–63. "Rock Carvings of
Chariots in Transcaucasia, Central Asia and Outer Mongolia" Mary Littauer: 53
R. Piper GmBH & Co. KG, Munchen 1975. Irenaus Eibl-Eibesfeldt. *Krieg und
Frieden aus der Sicht der Verhaltensforschung:* 14
Robert Conrad: 55

Manufactured in the United States of America

1 3 5 7 9 10 8 6 4 2

Library of Congress Cataloging-in-Publication Data
is available.

ISBN 0-684-84407-9

Acknowledgments

No book is ever written alone, least of all this one. It is filled with the thoughts and wisdom of those encountered during a long career—more than I can possibly name. I do, however, want to thank in particular Rob Cowley and Rod Paschall, the past and present editors of *MHQ: The Quarterly Journal of Military History*, for the support they have provided me over the years in my studies of arms and men. Also, I am greatly in the debt of John Batchelor for the wonderful illustrations in this book and of him and his wife, Liz, for the hospitality they provided me during my frantic sojourn at their home. Finally, I am very grateful to my editor at The Free Press, Bruce Nichols, whose patience and skill I can only marvel at.

To Jessica and Lucy

Contents

Chapter 1

A MOST ANCIENT
TALISMAN

It happened long ago, well before we became human. Perhaps the first was *Homo habilis,* the revolutionary successor to Lucy and her Australopithecine clan. Motivated more by fear than ingenuity, the little hominid might have raised a sharpened stick—heretofore used only to pry meat from the bone—to defend a carcass temporarily abandoned by its real killer. In that instant the tool became a weapon, and the partnership between arms and men was forged. For a creature without even a decent set of canines, this was a monumental step. Suddenly, the predator's apprentice could aspire to more than just scavenging: he might join the ranks of big cats and other carnivores as a full-fledged hunter. As this came to pass the veil of terror—woven out of the possibility of becoming the meal of something else—would gradually lift. This was the reward that sticks and stones and the intelligence to use them aggressively held out to the little ape-man . . . the mantle of top predator.

Even the most radical change is seldom without precedent: *Homo habilis* was not the first ape to take up arms. Chimps, for example, have been documented hurling rocks and brandishing sticks. The central difference is that hominids walked upright (there are fossilized footprints to prove it), imparting balance and leaving hands well shaped for grasping and throwing and bashing. With a little imagination we can spot the course of arms winding far deeper into the past.

Weapons, viewed in the broadest sense, have been around nearly as long as life itself. The key difference is that the earliest weapons were attached to their users. Nature is about competition, and competition demands aggressive behavior and the means to accomplish it. Most animals have come armed—both to hunt and also to compete among themselves. The distinction is a useful one. Aggression is highly complex. Yet it exhibits basic trends, classes of behavior that give every impression of transcending individual animal types.

Hunting or predation is based on the simple proposition of killing and eating a suitable victim, or avoiding such a fate. Beyond these stark alternatives there are few rules. The successful hunter earns his meal in the easiest and most risk-free manner possible—stealth and surprise are favored, the youngest, oldest, or simply weakest are preferred victims, every vulnerability is exploited, and the killing may grow wanton if it is convenient. The ruthlessness of the attacker is reciprocated by the survival mechanisms of the prey, at times resorting to the abandonment of offspring or herding together—not out of social solidarity, but simply the hope another will be chosen—or patterned behavior such as freezing, reliance on camouflage, or on hardened body parts like shells.

Kill or be killed contrasts sharply with the behavior surrounding competition among members of the same species—a phenomenon known to specialists as intraspecific aggression. Here, motivations are more complex. Although hunting frequently engages not only members of both sexes but also groups, intraspecific aggression normally takes place between just males paired off for individual combat. The struggle is frequently rough and even lethal, but it is not concerned essentially with killing. It is about perpetuating life, not ending it—achieving access to females, the territory to support them, and the dominant profile necessary to keep them. Violence is more ritualistic. The fighting frequently takes on the dimensions of a ceremony, with challenges elicited, a specific order of engagement followed, and mutually recognizable acts of dominance and submission ultimately exchanged.

Plainly aggression is not simply a matter of behavior, but also of teeth and talons, antlers and armor. If a weapon can be defined as "any instrument used primarily for attack and defense," it is apparent that they have an ancient and fundamental role in life. Seen in this light, the world of nature becomes an armed camp, the scene of countless weapons competitions stretching over millions, and, in a few cases, billions of years. Amidst this bristling tableau it is even possible to draw certain distinctions among the basic types of natural weapons that roughly parallel the characteristics of the two forms of aggression.

In the case of predation—especially among vertebrates—the simple matter-of-fact nature of the process is mirrored by the conservative, even prosaic character of the instruments employed: really only penetration, and to a lesser degree, poison, have been exploited. The resulting implements— basically teeth, fangs, and claws—have remained both stable over time and notable for their unobtrusive and generalized nature. Not only are the killing instruments seldom out of proportion to the user's body, but they are frequently hidden, or at least covered—teeth behind jowls, retractable claws. Defenses are conservative as well. Although there have been innumerable experiments with noxious taste and smell during the course of evolution, a basic reliance on analogous shell forms and body armor has remained a mainstay for those under attack. The chitin of the beetle, the shell of the tortoise, and the horny layering of the armadillo are all variations on the theme.

Obviously there are exceptions to this conservatism—spectacular adaptations to the possibilities of attack from above, the galaxy of variation in the class Insecta—but the basic sameness of predatory weapons remains notable when it is matched against the uniqueness, complexity, and diversity of animal arms specialized for intraspecific aggression. Here the passion of combat displays itself with an astonishing throng of utensils that would stretch the imagination were they not attached to the contestants: customized horns, antlers, tusks, bony plates and protrusions, many specialized as much to impress as to overpower. As befits their ultimate purpose, such instruments often appear as secondary sex characteristics, primarily among males.

Moreover, many of these intraspecific adaptations are used carefully and selectively. Deer and a number of other horned creatures use their antlers only against each other, while relying on their hooves to ward off predators. Similarly, piranhas, much feared for their teeth, never bite other piranhas, but strike at them with their tailfins. Rattlesnakes engage in wrestling matches rather than biting it out to determine dominance. Northern elephant seals and wild boars do attack their seal and boar rivals with their deadly tusks, but the blows are delivered on the shoulders and chest, areas heavily padded with layers of skin.

These are not aberrations. Unlike the violence associated with predation, fighting between pairs of the same species is often orchestrated so that the implements of combat are matched and applied alike, or symmetrically.

Frequently the instruments themselves are formed to serve the less-than-lethal purposes of the contest. Hence the antlers of stags and moose are designed for locking and pushing, not penetration, while the horns of bighorn sheep are ideal for their spectacular butting matches.

Utility is relative here. Unlike predatory instruments, the natural armaments of noncarnivores often give the impression of being oversized and unwieldy. But they are formidable-looking. Red deer frequently settle dominance issues not by fighting, but through the mere size of their respective antlers. The horn of the Hercules beetle—a tusk so long it exceeds the length of the creature's body—is used for fighting but actually protects the combatants against injury. In these cases and others, threat and its collaborator, size, are actually important forces in the evolution of natural armaments.

Thus evolution in the animal world has built arsenals in two separate stockpiles: one, like the teeth of a shark, simple and straightforwardly devoted to lethality; and the other, like a stag's horns, festooned with instruments of more ambiguous intent, elaborated in ways that serve the ends of what could be said to border on rough play.

This was the endowment awaiting hominids when they first took up arms some two and a half million years ago. As they crafted their own panoply there would always be nature's examples, providing mechanical suggestions along with archetypes and symbols to be transmuted in their growing imaginations into powerful underlying conceptions of what a weapon should and should not be. There was also a largess of available spare parts—horns and bones that might be harvested and used or further modified—as well as the compelling logic that came with wielding instruments of lethal potential. Just as with other predators, once hominids became hunters, they were faced with the same dilemma that drove rattlesnakes to wrestling and boars to the prudent and courteous use of their tusks. The implements they required to bring down dinner were, without modification or modulation, much more lethal than needed to clarify just who was the Alpha anthropoid.

But still, it was hunting, not competition, that molded our first weapons and the manner in which they were used. The first armaments must have been simple to the extreme but still deadly: sharp sticks and rocks. Yet even at this primitive juncture there must have been a basic dichotomy in how implements were used: close in or from afar.

Launching was plainly the safest approach, but also the least effective, with accuracy and damage potential diminishing as range increased. The famed anthropologist Louis Leakey believed that *Homo habilis* advanced to the point of fashioning a bola. If he is correct, the bola constitutes the first truly invented weapon, since there are no examples in nature of stones launched through centrifugal force. Bolas boosted both power and standoff capability; yet as with other such instruments that relied on mechanical advantage to cast objects, killing power was limited.

Dispatching really big thick-skinned animals required either penetration or heavy blows to the head—acts demanding handheld spears and also rocks. Stoneworking has been dated back as far as *Australopithecus garhi* 2.6 million years ago. Since the products were made of rock, they are still around to be studied. Sharp flakes allowed hominids to butcher kills far faster than other animals. Faceted hand axes were shaped for hammering and bashing—unquestionably handy for finishing off a prey in extremis, but contact this close was extremely risky. Spears, however, provided a modicum of breathing space. Stout wooden shafts could have been shaped by hand axes, and once fire was first employed (probably by *Homo erectus*) tips could have been char-hardened. These two implements would have constituted a sort of starter kit for big-game hunters—crude but effective. Later pioneering craftsmen took to lashing hand axes and faceted points to wooden extensions, thereby generating spears capable of really deep penetration and axes with true leverage and cutting power.

Such close-in weapons would have been objects of extreme respect. They and their users were involved in an exciting and rewarding pursuit; but one that demanded confronting at close quarters something very large

Faceted flint hand axes and spear points.

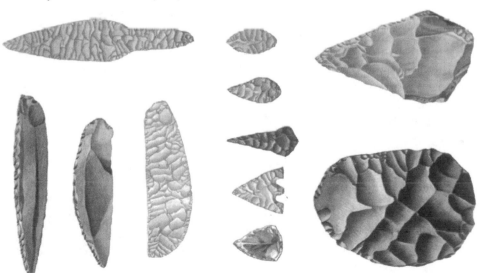

and lethally aroused. It follows that those involved were mostly, or even exclusively, male, forged into close-knit teams specialized to face danger, bonded individually to risk everything for the mutual objective and to protect other members in peril. What gradually emerged was a sort of brotherhood of killers—raw material for the cohesive small units upon which all armies would one day be built.

By contrast, devices and behaviors employed for killing game from longer ranges, while safe and effective, could not match close-in hunting in its capacity to confront and bring down even the largest beasts. It is quite possible that the prestige and traditions associated with close-in weaponry were carried forward in myth and memory, coloring the outlook of the earliest warriors as to what constituted courage and honor and the proper arms with which to fight.

This brings us back to the dilemma of the rattlesnake and the piranha. How did early members of our line handle implements so lethal against other species when they participated in disputes among themselves? This is a controversial question, since it relates to issues as vital and pressing as gun control, the root causes of violence, and even the possibility of nuclear annihilation. Moreover, the physical evidence—essentially a small sample of fractured skulls and other damaged skeletal remains—is ambiguous. The best that can be done is a logical extrapolation from what is known about intraspecific aggression.

With the possible exception of the eastern mountain gorilla, intraspecific aggression in some form has been found to exist throughout the primate order. Dominance-related combat is largely a matter of males and highly individualized. It is also marked by a considerable measure of threat and posturing, along with a degree of ritualization. In the case of weapon-bearing hominids this would have, in one way or another, served to mitigate the lethality of the weapons themselves. But there is another factor. Since the most lethal implements would have been the most prestigious, they were also the most likely to be used in this kind of combat. Rules governing their use would have been essential, though not necessarily successful. It is not unreasonable to conclude that some dead-ended hominids met extinction at the hands of their own weapons. Nevertheless, the emergence of *Homo sapiens* gives us reason to believe our line's solutions were more successful.

Yet they were still rooted in biological precedent. Weapons use among humans remains highly differentiated by sex, becoming (with the interesting exception of archery) almost exclusively the province of males. Resort to a champion—a single superbly equipped warrior—is also a phenomenon that has reappeared throughout military history. Despite an obvious willingness to employ weapons as a group activity, the enduring focus on the individual combatant and the continuing fascination with his accoutrements are key features of the course of arms. So is threat. Even a cursory tour of a military museum or a modern arsenal leaves the distinct impression that weapons are fashioned not just to kill, but to intimidate, whether a bigger spear, a terrifying monster depicted on the face of a shield, or the tortured logic of nuclear deterrence.

Ritualization is similarly apparent. Contrasted against the more laissez-faire predatory approach, the predilection to fight by the rules and employ like armaments in a prescribed manner would remain a continuing feature of human combat. It makes sense to associate such behavioral adaptations with the more general characteristics of intraspecific aggression. True enough, the violence reflective of the hunt—the inclination to slaughter, the willingness to include both sexes and the young as victims—periodically made a mockery of these standards of combat. Yet it is difficult to deny their existence. The two poles of aggression—ritual and predation—exist in a dynamic relationship, a tug-of-war that serves to bound the human potential for armed assault.

But caution is required. While this dialectic is useful in understanding the wars and weapons that became so much a part of our history, humans are motivated primarily by learned rather than imprinted responses, and these concepts should be seen in this context rather than as strict models of behavior.

The fate of other hominid species notwithstanding, it was *Homo erectus* alone who first marched out of Africa around 1.8 million years ago to spread across Eurasia as far as Indonesia. Then once again evolution took a creative turn, and began generating the antecedents of modern humanity. The process resulted in two basic strains—*Homo neanderthalensis* and *Homo sapiens*. Both had very big brains of approximately the same size, buried their dead, used weapons to hunt the animals of the glacial era, and coexisted for at least a thousand centuries.

Yet Neanderthals were strange from the beginning. Thick of trunk and heavy of limb, somewhat stooped, their large brains nestled in apelike skulls, Neanderthals left the impression of career underachievers. They may have scattered flowers on their dead before interring them, but there are few other signs of religious or aesthetic sensibilities. More to the point, their tool set and weaponry—basically indifferently formed hand axes—changed little over time. Just why this was the case has led to explanations as varied

Dawn of the Bow

Judged by longevity and accumulated victims, the bow stands as one of the most devastating weapons in history, not to be supplanted until the proliferation of firearms. A design solution seldom surpassed in elegance and simplicity, the bow capitalized on a number of mechanical principles to maximize lethality, while at the same time retaining remarkable lightness and portability. Combining efficient storage with sudden and precise release of energy, the simple bow generated significant range (in excess of 150 feet) and velocity (up to 200 feet per second), while taking advantage of projectile shape (long and thin) to convert this energy into a high degree of penetrating power. By the late Mesolithic (as testified to by a remarkable specimen pulled from a bog in Holmegaard, Denmark), certain bows had reached a point of development approximating the size and power of the English longbow—employing yew, carefully cut to take advantage of the tension resistance of the sapwood positioned next to the compression-resistant heartwood, to produce what amounted to a natural or "self-composite" weapon. Bows did require skilled and experienced archers to attain anything like their full capabilities. But in such hands, a bow could be highly accurate and fast firing—up to six shots a minute. As a hunting instrument and later a weapon of war it did continue to lack the prestige and sheer lethality of arms specialized for close quarters. But it could certainly bring down a human.

as hereditary disease, speechlessness, overspecialization for the glacial climate, and simple misunderstanding; but the enduring image of Neanderthal is that of the archetypical oaf.

The same cannot be said for the other branch of the family, the Sapiens. After skating through the Lower and Middle Pleistocene, around 40,000 years ago our immediate ancestors underwent a remarkable cultural takeoff, evidenced by a spectacular series of cave paintings depicting with consummate skill and artistry the profusion of game animals off of which they made their living. As with the Neanderthal's torpor, there are no definitive explanations of Sapiens's turning point.

Yet cave paintings do indicate that Sapiens became a far more efficient predator. Also, during this time frame, known collectively as the Upper Paleolithic, bone piles in certain spots became monumental, the graves of thousands of beasts ranging from mammoths and mastodons to rhinos and aurochs. And this was just the big game; the kills extended to much more elusive quarry—the swiftest ungulates and even lacustrine and marine creatures.

Evidence of the takeoff also includes fundamental improvements in weapons—particularly those useful against hard-to-get-at prey. Thus, a group of arms that took clever advantage of leverage, aerodynamics, or other quirks of physics to dramatically advance standoff performance sprang into existence—the boomerang, spear thrower or alatl, the harpoon, the sling, and the bow.

Around 30,000 years ago the last of the dwindling population of Neanderthals died. Sapiens stood alone in the unforgiving Darwinian spotlight, the final survivor of the hominid line. Our ancestors were by definition fittest, simply because they had survived while others had not. But our hominid cousins, especially Neanderthal, were far from evolutionary chumps. Numerous factors may have contributed to their demise, but there remains a persistent suspicion we pushed them over the edge. Essentially, this would have meant war. Not individualized violence over issues of momentary importance, but premeditated large-scale aggression with permanent results, the first example of genocide successfully conducted.

The implications are profound. If it were to become clear that we, rather than circumstances, eliminated the Neanderthals, then it could be said we were warlike from the beginning. For a species harboring thousands of nuclear weapons, this is not a pleasant prospect to contemplate. There are no definitive answers. But the weight of evidence seems to argue against such a sanguinary conclusion.

From nearly the beginning we were hunters and gatherers; that was the essential evolutionary context. Careful observations of the remnants of such societies by pioneering anthropologists in the late nineteenth and early twentieth centuries provide us with a reasonable idea how we might have lived—bonded together in highly social groups of from twenty-five to fifty. Frequency and intensity of contact demanded that daily existence be configured to preclude tension. Sharing was probably the rule of thumb, and—since everything had to be portable—personal property would have been kept to a minimum. Decision making was likely a matter of consensus, with all participants having their say. Human relations in such societies are inevitably accompanied by a substantial measure of equality. Hence interaction between the sexes, despite the males' superior strength and control of weaponry, probably took place on basically a level plain.

Yet sex and reproduction inevitably must have posed challenges for such societies. For one thing they were too small to breed internally. If continual birth defects were to be avoided, fresh genes had to be brought in from other bands. Mates, of course, could be stolen; but exchanges were safer. Meanwhile, disputes over sex and other similarly personal matters within the group—inherently parlous when arms were around—were best short-circuited by allowing members the option of leaving and joining another band. Yet such a safety valve presupposed at least tolerable external relations.

The land was basically open to roving, and survival simply meant following wandering herd animals or the life cycles of ripening plants. Thus in a world that amounted to a movable feast, territory must have meant little. Meanwhile, human populations remained very low (probably never exceeding 20,000 in Upper Paleolithic France, for instance) and contacts between bands were likely to have been infrequent. For the most part there was plenty for everybody, and should a situation turn ugly the easiest thing to do was retreat to one of earth's many unoccupied spaces.

Territoriality, property, and labor—the real drivers of war—were simply not present in sufficient measure in the Upper Paleolithic or among earlier hunter-gatherer bands of humans and their ancestors to sustain true warfare as a useful and self-perpetuating activity. It is possible that humans viewed Neanderthals as prey; but in a world teeming with game this too seems a bit lurid and far-fetched from a cost-benefit perspective. It is likely that there was violence within and between bands and species, but it also stands to reason that it was highly personalized, lacking sustained economic and political inducement, and focusing instead on revenge and females. The raw materials for genocide and war were present among humans—a capacity for both intraspecific and predatory aggression, small teams bonded to face danger, and weapons—but the motivation probably was not. For us war was a thing of the future. For others, however, war was a thing of the past . . . the deep past.

As the glaciers retreated and human habitation spread, the really big animals began to disappear. In North America this was particularly abrupt and dramatic, coincident with the arrival of the two-legged top predator. In the Old World, where the great beasts had much longer to adjust to evolving human hunting skills, the decline was less complete. But in either case the monster mammals, whose big bodies were specialized for very cold climates, became fewer and fewer in number, either following the ice sheets north or simply going extinct.

Meanwhile there were more human mouths to feed. This was not a result of growing band size, since the supply dynamics of hunting and gathering imposed inelastic limits on the number of people who could live together in one group. Instead, the circumstances encouraged social fissioning, with elements breaking off from a main group and forming new pods, which in turn expanded to band size. But appropriate real estate was also

Tiny Warriors

War is not simply mass violence. It is about societal, not individual, issues; it implies direction by some form of rulership and military organization; it has palpable economic and political goals, and it is aimed at permanent results. Human violence, so long as we remained hunters and gatherers, does not seem to have measured up to these criteria. But ants are a different story. For as long as 50 million years, certain species not only lived in tightly organized and stratified societies, but engaged in mass aggression featuring specialized warriors (bigger and better protected), the equivalent of tactics and strategy, the taking of slaves, prolonged battles over territory, and even attacks aimed at annihilation. Ants can do this because they are genetically specialized for self-sacrifice, sharing three quarters of their genes with other members of the colony. But if looked at in general terms, the manner and apparent reasons for which they fight are highly suggestive.

Warlike ant societies come in two forms, sedentary and nomadic, and both characteristically have large populations. In the former case, control of areas around the nest looms large, with organized fighting acting as a key determinant of colony spacing and, ultimately, success. Essentially, war here works not only to regulate competition between colonies, but also as an arbiter of factors operating within these societies. Slaves, for instance, are taken as a sort of energy bonus, while colony population corresponds to the extent of territory that can be defended.

The equation is entirely different among nomadic or army ants. While their depredations are actually based on voracious predation, their mass migrations are so reminiscent of the movements of human pastoral nomads that W. M. Wheeler, the founder of modern American ants studies, called them "the Huns and Tartars of the insect world." The ruthless character of raiding in both ants and humans is plainly dictated by life on the move—territory, slaves, and long-term dominance are meaningless—victims are immediately exploited, and there is little incentive to keep them alive. Nomads are wandering masses of mouths—the ants being driven by their own hunger, the pastoralist humans by the capricious population dynamics of their flocks.

These phenomena almost certainly say less about ant or human nature than they do about war as an adaptive mechanism. Given a certain way of life, war served an important purpose. When certain solitary wasps turned social and became ants, their genes invented war. And when humans domesticated plants and then animals, their minds came up with the same solution. The difference was that the insects were caught by their own genes and marooned as perpetual warriors. We could change. But then ants had little in the way of long-term memory to prod them to seek revenge. We did.

finite. At some point the distance between bands must have started to narrow, and the concept of territoriality began inching away from the loose and permissive to the much more exclusive. This was not yet war, but it was pointing in that direction.

Outside of shrinking elbow room and a diminishing supply of lumbering meat bonanzas, there was another basic bane of life on the move at the end of the Paleolithic: too many babies. The average human female was and is capable of bearing around eight healthy offspring during her life span. Birth-related deaths among both mothers and children were undoubtedly high, and prolonged nursing, high-protein diets, and continual movement would have kept body fat low and stretched the gap between pregnancies. There may also have been some knowledge of plants with birth-controlling properties. Nonetheless, it has been calculated that live births exceeded by 50 percent the levels needed to sustain populations at Upper Paleolithic levels. This probably meant infanticide. Yet there was an alternative. When a band accustomed to roaming lucked across a food supply bountiful enough to support larger populations, it was to their advantage to settle down around it, multiply—and defend it to the death from other bands.

Until recently the Mesolithic has not drawn much interest among archaeologists and anthropologists. It was viewed as a kind of evolutionary intermission between the full flowering of the Upper Paleolithic and the coming of agriculture and pastoralism that made the Neolithic possible. In part this was because there was not very much evidence. The most interesting sites were located along riverbanks, lakes, and shorelines now covered by up to several hundred feet of water due to the great glacial meltdown.

Lately, however, the Mesolithic picture has clarified and grown more interesting. Thanks to new techniques and insights gleaned from modern peoples living under similar circumstances, a consensus is forming that complex hunter-gatherer societies—those based on the intensive exploitation of stationary and abundant food sources, rich mollusk beds or fish runs, for example—had begun to evince most of the same social characteristics previously attributed only to the arrival of agriculture: a settled existence, communal population growth, productivity-enhancing technology, the start of social stratification and formalized leadership, the expansion of trade networks, increased competition for resources, intensified territoriality, and the appearance of what looked very much like warfare. The last of these is supported by pictorial representations of combat with clear signs of violent and simultaneous death.

There were not yet armies, set-piece battles, and stable military institutions. These Mesolithic societies and their institutional manifestations were by nature transitory. The food sources on which they were based acted as a limiting factor, being resistant to storage, impossible to intensify, and prone to run out. A band would settle around a good salmon run or shellfish bed and the various institutional drivers would start grinding away. But before much of a social superstructure was built, the food supply would collapse beneath its weight and the process would reverse. So everything remained immature, including the fighting.

Violence was likely inchoate and fairly spontaneous. It reflected the underlying causes of warfare as we understand it today, but its conscious motivations were different. Physical aggression would have been driven by the urge for revenge, and taken the form of sporadic but still brutal attacks primarily focused on individuals or those close to them—blood feuds rather than societal warfare. This judgment is supported by the apparent absence of stockades or walls protecting Mesolithic communities, a feature that would one day be among the clearest signs that true war had arrived. Instead of going to the trouble of building fixed defenses, a more viable option in the face of continued or serious attacks was simply to leave. The less dangerous but more rigorous life as classic wanderers remained open. One day this would no longer be true; a time would come when farmers became too numerous and their skills too specialized for them to do anything else but defend their territory. Yet this was still thousands of years in the future. So we are left in the Mesolithic with a kind of proto-war fueled by competition for short-lived food supplies and the emotional desire to get even.

Vengeance is worth considering. If it does not in itself constitute war, it remains an important factor even to this day, being embedded in sophisticated concepts such as nuclear deterrence. While it is often seen as primitive and somehow inhuman, it is nothing of the sort.

Arguably human institutions are held together by networks of reciprocation, the basis of guilt, obligation, and even justice. We are unique in that our powerful capacity for reasoning allows us to make stunningly accurate judgments as to the quality of reciprocation, and our prodigious long-term memory makes sure that debts are not forgotten. The urge to reciprocate and sanctions against not reciprocating are very strong . . . they have to be or the entire social structure would collapse.

Vengeance is simply its punitive equivalent. At the interpersonal level it is a primary cause of violence, its strength testified to by the number of institutions designed to curb retribution—the law's depersonalization of punishment, for example. Like reciprocation, retribution tends to perpetuate itself through chains of paybacks. This is a key reason why wars, even today, stretch on beyond all political advantage. Certain types of war, civil conflicts for instance, sometimes come to be driven by little else, and are therefore particularly hard to stop.

As always historical evidence is quirky. Although the relative roles of vengeance and competition for resources in Mesolithic conflict can only be approximated through mental models, good fortune has preserved for us pictorial evidence of what the fighting actually looked like. Of those Mesolithic cave paintings that deal with lethal attacks, the four best known come from the hills above the eastern Spanish coast. The first two depict a single victim hit with multiple arrows, in one case by ten archers—most plausibly the aftermath of an ambush reflecting the emphasis on individual targets. The final two—a small scene of only seven warriors from Morella la Vella and a much larger grouping of at least twenty-nine figures from Les Dogues shelter in Castellón—are obvious representations of massed combat. In both the action is confused, a group of swirling, darting figures clearly on the run, perhaps hoping to rip off a few quick shots before retreating. This is the same chaotic tableau reflected in other Mesolithic paintings. In every single image of combat or violence the warriors are

*This Mesolithic cave painting
at Morella la Vella, Spain, is
the earliest depiction of
humans fighting.*

shown fighting solely with the bow, or, in the case of recently discovered Australian fighting scenes, javelins and boomerangs.

This is telling, since these long-range weapons are particularly suited for hit-and-run attacks and combatants lacking the willingness to close and fight stubbornly to the death. Their use here points to the tenuous and indecisive nature of Mesolithic fighting. While the bow and similar devices were likely to have been the primary implements used in hunting the smaller, more elusive prey available at this time, it is implausible that humans would have forgotten close-in weapons and what they stood for. They may have been used in contests over leadership or other issues where prestige was critical, but their apparent absence in group combat strongly suggests that the fighting had not reached the all-or-nothing stage.

Unlike war, weapons reach back deep into the history of our line. In fact they preceded us as a species. Similarly, the subsequent course of arms drew on some consistent themes that arguably have their roots very deep in our history. Since hunting, especially big-game hunting, was largely the province of males, it was they who came to fashion, possess, and use weapons pretty much exclusively. Weapons employed in mass killing tended to mirror the implements of predation in being simple, generalized, and influenced by little other than utility. Those arms used primarily in activities associated with intraspecific aggression were subject to competitive elaboration in ways that made them more impressive and intimidating, at times actually at the cost of usefulness.

Generally, the course of arms would remain stable, with several very ancient implements perpetuating themselves, up through most of human history. When change did arrive, it came in an all-consuming fashion analogous to the evolutionary concept of punctuated equilibrium, and then settled down to stasis—a pattern anticipated by the long hominid/human experience with a narrow range of weaponry, interrupted by the virtual eruption of the bow, harpoon, boomerang, alatl, and sling, followed by another stretch without major innovation.

Weapons use also drew on earlier themes. Hence, in wars where dominance rather than complete overthrow was sought, arms would tend to be matched symmetrically in ways that reflected shared rules of engagement—patterns clearly analogous to fighting behavior within other species. Also, in environments where combat at close quarters was firmly established, courage would come to be equated with a willingness to fight this way using appropriate weaponry—an outlook that is logically tied to our experience with prehistoric big-game hunting. Conversely, although long-range arms eventually were fit into a heroic context, this required elaborate cultural accommodations.

We arrived upon the scene, not just with a weapon in our hand, but with a headful of lore about this most ancient talisman. Military history would be shaped accordingly.

Chapter 2

WAR'S ARRIVAL

The first victims of war would have perceived their plight in the same way Mesolithic villagers viewed aggression long before. The same drivers were present—premeditated violence by groups and an evolving social structure with issues of inclusion, exclusion, and personal property. But this time the economic underpinnings were entirely more robust—plants that could be cultivated and stored, along with animals that could be bred and accumulated to create a self sustaining, permanent base. With this came a perpetuation of organized violence: the arrival of true war and military architecture specifically designed for the purposes of armed conflict. War and the evolution of arms were joined at the hip, the possibilities and limitations of arms defined by the way people lived.

War would spread to become a central part of a variety of Old World societies, and then reinvent itself in the New World, under somewhat different circumstances, but still serving analogous functions. In all cases war survived because it addressed a vital function for each type of society that employed it. And at the root of everything was domestication, particularly agriculture.

Most specialists believe that the transition to agriculture first took place in the Middle East, with the initial steps taken by a group known as the Natufians around 10,500 B.C. The Natufians, up to that time hunter-gatherers specializing in gazelle, came upon a stand of wild cereal rich enough to sustain temporary settlement. When it was exhausted the humans moved on, not returning until the next year. But over time something remarkable happened, the yields—as if by magic—started to increase. Actually, it was natural selection at work. The very act of harvesting greatly increased the chance of a cereal plant being reseeded. Since the shoots with the largest edible portions were the most likely to be cut by the humans, it was these that were most likely to be reseeded, allowing them to dominate the fields and further increasing the yields. From this point it was just a matter of time before Natufians and other groups settled down more permanently around these fields and learned to plant seed grain in the formerly self-germinating fields. What emerged was a true symbiosis—opportunistic plants thriving off human labor and humans living off the plants. Just as we domesticated the plants, the plants domesticated us.

Meanwhile, the steady, storable, renewable source of nutritious plants offered humans the chance to fulfill their reproductive potential. Skeletal evidence indicates that living off one or two carbohydrate-rich food crops and much less meat left humans a head shorter than their Paleolithic ancestors and in some cases seriously malnourished. But it also kept women consistently above the 20 to 25 percent body fat threshold necessary for ovulation and pregnancy. Soon enough there were a lot more of us—approximately a forty-fold increase in the Middle East from 8000 to 4000 B.C. It amounted to a veritable plant trap, since such numbers could never be supported by hunting and gathering.

The human inheritors of what was once called the Neolithic Revolution found themselves stuck in fields surrounding an expanding number of farming villages, working long hours in the hot sun fighting weeds and doing everything possible to nurture plants that once took care of themselves. There was always the possibility that the crops might fail, leaving little alternative but to seek new, more fertile fields to cultivate the same food crops, spreading the system until basically all suitable land was brought under cultivation. At this point, the sole avenue of agricultural improvement became further intensification through irrigation in those areas where it was possible. And with this came even greater population and dependency. So the stalks of the plant trap coiled still tighter around almost all of those who once had roamed free. Yet a few would escape to wander again, exploiting another renewable food resource that moved with them. But this would require domestication of a fundamentally different sort.

Like agriculture, animal husbandry is coming to be seen not as an invention but as an example of what is termed coevolution. In this case it amounted to a true biological partnership. Unlike crops, whose drift toward domestication was pushed by the random influence of natural selection, the mammals involved were self-directing entities capable of making what amounts to a decision to join up with us. Wolves were first, around 10,000 B.C., during the Mesolithic. The time was right, since they were fast predators with acute noses capable of complementing our superior cunning and weapons, to give both species the aid of the other in hunting the agile smaller prey that now predominated. So the wolves, who were genetically plastic and opportunistic enough to forge an alliance with humans, eventually became dogs.

It is probably no accident that the second great wave of creatures to join up were herbivores. Humans had become, rather suddenly, the masters of nutritious crops and food stores that drew opportunistic plant eaters like a siren's song. We ate some, but we kept others, selectively breeding until they became more dependent, docile, and useful. By 6500 B.C. there is clear skeletal evidence that both sheep and goats had been domesticated. Cattle came somewhat later. Yet these herd animals only gradually revealed their full utility—sheep developing their woolly fleece over time during the Neolithic, and goats and cows awaiting the spread of lactose tolerance among adult humans and the invention of more digestible dairy products like yogurt and cheese.

The growing numbers of ruminants in Neolithic villages were bound to lead to problems of crop damage, either through eating or trampling. The solution was obvious: move the herds to the outskirts of the settlement. There they would be free to graze under the watchful eyes of the relatively few humans and dogs who fell into the role of shepherds. As the herds grew larger so did the incentive to range still further afield. By around 6000 B.C. Neolithic communities had begun to split, as two separate ways of life formed around first plants, then animals. Most likely the shepherds first established base camps out in the areas of best pasturage, still returning home at regular intervals. Independence had not yet been declared, but everything pointed in that direction—the possibility of pastoral self-sufficiency and a return, albeit for a relatively few humans, to a life of wandering, freedom, and at least an approximation of the old ways.

For shepherds, the basic combination of sheep, goats, and cattle offered a flexible economic mechanism, with the relative numbers of each species readily adjustable to take account of variance in terrain,

vegetation, water . . . and also risk, since these animals—especially the
sheep and goats—were prone to very rapid die-offs in bad weather or
when deprived of sustenance for even short periods. In such circumstances, there was little recourse but to look to the farming communities
for help—though not necessarily as supplicants or traders. Pastoralists in
such straits would have had little to exchange and little inclination to do
so, since they characteristically viewed farmers as drudges and historically brutalized them. The stage was set for true hostility, not simply
individualized violence, but a clash between two very different ways of
life . . . war.

Down on the farm the social dynamic was also ripening, and one of the
fruits was a growing proclivity for violence. Archaeological evidence
points to much the same pattern of evolving and self-reinforcing social
complexity seen earlier in the Mesolithic. The dramatic increase in agriculture and growth in populations in the Middle East appear to have led
to the beginnings of true societal organization and economic specialization. Housing, which started as loosely clustered circular huts, progressed
steadily toward ever more tightly grouped rectangular mud-brick structures. By around 6200 B.C. places like Oal'at Jarmo, Alikosh, and Tepe
Guran were showing signs of a considerably broader inventory of tools
and artifacts, including that hallmark of sedentary existence, pottery.
Also feeding the stock of valued things were ever expanding webs of
trade, which by 6000 B.C. were transporting goods such as turquoise and
decorative shells several hundred miles. Finally, graves started reflecting
not only increased possessions, but also differentials of wealth.

All of this reached early fruition in Çatal Hüyük, a Neolithic community that thrived in what is now southern Turkey between about 6250
and 5400 B.C. Made up of about 1,000 houses and a population of as
many as 6,000, Çatal Hüyük is one of the largest and wealthiest
Neolithic sites ever unearthed. But in spite of this and the town's obvious capacity to grow a wide variety of food crops—something like fourteen edible plants were cultivated—there are clear skeletal indications of
dietary deficiencies among elements of the population. Rich burial sites
for a minority also point to emergent class differences and something
approximating leadership.

It is also notable that Çatal Hüyük's dwellings were packed in together
pueblo-style so that the outer edges of the town presented a single face

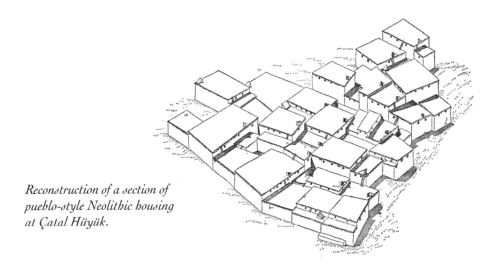

*Reconstruction of a section of
pueblo-style Neolithic housing
at Çatal Hüyük.*

without a break. The town's excavator, archaeologist James Mellaart, does
not believe Çatal Hüyük was specifically designed to be fortified, but
common sense points to at least some increasing outside danger. Then
there is the matter of weapons. Mixed among the remnants of devices typically associated with the hunt at Çatal Hüyük were arms indicating something more ominous—ceremonial flint daggers, a profusion of baked clay
sling balls, and mace heads.

The mace was not the first weapon, but it was the first weapon specifically crafted to be used against humans. It could break bones. But more
importantly, it could break heads.

At first glance it might seem that a heavy blow to the head would be
equally devastating to man or beast. This is not the case. Large animals
have brain caps reinforced with a raised bone suture, and the back of their
heads (a particularly vulnerable area in humans) is generally fused with the
thick first vertebra. In the case of ruminants, horns and antlers not only
thicken the top of the skull, but act to deflect that blow. And all of this
presumes the animal would simply stand and face its assailant. More likely,
it would flee, presenting its hindquarters and making an effective blow to
the head all the more difficult to deliver. Most big prey were best attacked
from the side. Our careers as hunters reinforced this conclusion, resulting
in an emphasis on a series of weapons designed for penetration, not
concussion. Such arms would enjoy a long span of usefulness against other
humans; but it was the mace that proved singularly—though temporarily—
effective when applied to us. Our heads were bigger, more fragile, and, due
to our upright posture, similarly exposed whether assaulted from the side,

the front, or the back—a vulnerability that led eventually to what must have amounted to hundreds of thousands of mace-inflicted skull fractures. By the time we began memorializing and depicting our leadership, the mace was clearly the primary weapon of hand-to-hand combat. Thus we would be introduced to Egyptian king Narmer exercising his sovereignty with a mace, bashing the brains out of a captive in a stone carving around 3000 B.C. Barring an excess of royal enthusiasm, a single blow would have done the job.

Despite the appearance of maces and other signs of violence at Çatal Hüyük, especially during its latter stages, the picture does not justify the label warrior state or anything like it. Rather the evidence points to Neolithic communities in the Middle East remaining relatively free of endemic hostilities for upward of two millennia after the domestication of food plants and the coming of sedentism. Social complexity did seem to be gradually pushing toward war; but before this stage was reached things changed suddenly. Around 5500 B.C. it is generally conceded that the evolution of Old World agricultural communities was violently, even catastrophically affected by outside forces.

Natural disaster might have played a role. Right at this point, 7,500 years ago, melting glaciers at last raised the sea level sufficiently that the salt water of the Mediterranean breached the natural dam at what is now the Bosporus and poured into the freshwater lake occupying the Black Sea basin, resulting in a flood of biblical proportions. Some have hypothesized that this set off a series of migrations, which could have resulted in a general climate of turmoil and usurpation. But this remains a long shot, since the populations most affected would have been small groups of advanced hunter-gatherers, living along coasts far from sedentary habitation. Meanwhile, there was a more plausible candidate for aggression.

The slate Palette of King Narmer, Hierakonpolis.

Among pastoralists the process of separation occurred unevenly across the Middle East and the areas to the north. Herds and those who took care of them continued to hang around the outer edges of agricultural centers. But as numbers of people and beasts increased, a kind of critical mass was reached, accelerating the development of a distinctive and independent shepherd culture.

There remained several impediments to its full flowering. Without effective pack animals, possessions must have been kept to such an absolute minimum that even the evolution

Mechanics of Maces

I t could hardly have been more simple, essentially a rock attached to the end of a stick; but it possessed remarkable mechanical efficiency. In recent tests a mace weighing 1.8 pounds generated 101 foot-pounds of energy at its point of impact—the greatest among ten of the most common ancient arms and enough to seriously damage the cranial structure with even a glancing blow.

Yet the mace's power presented structural challenges. The key problem in manufacture was the joining of the head to the two-foot-long handle in a way that prevented it from flying off.

The mace—the first weapon specifically designed against humans.

By the time of Narmer, the problem was solved. Mace heads were drilled through their whole length, with handles protruding slightly at the top, allowing the business end to be wedged tightly in place. Contemporary illustrations also show the handle bound with cord for reinforcement against shattering (another common point of failure) and thickened slightly at the bottom to lower the chances of it slipping out of the hand when swung.

Once perfected, the effectiveness of the weapon was amply attested to by Egyptian and Sumerian medical texts, which reveal a marked familiarity with skull fractures and considerable skill in treating them, especially by trepanning (removing circular sections of the skull). But whatever the mace's contribution to medicine, it was not destined to last.

The first sign of the mace's decline came around the middle of the second millennium B.C. In Mesopotamia mace heads began to be made with fluted edges to better cleave the skull. Later they would come to be molded from solid copper to add mass. Skulls weren't growing any thicker, but they were better protected. For the two most important Sumerian military relics of the time, the Standard of Ur and the Stele of Vultures, both show massed infantrymen wearing helmets, some even with nose guards. The mace had met its match. Tests reveal that a helmet made of four millimeters of leather under two millimeters of copper is sufficient to attenuate the mace's killing power. The helmet spread the force of a blow over an area large enough to render it far less dangerous. Maces would remain symbols of authority, and even reappear on European battlefields during the late Middle Ages and also in China and India, but never again as dominant weapons. This confrontation between offense and defense at once delineated the basic routines of arms racing, and led to history's first obsolescent weapon.

of a very austere lifestyle was inhibited. Still more basic was the fact that nascent pastoralists remained pedestrians, living a life based on mobility. Later, this would change rather dramatically with the domestication of the horse, enabling the exploitation of the vast steppe lands to the east. But until that time, these newly minted pastoralists would be confined to roughly the same environs as the agriculturalists and still subject to the possibility of rapid animal die-offs that would always plague shepherd peoples.

The solution in the event of such a disaster was to prey upon the sole source of stored food, the farming villages that dotted the landscape. Even without horses, these shepherds had a very significant military advantage— their mobility. They lived nowhere in particular, and had nothing in the way of territory to defend. Their targets were stationary, while their own assets were movable. They could plan an assault against a known location, while the farmers could not know where to seek vengeance—just as long as the raiders could get away cleanly.

This was a problem. While pedestrian attacks, presuming the advantage of surprise, were plausible enough, successful escape was another matter entirely, particularly if the aggressors were weighed down with foodstuffs and other booty. Prudence then would have pointed in the direction of brutality and terror—to kill wantonly all who might follow, to reduce the victim community to helplessness, and to be violent in ways that would induce shock and paralysis. It also stands to reason that cruelty would have been reinforced by inclination, since pastoralists historically viewed their agricultural cousins with contempt as slaves to their fields. Farmers occupied a lower moral rung in the eyes of the nomads; being plant eaters, like their own herds, they were seen as resources to be exploited. There was little reason to be merciful.

This pattern of aggression can be traced all the way back to this initial confrontation between raiding wanderers and settled farmers. Archaeologically the evidence points to this having taken place from the Late Neolithic through the Early Bronze Age in Anatolia, the Levant, and the areas east of the Zagros Mountains extending into the northern portions of Mesopotamia—a hit-and-miss pattern attributable to the mobility of the attackers.

It is possible to speculate how weapons were used in such attacks. Archaeological remains contain mostly stone mace heads and baked clay sling balls, since both weather very well. But arrowheads and spear tips would have also been fashioned of stone; that the stratigraphy of the period does not appear to include large numbers of them is potentially quite significant. It may well be that combat between raiders and agricultural villages was actually dominated by maces and slings.

Since these weapons are exceedingly simple and require only commonly available material, it is possible that each side would have had access to both. But it is likely that the farmers might have placed greater reliance on the mace. Attacks must have been sudden, and the defenders faced with an enemy already upon him. The farmer needed something that was immediately available, easy to use, and optimized for hand-to-hand confrontations.

Alternately, shepherds likely favored the sling. It was their traditional weapon, thoroughly in keeping with their lifestyle. It demanded considerable skill. But since the shepherd needed a weapon to keep wolves at bay, or just as a reminder for strays to get back with the herd, its use constituted an occupational requirement. Used proficiently against humans it had the range and accuracy to begin an attack with lethal surprise . . . a sudden hail of sling balls punctuating a peaceful morning . . . along with the portability not to burden the equally critical escape. Like so many weapons of predators it was inconspicuous but exceedingly effective. Goliath was far from the only military juggernaut to fatally underestimate its power.

The invisibility of the sling as an ancient weapon was but one manifestation of the farmer-shepherd split. The net effect of the division had greater implications. It set off a sporadic pattern of aggression, sufficiently terrifying to result in the gradual abandonment of open village sites and the concentration of populations in fortified townships. In Central and South America this apparently never happened, as witnessed by the fact that populations continued to live largely in unfortified communities in spite of the obvious presence of warfare. The sudden change in Old World Neolithic development patterns reflected a deep economic and ideological schism and a pattern of hostility with genocidal overtones—a scheme destined to impose itself on the subsequent development of armed hostility all across Eurasia and ultimately around the globe. This is why it makes sense to pinpoint the origins of true human warfare in this initial confrontation between the sedentary and the nomadic. While this is a stark and categorical conclusion, a good deal of the reasoning is founded on a solid foundation of stone and mortar.

Weapons encompass the defensive as well as the offensive. The appearance of circumvallation—the building of defensive walls around towns—should be thought of as municipal armor, a stone shield against hostile outsiders. Throughout most of recorded history, walled fortifications were at the heart of the military-industrial complex, the most expensive and commonly employed defense-related projects.

Silent Sling

I t could not be simpler—a length of rope or leather with a pouch in the middle to hold a projectile. A long-range model typically weighed in at just over an ounce, and in a history that stretched between the Late Paleolithic and the Middle Ages its sole improvements came with better ammunition. With the exception of the composite bow, no projectile weapon exceeded its range and power. Yet it is among history's most unsung weapons.

We barely remember how it was used. Most have the image, probably from Hollywood biblical epics, of slingers whirling it round-and-round over their heads. This was nothing like the real delivery, which slinger-historian Foster Grunfeld describes: "Much faster than it takes to describe he would . . . anchor the loop to the base of the pinkie of his throwing hand, hold the knotted end between thumb and index finger, load a stone into the pouch with the other hand, take aim, swing forward underhand with force (rather like a fast-pitch softball), and release the knotted end after one full rotation." The result of this simple motion seems all out of proportion to the effort invested, sending a projectile whizzing along toward its target at up to 120 miles per hour—fast enough to kill a man with ease at fifty paces. At such a range Livy, the Roman historian, reports that Achaean slingers could hit "not merely the heads of their enemies but any part of the face at which they aimed."

For all this, however, the sling made very little impact on those who made it their business to record military exploits. David is, of course, celebrated for slinging Goliath into oblivion. But he is the only famous slinger in the Western military tradition, and even David quickly traded his trusty sling for more prestigious arms when opportunity beckoned. As armies evolved into complex entities employing a variety of arms, slingers were consigned to the role of auxiliaries, a lowly station above which they would never rise.

It was a sort of Rodney Dangerfield among weapons—if not exactly despised, then certainly ignored. In part this relates to its long-range character. But it also reflects who was doing the fighting. Military history is almost solely a product of settled societies, if for no other reason than they monopolized literacy. The kind of warfare in which they normally engaged was against similarly armed and motivated opponents—combat reflecting many of the characteristics of intraspecific aggression. Honorific weapons were very much a part of this, and size really did matter. Conversely a sling one ounce in weight and capable of doubling as a belt was not likely to inspire much respect.

More to the point, their massive and lasting nature left them the most obvious and convincing pieces of the archaeological puzzle as to how and when war arrived. Put simply, walls are hard to build and can be dated with a fair degree of accuracy. The sudden proliferation of walled communities after 5500 B.C. leads to the conclusion that not only were they a response to a specific danger, but that it was one that hadn't existed before.

There are complications. It has been maintained that human behavior is more cultural than utilitarian. Thus people in the Middle East might have built walls largely because they felt like it, and continued doing so because it became tradition. In the New World, on the other hand, the circumvallation habit might have simply never gotten started. The obvious rejoinder is that this is not dancing or table manners, walls require countless hours of valuable labor, and simple logic argues they would have some significant purpose. What then besides safety meets this criterion? Shade? A desire to live in an exclusive gated community? A rock-clad insurance policy is a better bet.

The farmer-shepherd split began a chain of catastrophe. Halicar was

The Mystery of Jericho

In terms of the physical evidence, there is a more troubling problem. The first walled fortifications in history were excavated at Jericho, reliably dated at around 8000 B.C., long before the split between agriculturalists and pastoralists—indeed, well before the taming of sheep and goats, and the first site where remnants of truly domesticated barley and emmer wheat were found. If walls were the key sign of the true coming of war, and this was catalyzed by raids on the part of shepherds, why was Jericho built?

There are several potential explanations. Archaeologist Ofer Bar-Yosef and a number of others argue that the walls of Jericho are no such thing. Instead, they are simply the results of a massive flood control project. The hypothesis is unconvincing. For one thing there is little evidence of deeply incised channels or other signs of flooding. Secondly, not only are the walls fully twelve feet high, seemingly indicating something other than hydraulic overkill, but the complex includes a massive twenty-five-foot-high stone tower that is inexplicable in terms of flood control.

There is an alternative line of reasoning. Jericho was a highly unusual site, in some ways well ahead of its time. It was far bigger than any known contemporary Neolithic community, occupying ten acres with a population estimated at between 2,000 and 3,500. It was also located on a well-watered site that was not only good for agriculture, but also positioned to exploit and even control trade in salt, asphalt, and sulfur from the nearby Dead Sea. So it is possible to conclude that Jericho was rich, a horn of plenty surrounded by a dry and hungry world, and therefore a magnet for have-nots. The walls then can be explained as a monument to fear, but a fear, unlike later, that was unique to Jericho itself.

burned repeatedly. Mersin was destroyed, leaving a heap of human remains. Arpachiya was smashed and looted by an unknown assailant. In Azerbaijan at the site of burned Hajji Firuz II, the corpses of twenty-eight massacred people were found. In Thrace the settlement at Tell Amak was torched. With growing frequency, between the Late Neolithic and Early Bronze Age, town after town in the Middle East was destroyed, built walls, or both. Whole cultures like the Halaf simply disappeared. And while this was happening, weapons-related evidence—particularly mace heads—increased dramatically among the ruins. The violence was accelerating so fast that it seems no longer attributable to just pastoralist raiding. Settled localities seem to have begun fighting among themselves.

How might this have taken place? Circumvallation, by concentrating populations, very probably accelerated and intensified the factors that made for social complexity, among them leadership, armed elites, and an accompanying propensity for organized violence. Meanwhile, external pressure was having its effect, for war tended to spread in a manner analogous to contagion, its key vector being fear. Once underway, hostilities often establish themselves at higher, not lower, levels of violence. Agriculture had filled up the land, and for those with no avenue of retreat, aggression could only be met by submission or equivalent countermeasures. Once the precedent of attacking whole communities was set, it was continued by the original victims—jumping from agricultural community to agricultural community until circumvallation and eventually siege warfare became the norm. War continued to spread until it ruled Eurasia, marching hand in hand with social development until it became very much a part of what we know as the dawn of civilization.

Chapter 3

HISTORY POISED
ON THE TIP OF A SPEAR

As the process of circumvallation continued it stimulated the rise of what are now called central places—networks of roughly evenly spaced focal points of communications, organization, and distribution. Safety in numbers encouraged the consolidation of people into larger settlements. Working together, these two trends began to form something that looked less like a village and more like a city.

Cities were the vessels of civilization, acting like reactors in a chemical factory, compressing and catalyzing societal development until eventually they took us to the next stage—a time when millions were ruled by a king and his bureaucrats, when religion thrived and huge temples and monuments were constructed, and when armies numbering in the thousands marched out to risk it all for those in charge.

Historically, the process first began among a group of communities that came to be know as Sumer, city-states that sprang up on the flatlands bordered by the Tigris and Euphrates rivers in what is now Iraq. Well watered by the two rivers, the plain was essentially rainless and subject to withering heat—a fragile but potentially productive environment. The Sumerians appear to have originated in the hill country to the north, but came down to the southern plain to practice irrigated agriculture, with all the digging and maintenance it took to keep a ditch network delivering life-giving water. The immediate result was a dramatic increase in crop yield. Populations shot up, and a fundamentally new type of society was born.

Living on flatlands naked of natural defenses and possessing relatively huge surpluses of grain, there was an obvious incentive to construct fortifications—not just village walls, but massive ramparts designed to protect whole cities. The result—life at close quarters in large numbers—meant dramatic changes in social patterns and organizations. Populations did not simply grow, they were transformed. Reproductive pools were for the first time large enough to support true inbreeding. Communities no longer had to cooperate with neighboring communities to avoid birth defects. This also tended to devalue the social safety valve inherent in the "escape option"—a development ratified by circumvallation that not only kept outsiders out, but insiders in. People now had every reason to stay home. And in these close quarters vast amounts of energy were liberated—and also, hostility.

Living packed together had other consequences. One of these was the rise of a lethal but invisible threat . . . epidemic disease. Also, Sumer's reliance on a few food crops—utterly dependent on a well-maintained ditch network—practically insured periodic crop failures along with the malnutrition and vulnerability to disease that were famine's handmaidens.

Sumer was in the midst of a revolution, one that eventually would lead to a self-reinforcing cycle of demographic instability. But even in its adolescence it brought forth a remarkable style of war—one that would be repeated under analogous conditions in the West up to the end of the Middle Ages. What happened in Sumer was emblematic of the human condition under agriculture. Massive irrigation required coordination, which encouraged a pyramid of power capped by a single ruler. On the other hand, our evolutionary heritage of working together in mutually dependent bands fostered rule based on consent of the governed. Out of this tension emerged a distinctive approach toward armies, weapons, and warfare. Key aspects of the process are captured for us in two remarkable artifacts—one pictorial and the other written—the latter being a chronicle of the life of an actual figure and quite literally humankind's initial literary hero, Gilgamesh of Uruk.

Among Gilgamesh's exploits is a particularly suggestive tale having to do with a war between Uruk and the rival city of Kish over water rights. The action opens after envoys from Kish have warned the men of Uruk to stop digging wells and irrigation ditches in disputed territory. Gilgamesh opts

Epic Bloodshed

Epics were not the product of any imperial mandate or the ancient equivalent of a gonzo Marine drill instructor. They were composed by illiterate bards to be sung for popular entertainment. The orally composed epic poem that forms the basis of so many national literatures is a phenomenon unique to the military histories of Indo-European peoples. Epics were the repositories of Western martial traditions long before the arrival of the gun. The heroic figures around which each poem is based—a rough bunch one and all—embody virtue on the battlefield.

The bruising, skull-splitting action not only instructed apprentice warriors on how and when to kill, but also on the courtesies of killing—the rich, intraspecific, aggression-laden code of battle ethics. Although these poems seldom fail to startle modern readers with their sheer gore and gleeful infliction of mayhem, they are also cautionary tales of those who fail to observe conventions, fight with the wrong weapons, or otherwise transgress the warrior ethic. At the core these poems are about courage, or at least what is perceived to be courage. The consequence of breaking the code, worse even than an excruciating death or disfavor from one or another god, is to be seen as less than brave.

The historic trail of epics begins with Sumer's *Gilgamesh,* moves to the *Iliad* and *Odyssey* of the Greeks, the *Mahabharata* of Vedic India, up through *Chanson de Roland* and *Beowulf.* All favor elite warriors willing to mix it up at close quarters. With the exception of the Vedic Indians, they favor the spear and sword over the bow. All don armor and helmets configured for close in fighting. Heroes are big, swift afoot (or at least possessors of fast chariots or horses), and loud, intimidating their opponents with their reverberating yelps of war. Given this basso profundo profile, it is not surprising that epic heroes typically disdain anything that might amount to trickery, preferring to do battle in the most confrontational manner possible. Together, this list of traits amounts to a sort of basic text on the etiquette of intraspecific aggression among humans. It also says a good deal about weapons. It is a singular fact that the course of arms would be dominated by lethal systems that were big, loud, fast, armor-plated, and in their essence confrontational; in this Achilles, the medieval knight, the dreadnought battleship, and the modern M-1 Abrams tank are united.

for war. But it quickly becomes apparent that, even in his dominant position, he lacks the power to make such a unilateral decision. Instead, he must go before a council of elders, where he is rebuffed, only managing to have the decision reversed by an assembly of all the city's fighting men. So, although Gilgamesh gets his way, it entailed the approval of those who would bear the brunt of the fighting.

Nor was war unusual in Uruk, Kish, or anywhere else in Mesopotamia. It

Stele of Vultures.

was a key element of intercity politics, enforcing a balance among the system's competitors. Each resorted to war opportunistically, but the net result across the region was a rough equilibrium—or put negatively, universal frustration. Military advantage could be quickly countered through alliance, imparting a characteristic cast of amorality and perceived futility to the entire process, or as one contemporary put it: "You go and carry off the enemy's land;/The enemy comes and carries off your land."[1]

The endless ebb and flow of violence would prove the hallmark of balance-of-power schemes. It also produced a characteristic kind of war—short, violent, and aimed at limited objectives. And in the case of city-states it generated a style of fighting that was a product of distinctive internal politics, and was destined to repeat itself over a vast stretch of time.

We know a good deal about the kind of army Gilgamesh led. Our knowledge is derived from the second of the era's remarkable relics—a victory monument carved around 2500 B.C. known as the Stele of Vultures. It is nothing less than a limestone snapshot of the Sumerian order of battle, and it reveals a basic split. Out front, armed for single combat is Eannatum, ruler of Lagash, symbolically looking forward to a day in Mesopotamia when battles would be won by elite warriors seeking out and defeating their coequals, while a mass of underlings essentially looked on. At this point, however, Eannatum was backed up by something entirely more lethal.

The infantry column to his rear—all wearing helmets, advancing shoulder to shoulder behind a barrier of locked rectangular shields reinforced with bronze disks, and presenting a hedgehog of spears protruding from several rows back—is a full-fledged phalanx. The implications of this very convincing bit of evidence have frequently been overlooked by military historians, who focus instead on the development of this presumably advanced formation by the culturally adroit Hellenic Greeks almost 2,000 years later. Actually, the tactical and technical requirements for a phalanx are simple. It demands mainly a willingness to confront adversaries at close quarters and face danger in a cooperative fashion. This is a commitment virtually impossible to draw from any but highly motivated troops, those who view themselves as having a true stake in not just the fighting, but charting the course of their polities.

The reason is simple. In this formation, battle is a short but vicious spasm of fear and brutality, a collision between two compressed human hedgehogs bound to leave many in the front ranks skewered by the spears of their opposites. This was combat for amateurs, more demanding of courage and determination to push ever forward than the skills of professionals—a style of war guaranteed to deliver the participants back home in short order, either alive or dead. And there were plenty of the latter. The Stele of Vultures records over 3,000 Ummaite battle deaths in a single episode.

There are also economic and social overtones. It is notable that the Stele of Vultures depicts phalangites and Eannatum as similarly dressed and protected. The phalangites' adoption of what amounted to aristocratic arms, a phenomenon to be repeated in ancient Greece, was not only indicative of the community's ability to accumulate metal, but also of the economic independence of its fighting members.

Neither this independence nor the phalanx was fated to last in Sumer. The great wheel of social complexity continued to turn, and in a direction that moved away from relative social equality and its associated fighting qualities. Just 200 years later we find Mesopotamia ruled by a single tyrant and a small Semitic warrior class. Gone was the determined human hedgehog, replaced by a cadre of elite warriors and a mass of peasants armed with bows. It was less effective, but it fit the times.

The story of the phalanx continued in Hellenic Greece. Here there was an economic and political structure uniquely suited to its perpetuation. In Hellas the phalanx would dominate war for 350 years, and turn the hoplites who manned it into the most feared and respected soldiers of the Mediterranean world. Greek military history was the Sumerian experience in reverse—first individualized and only later corporate. We pick up the thread amongst the palace-dominated kinglets of the Achaean Greeks—a

society that thrived between 1600 and 1200 B.C. and subsequently came to
be known as Mycenaean, after its most famous archaeological site. Here, a
class of warriors under a hegemon ruled an underclass of agricultural peasants. War was frequent and in battle the elite dominated the action. These
Achaeans were great sailors, and therefore extremely mobile. But, this
aside, there was not a lot to distinguish them from any other Middle Eastern or Mediterranean Bronze Age society . . . had it not been for one man.
Because of him we have a vivid picture of the values and attitudes that
would form the core of the Greek martial tradition and ultimately the
Western way of war.

The palace economies of Mycenae collapsed suddenly around 1150 B.C.,
apparently weakened by the conquest of Troy, and made vulnerable to
another band of Greeks, the Dorians, who moved in behind them and then
swept over the entire peninsula. The ensuing period is known as the Dark
Ages (c. 1100–800 B.C.). But its legacy was hardly dark.

The society that emerged remained essentially aristocratic, but the
Mycenaean state structure had been replaced by other principles of social
solidarity—still nascent but full of promise. A radical change in agriculture
was taking place. Under population pressure the Greeks turned to a
network of small, family-run, and privately owned farms, and around this
crystallized the polis, or city-state, which would form the basis for the
remarkable evolution of Hellenic politics.

There were military implications. Through most of the Dark Ages aristocratic power appears to have been enforced by heavy cavalry. Around 700
B.C., however, a new scheme of fighting emerged—massed heavy infantry, or
hoplites (from the Greek *hoplon*—"tool, weapon, piece of armor"), armed
with spears and lined up in rows and columns, a formation that cavalry
lacking stirrups could not hope to overrun, so long as they stuck together.
So the phalanx reappeared, sweeping all before it.

On one level this was a matter of sheer lethality; but it was also based on
economics and politics. The ranks of phalanxes were filled with farmers—
warrior-landholders whose style of fighting reflected their way of life.
Soldier-historian Xenophon would write in 400 B.C.: "Farming teaches one
to help others. So, too, in fighting one's enemies" (*Oeconomicus*). Strength in
numbers also had a political dimension, because the privileged ranks of
yeoman became the voting heart of the broadly based oligarchies of the
thousand or so city-states that came to populate the Greek-speaking world.

Blind Visionary

The man most essential to the modern conception of the Achaean Greeks was no soldier or king; he was a blind poet describing events 450 years after Mycenaean society had collapsed. Homer created images of combat that would never fade. His *Iliad,* the archetypical martial epic, was destined to become the most influential tale of war ever composed. To say the *Iliad* was the Greeks' favorite story entirely understates the matter. In their minds it conveyed the essence of everything Greek. But above all it instructed Greeks on how to act when they fought and killed each other.

The action in the *Iliad* spins out over several weeks or months of the ninth year of the siege of Troy in Asia Minor by a confederation of mainland Greeks, intent on avenging the abduction of Helen, the beautiful wife of one of the leading invaders, by Paris, a Trojan prince. (Calling the city besieged is a stretch, since the invaders have no siege machinery and the residents of Troy come and go as they please, the major battle action occurring when they sally forth on the plain below the city.)

Homer's characters are among the very most violent in history, but as with so many other epic heroes they are polite killers. They do their fighting on neutral ground away from women and children and according to a prescribed code of conduct reflective of intraspecific aggression in other species. Key combatants on both sides are less members of an army than a collection of individuals seeking glory through single combat with social equals. Fighting takes place in a standard prescribed sequence. After first casting spears from relatively long range, combatants move in to stab with an extra spear, and, if there is still no lethal result, settle the issue with their swords—"he who fights at close quarters" was a favorite Homeric epithet.

Homer's Greek and Trojan elite are armed for close combat, and on most occasions virtually alike—helmet, breastplates, greaves, and shield, all made of (or at least reinforced with) metal. Even though the *Iliad* is filled with metallic imagery, the most frequent being "bronze-clad Achaeans," the material was clearly in short supply. Only the heroic minority can afford a full set of armor. Arms and armor are not only intrinsically valuable, but are also considered trophies to be stripped from a fallen opponent, an act that frequently leads to further combat.

By contrast archers are portrayed as basically ineffectual, or worse in the case of the adulterous bowman Paris, whose abduction of Helen was the *casus belli.* "You archer, foul fighter," Diomedes tells him. "Lovely in your locks, eyer of young girls./If you were to make trial with me in strong combat with weapons,/your bow would do you no good at all."[2]

The *Iliad* was more than just a story. Modern scholarship, beginning with the excavation of Troy in 1871 by Heinrich Schliemann, has supported the accuracy of the information in the poem to a degree undreamed of by earlier generations. The attitudes of the characters in the *Iliad* represent the outlook of actual Achaean warriors around 1240 B.C., passed down to Homer relatively intact through the oral tradition.

*Early Hellenic bronze
body armor.*

Like their Sumerian forebears, these were men who had a stake in their
societies, but this time it was less transitory. These Greeks, unlike the
Sumerians, were blessed with conditions that favored dry farming,
produced food surpluses without bureaucratic interference, and perpetu-
ated individual land ownership . . . the linchpin of the whole system.

Property had another manifestation. Each member of the Greek phalanx
brought his own weapons and armor, an expensive and weighty proposition
made largely of rust-free and easy-to-cast bronze—a quarter-inch-thick
breastplate and helmet (thirty and twenty pounds respectively), greaves to
protect the lower leg (three pounds apiece), a round wooden shield three
feet in diameter (twenty pounds), an eight-foot thrusting spear, and a short
secondary sword—a total of about seventy-five pounds, far more burdensome than the Sumerian equivalent. This was heavy infantry with a
vengeance, so heavy that the most common cause of death in battle was
getting knocked down and trampled. The very weight and imperviousness
of this armor conditioned the whole nature of Greek phalanx warfare, slow-
ing it down to a crawl and insuring that victory would come not through
tricky maneuvers but sheer stubborn pushing.

Why did this expensive, burdensome armament flourish? Successful
phalanx warfare is perfectly possible with much less armor, even virtually
none at all—the Macedonians did it and so did Swiss pikemen. There was
no overwhelming necessity for these Greeks to be so heavily armed; it was a
matter of choice, even irrational choice.

Homer provides the answer. These hoplites were committed to fighting
as a group, but they went into battle armed as individual warriors in accoutrements functionally identical to those of the heroes of their favorite war
story. It was at once a testimony to upward mobility and the power of the

Iliad. But they were more than poseurs dressing up for battle; the whole way they fought was suffused with the lessons preached by their blind visionary.

War was virtually a matter of Homeric ritual. Battles were staged on familiar ground well away from noncombatants. Sieges were seldom attempted. Instead, one side invaded the other's borderlands and, by doing a bit of damage, provoked the opponent to come out and accept the challenge. Animals were sacrificed, prayers uttered, hymns sung, and then battle commenced, fought by two virtually identical forces in the most confrontational manner possible. Like Achilles, Diomedes, and the rest, each side was intent on a decisive victory through a single monumental effort. When it was over, armor was stripped, a truce was declared to exchange bodies, and a trophy erected. Homer's King Priam, intent on retrieving his son, Hector's, corpse, would have felt right at home. Paris, on the other hand, would have been unwelcome. Take, for example, the complaint of a mortally wounded Spartan hoplite that "death was of no concern, except that it was caused by a cowardly archer."[3] Strabo, the first-century A.D. geographer, even claims to have seen a very old inscription in Greece that forbade the use of missile weapons.

But if Homer was interpreter-in-chief, other factors lay further beneath the Greek military tradition. Phalanx warfare was undeniably lethal, but it was conducted far more as a game or contest than as a people hunt. This was perfectly appropriate, since its ends were limited. Just as in Sumeria, the role of organized fighting was to maintain the balance of power among many small centers. The aim was not conquest but adjustment. And the Greek phalanx proved an efficient and even parsimonious mechanism. The march through the hills leading to the opponent's territory, the battle, and the return usually could be completed in under a week. This short time frame together with the absence of a heavy siege train made elaborate and expensive logistical preparations largely unnecessary. This was war tailored to the needs of a small farmer, delivering him quickly back to his fields where he could get on with the real business of his life. The phalanx was ideal for the amateur soldier. Little if any drilling was necessary. He just stood in line, probably surrounded by relatives, and summoned the nerve to follow the fellow in front into the abyss of combat where he pushed with all his strength and fought for his life over the course of an hour or so. If he was still standing he could go home—a true weekend warrior.

What was it like to fight in a Greek phalanx? Until about a decade ago, nobody had given it much thought. Then Victor Davis Hanson published *The Western Way of War,* and classicists and historians realized that the nature of the experience was a significant piece of the puzzle of how and why this style of war was fought not just in Hellenic Greece but elsewhere.

The Pros from Sparta

There was one bastion of dissent from the amateur spirit in Hellenic warfare, and that was Sparta. Located in the southern portions of the Peloponessian peninsula, Spartans controlled around 3,500 square miles of territory, as much of it as possible devoted to agriculture. But they were not farmers, they were soldiers . . . and parasites. They lived off the labor of an enormous class of serfs, or helots, made up primarily of local peoples subdued in wars of conquest during the eighth and seventh centuries B.C. Henceforth, Sparta became a mechanism to keep the helots subdued, resorting to ritual humiliation, secret police, and even organized assassination. And in doing so the Spartans created history's first totalitarian state . . . a society on autopilot, so regulated and oblivious to individuality that it better resembles a colony of ants than anything human.

Ironically, it was toughest on the ruling class, the warrior elite. To insure supremacy, Sparta professionalized its hoplites, freeing them from the farm but putting them through a lifelong training program more onerous than anything suffered by the helots. Separated from their mothers at age seven, they began their indoctrination in "herds" of youths supervised closely and sadistically by older boys. Described by one historian as "Boy Scout troops in hell," they engaged in endless physical conditioning, military drill, and war games, while they learned to steal most of their food. Recreation largely consisted of the opportunity to engage in a homosexual relationship with an older boy—encouraged by the elders as a means of bonding future Spartiates for battlefield purposes. All of this culminated at around age twenty, when a trainee was deemed transformed into a "Similar," an identical warrior module capable of being fit neatly and homogeneously into the ever victorious Spartan phalanx. He then joined a mess of about fifteen other Similars where he lived until marriage, forever being responsible to his messmates for dinner on a regular rotation. Should he miss a meal contribution, ever show the slightest hesitation in battle, or in other significant ways violate the code Spartiate, it meant demotion to the ranks of "Inferiors," permanent exile to a shadow class not only for himself but also for his male children. There was social mobility in Sparta, but most of it was down.

Militarily it worked. Down to 371 B.C. and its defeat at the battle of Leuctra, the Spartan phalanx was the best in Greece. Not only were they better disciplined and less likely to break and run in a tough spot, but they had a singular tactical advantage. Other phalanxes tended to lurch to the right as they moved into battle, each member seeking to cover his exposed side with his neighbor's shield. Spartans went straight ahead, creating a kind of automatic flanking maneuver, which sometimes allowed them to roll up their enemies from the side. They won victory after victory, and built up an awesome military reputation. So awesome that Sparta was the only major Greek polis not circumvallated. Ranks of Similars were the only walls.

Spartan and Theban phalanxes clash at the second battle of Coronea.

Yet their military advantage was probably more marginal than absolute. Phalanx warfare was not by its nature decisive, nor was it meant to be. The very weight of equipment prevented effective pursuit of a vanquished foe. The field might be taken, but the enemy army was seldom permanently destroyed. Basically any able-bodied farmer could fight in a phalanx. Spartiates were better, but only somewhat better. And the societal price for turning themselves into professionals was excruciatingly high. In the end it was only the Similars who had any real stake in society, and their life was miserable and dangerous. Ants behave the way they do because their genes command it. Humans always consider their options, and will act to throw off their yoke if opportunity arises. In the case of the Spartans it came down to a matter of numbers. It was just too hard to become and then live the life of a Similar, so eventually there were not enough of them.

Other extremely militaristic societies—the Assyrians and Aztecs—ran into similar problems and were destroyed. Defense is an open-ended proposition, and when war and warriors are allowed to become ends in themselves, the numbers willing to support such a proposition inevitably dwindle.

Basically it was horrible—combat reduced to its most elemental terms. After prayers were said and rituals practiced, the two formations advanced toward each other, at first marching and then jogging. Colliding at a combined speed of perhaps ten miles per hour, the first three ranks grasped their spears with an underhand grip and went for the groins and upper thighs of their opposites, seeking to inflict a mortal wound. After only a few seconds, though, the remaining five ranks came crashing into their backs, compressing the whole mass into a knife-edged rugby scrum. Hoplites in the front then switched to an overhand grip, and sought to stab downward over their opponents' shields and into their faces and necks. If spears were splintered or dropped, combatants might also use their short swords, but it was hard to strike a clean blow in this hoplite sandwich. A misstep or a trip was probably lethal, since there was no getting up wearing seventy-five pounds of armor. Forward was the only direction that mattered. Eventually several at the front would fall or two columns would be split apart, creating room for a human wedge, who would then split the formation. That was usually the end, and the losers would fall back and run.

Left behind were the casualties—around 5 percent for the victors, double that for the vanquished—most of them fatal. Any penetration of the gut, or even a gaping flesh wound, was almost certain to lead to infection, and within a day or two, an excruciating death. Separation of a major artery left many to bleed to death within minutes. But most prevalent were bodies asphyxiated or simply broken and crushed by the feet of both comrades and enemies. Despite this carnage, over 90 percent probably came through unscathed—exhausted, mortified, but assured that their time of trial had passed.

This was the beauty of classic Hellenic phalanx warfare: it was awful while it lasted, but when it was over it was over. As such it could be repeated again and again, essentially without change for hundreds of years. Unfortunately, this and other highly ritualized forms of warfare not only depended upon a shared set of values among combatants, but were subject to outside influences. Eventually, both would prove the undoing of the Greeks and their spear-studded balance mechanism.

The first cracks appeared during the second Persian invasion of Greece by King Xerxes in 490 B.C. While the Greeks, led by Athens and Sparta, gloriously repelled the invaders, the campaign revealed the usefulness of naval forces, light troops, or peltasts, and cavalry, along with the inadequacies of the phalanx faced with an adversary unwilling to meet it head-to-head. The next blow came during the disastrous Peloponnesian War, which stretched from 431 to 404 B.C., a nearly thirty-year time span during which not more than three or four traditional hoplite battles were fought. Indeed, chief protagonist Athen's entire strategy was based on not fighting the Spartan Alliance when they invaded. Instead, Athenians stayed behind their walls and relied on their massive navy to preserve lines of supply and communications, while pressing the issue elsewhere. Athens eventually succumbed, but the lessons of the war were plain. The succeeding fifty years saw a much greater reliance on peltasts throwing javelins, slingers, and other light troops. Even phalangites trimmed down their accoutrements in an effort to gain better mobility. But this was nothing compared to the revolution in arms going on in the north.

Macedonia was conceded to be part of Hellas, but was thought to be the land of the bumpkin, a rural enclave that missed the political revolution that shaped the rest of Greece. Instead of city-states, Macedonia was organized around a tribal structure and ruled by a hereditary line of monarchs, the Argeadae—a mediocre bunch who had great trouble imposing their will on their fractious and independent-minded subjects. Plagued by succession quarrels, petty revolts, and hereditary squabbles Macedonia limped along as a backwater, prey to Greek and barbarian alike.

Then, without warning, the line began producing extraordinary rulers. First two of them, a father, then a son, followed by a host of successors, some only a few degrees less talented. These Macedonians would conquer half of the known world and rule the Eastern Mediterranean basin for upward of two centuries. Philip II, elected regent in 359 B.C., was the first great Macedonian leader. At twenty-one he was already a young man of considerable experience and determination, with an excellent understanding of the latest Greek military developments, having spent three years in Thebes, the hotbed of such change. Philip wanted to take it to the next stage, creating a force based on a number of different weapons specialists all working together—the Hellenic equivalent of combined arms. He joined traditional Macedonian heavy cavalry with mercenary peltasts and slingers, and then anchored them with a phalanx like none the world had yet seen. Although it possessed the one imperative for such a force—men willing to fight at close range—this phalanx was not a natural development of Macedonian society, but a contrivance barely ten years old when Philip took this still pliable mass and created a military implement specifically aimed at trumping its Greek equivalent.

Designed around a *syntagma* of 256 men, lined up sixteen deep and sixteen across, the modified phalanx was not only uniformly thicker and more massive, but projected considerably further forward than Hellenic counterparts. Philip had replaced the conventional eight-foot spear with a sixteen to eighteen-foot-long *sarissa,* pointed at both ends, and designed to be wielded with two hands—precluding any shield beyond a wicker target hung from the neck. Although helmets and greaves were worn, body armor was reduced to a light leather tunic, the combined effect making the Macedonian both more agile and mobile than his Greek counterparts.

The new-model phalanx could assemble and move into battle with unprecedented speed, and when it hit, the effects were devastating. The first five rows of *sarissas* created a wall of pikes that at once impaled those in front and absorbed, then rebounded, the shock of the enemy advance. This was followed by a continuous stabbing action, taking full advantage of the force of both hands and lighter equipment to generate a multitude of very sharp battering rams. Meanwhile, the men in the middle and rear pushed those in front, while simultaneously providing cover from javelins and arrows with their raised pikes and stabbing fallen adversaries with the butt spikes. The net effect was what Greek historian Polybius called "a storm of spears."

Philip's new symphony of arms chewed through Greece. Sluggish and beleaguered Greek phalangites staggered into Philip's forest of pikes. Hopelessly out-spear-ranged—facing up to ten spikes per man—the Hellenes were either "processed" through its points and marching feet and butt spikes, or ran for their lives. Most chose the latter and Greece was conquered, the climactic engagement coming at Chaeroneia in 338 B.C.

The mobile and agile Macedonian phalanx — "a storm of spears."

Two years later Philip fell to an assassin's blade, and Alexander took his father's band on the road with great success. Before his short life was over he had conquered Egypt, the Persian Empire, and made inroads as far as modern Afghanistan, Pakistan, and India. Although cavalry was his primary strike force in Asia, 70 percent of his troops were phalangites, the bedrock of his army. In the subsequent 150 years this changed only moderately. In the successor kingdoms of the Hellenistic east, phalanxes were expensive, since in the absence of willing natives, they required Greek mercenaries, who were less than optimally effective. Nevertheless, in 217 B.C. at Raphia, after the cavalries of Ptolemy IV and Antiochus III had canceled each other out, the heavy infantry settled the issue. Back in Macedonia, where patriotic troops were available, the phalanx continued to thrive. In an effort to increase still further the number of protruding spear tips, the *sarissa* continued to grow, first to twenty-one feet, later to twenty-five. This was too much like wielding a telephone pole, so the length was quickly dropped back to twenty-one, which was still too long.

"Nothing can stand up to the phalanx," thought Polybius. He was wrong. In all its manifestations the phalanx remained a fairly clumsy and brittle instrument with only a limited capacity to adjust to technical or tactical challenges. The truncated career of the gastrophetes, a large, but man-portable crossbow, illustrates how technical threats to the system were marginalized. An enormously powerful weapon that could shoot through any armor, it was apparently deliberately suppressed by the Greeks, a significant act, since the roughly simultaneous appearance of the crossbow in China would have a major effect in undermining close combat there.

The phalanx's tactical vulnerability were less easily circumvented. If it could be penetrated or taken at the flank or rear the consequences were bound to be horrific. Once within its ranks swordsmen would be free to hack apart its occupants with little fear of retribution. The phalangite's dagger was as small as his pike was big, and his pike was useless against an opponent two feet away. A pikeman facing an opponent at his side faced a nightmare choice—hold on and be disemboweled or drop the spear and try to somehow escape through the press of humanity. This awful choice remained for the most part an apparition so long as phalangites faced their own kind, or the undisciplined hordes of the east. Then one day in the misty hills of Cynoscephalae the Macedonian pikemen of Philip V found themselves facing a new kind of soldier armed with Spanish short swords and the tactical flexibility to make their worst nightmares come true. They were Romans and for the next millennium and a half the phalanx would be a thing of the past.

Swiss Sequel

Late in the thirteenth century A.D. the phalanx arose anew. In the solitude of the Swiss Alps the urge for independence joined an egalitarian streak among free peasants and burghers, leading them to gravitate into close battlefield formations, which proved astonishingly effective against the arms of the day.

In their early confrontations with Habsburgs and Burgundian nobility, Swiss confederates relied almost exclusively on the halberd—an eight-foot-long man-cleaver capable of paring a man's face from his skull. Then, because the need to repel cavalry charges had become painfully apparent, the primary arm gradually shifted to an eighteen-foot ashen pike topped by a ten-inch steel head . . . a near reincarnation of the *sarissa*. Like the Macedonians, the Swiss chose agility over personal protection, paring down their equipment to a steel cap and breastplate, itself frequently sacrificed for a leather jerkin or buff coat. As a consequence, the phalanx that emerged was as much feared for its suddenness of attack as for its lethality when contact came. The record is full of heavily armored victims who first spotted a body of Swiss as it crested a hill or emerged from a woods, only to find their opponents upon them before they had time for much in the way of preparation beyond a farewell prayer.

The political and military parallels with earlier times were hard to miss. Like the Greeks, the Swiss placed near total reliance on infantry, with cavalry or missile weaponry basically ignored. As with earlier phalanxes, it was the spirit of equal participation not obedience that drove Swiss pikemen into the fray. Also, like the Greeks, the Swiss were quick to grasp the possibilities for profit in tactical ascendancy and found a niche as mercenaries—"pas d'argent, pas de Swisses" becoming virtually a national motto. But as cash drew them to battlefields all over Central Europe, exposure brought forth predictable reactions. In Germany, Emperor Maximilian built his own phalanx of pikemen and halberdiers and trained them to fight just like the Swiss. The Confederates more than held their own on many a blood-soaked battlefield, but their lethal style of fighting was no longer unique or unchallenged. Worse yet, the Swiss eventually insisted on charging straight into the face of murderous artillery fire capable of knocking down whole files with a single shot. Their days on the battlefield were effectively over. Even so the pike lingered as a defensive measure for slow-loading infantry formations until the invention of the bayonet, which, when attached to a musket, performed an analogous role until at least the first decades of the nineteenth century . . . a total life span of over 4,300 years.

From pure halberds to halberd-pike combinations.

Chapter 4

GHOST RIDERS

Horses were the missing piece in the pastoralist puzzle—the last basic act of domestication that set shepherd peoples on their way across Inner Asia, and on a 5,000-year rampage through military history.

As with other animals who joined humans, the relationship probably began with curiosity on one side and hunting and capture on the other. Horses were big and could be bred, but were not at first useful except for their meat. Their real assets, strength and speed, were hard to exploit. Yet it must have been obvious from the beginning that they had great potential for improving human mobility. Just how and when this was realized is a subject of controversy.

Opinions on the origins of equine domestication were once split between those who believed that asses, onagers, and then horses were first used as draft animals by farmers in the Middle East, and those who thought they were first ridden in Central Asia and later hitched to vehicles in Mesopotamia. Whether the cart was placed before the horseman or vice versa, the initial act was not dated much before 1500 B.C.

This has changed dramatically. The bickering is far from over, and there have been several false starts in terms of evidence. But gradually data is being accumulated by archaeologists working at sites on both sides of the Urals that should catapult the initial date of horse riding up to twenty-five centuries further into the past, and place it firmly on the western margins of the Inner Asian steppe. What they are finding is concentrations of horse remains, sometimes thousands of them. Some were hunted and butchered for food, others were bred. But the teeth of certain horses show unmistakable signs of "bit wear," characteristic spalling of the enamel caused by repeated partial dislodging of these control devices—"taking the bit between the teeth." There are two uses for such a bit: either the animal was employed in pulling a vehicle or it was ridden. Since the wheel was not invented until about 3000 B.C., riding is likely. Although penetration of the deep steppe took some time, riding made subsistence along a broad swath of rim lands possible almost immediately. Once out there the lives of mounted shepherds would have been transformed nearly as quickly. Using the analogy of American Indians, whose exploitation of horses transpired largely independent of the Europeans who introduced them, it is logical that the combination of mobility and a new environment was sufficient to produce a fundamentally different lifestyle within a century or two: true nomadism.

A horse culture in Ukraine known as the Yamna provides a good example of what happened. Suddenly people who were once anchored to a specific place began to roam. Significantly, their graves grew not only richer, but yielded more weapons. Even more suggestive are a series of mace heads carved to represent horses. But the most intriguing evidence that something fundamental had happened is indirect. A sedentary river valley culture known as the Tripolye C-1 began to form large, aberrant settlements of almost a thousand dwellings toward the end of the fourth millennium B.C., only to be replaced by the Yamna. The best explanation is that these macro-villages comprised defensive concentrations of peoples forced together by outside pressure—pressure that had arrived on horseback.

The threat did not remain localized in Ukraine. Until the 1990s it was assumed that there were no horses in the Middle East before around 2000 B.C. Since then, domesticated horse bones have been found on both sides of the Euphrates River in southeastern Turkey and in the Negev region in Israel, all of them dated to the fourth millennium B.C. While their presence can be explained by trading, the military advantage of those who introduced them must have become almost immediately apparent. And it is entirely possible that this set off a pattern of raiding—this time by a far more dangerous and frightening mounted predator.

In the same time frame two important developments took place: the acquisition of wheeled vehicles, which, along with the domestication of camels, enabled horse nomads to move into the depths of the Inner Asian plateau; and the rise of the state and the consolidation of much larger walled urban concentrations in Mesopotamia. Keeping in mind the example of Tripolye C-1, it is quite possible that this urban consolidation was hastened by nomadic horse raiders.

Bow of Bows

The composite bow was a weapon of stunning efficiency, unsurpassed until the evolution of effective firearms. Unlike the natural sapwood-heartwood amalgam of yew, the composite bow was carefully laminated out of an elegant combination of different materials. Indeed, its very sophistication may well have thrown us off the trail leading to its origins.

Horse nomads probably moved out on the steppe armed with a combination of maces and slings. Neither was optimal on the back of a galloping steed—one lacked reach and the other was simply too difficult to use from such a perch. Bows offered an alternative. To gain sufficient power, however, bows shaped from a single slat of wood, even those fashioned from yew, had to be nearly six feet long. This was much too lengthy to shoot comfortably from horseback, and besides the steppe was a vast grassland with few trees and less yew. So there was ample reason to experiment.

The power of a bow is a function of three factors: resistance (draw weight), the efficiency with which energy is delivered to the arrow, and the distance over which force is applied (draw length). Unitary, or self, bows have two key shortcomings. First: even yew can only be bent so far before it breaks; hence, increases in draw length mean equivalent extensions of the bow's arms—ergo longbows. Second, the long, heavy arms of the self bow consumed a good deal of force as they sprang forward, so they were not particularly efficient in transferring energy to the arrow.

American Indians again provide an example of how the evolution of the bow might have occurred. In the late sixteenth century A.D., shortly after nomadic tribes began riding horses on the Great Plains, warriors started experimenting with shorter bows, reinforcing them with longitudinal layers of sinew, using a glue derived from hides. (Sinew was extremely resilient and had a tensile strength four times that of wood.) Soon some braves removed wood completely, substituting elk antler, which bore compressive loads better. The result was a weapon both shorter and considerably more powerful. The arrival of the gun ended the process before a true composite bow could be perfected, but a great deal of progress was made in barely a century or two.

Tests recently performed on a fully developed Asian composite bow demonstrate that it roughly doubled the power of an equivalent self bow. It could send an arrow 300 yards with great accuracy and was capable of penetrating armor at 100 yards. The key is the lightness and flexibility of its arms, which unstrung are actually reflexed, or curved forward. When strung, this is translated into much extended draw length, despite being only around three feet across. Combined with the capacity to shoot up to six arrows a minute, it constituted a true technological tour de force.

That mounted shepherds devised such an elegant solution remains implausible to many modern scholars. Just as with horse domestication, they tend to look to the settled river valley civilizations of the Tigris-Euphrates, Nile, and Yellow rivers for the composite bow's origins—beginning around 2400 B.C. in Mesopotamia. But this interpretation ignores the fact that in each case the bow's appearance followed shortly after the first sustained contact with large numbers of steppe nomads bent on conquest and systematic plunder. It also ignores the steppe nomads' incentives and opportunities. Unlike the other cultures, these horsemen had a pressing need for such a bow, and the materials used were appropriate to their culture and environment. It also follows that, like the American Indians, they did so quickly, and in short order found themselves with an unparalleled military capacity. Once mounted and armed, all it took was practice.

Meanwhile career horse riding and life on the steppe were putting the finishing touches on one of history's most fearsome military instruments— the closest the ancient world came to an ultimate weapon. (One very persistent definition of such a weapon is a system that combines the speed to outrun any contemplated foe with the capacity to outrange him. Under such circumstances the adversary is helpless to inflict injury, and can be destroyed at leisure. If there is such a thing as a grail of arms, this is it.) Horses provided shepherds with the first prerequisite, his own ingenuity likely supplied the second.

The horse-riding culture of the steppe emerged suddenly, then changed very little—it was a social organism so elegantly adapted to its venue that pressure to evolve dwindled to the point of evaporation. This was a society so deeply and irresistibly shaped by environment and psychology that as centuries passed and ethnicity replaced ethnicity on the Inner Asian plateau, each would end up living exactly like its predecessor.

If the horse was the key that opened the steppe, the wagon was the conveyer. Shortly after the invention of the wheel, pastoralists would begin elaborating on the cart until they generated a sort of primitive prairie schooner, a massive contrivance rolling on four solid wheels and drawn by teams of oxen, or in the desert, camels. Although material possessions remained austere, vehicular transport enabled pastoralists to carry the necessities for survival—including women and children—in one of the planet's harshest environments.

Men lived on horseback, not only riding from dawn to dusk, but virtually from infancy. Before little boys could walk they were placed on the backs of sheep as a first step toward riding. An older lad would vault on a horse rather than run to complete even the shortest errand. Adult males ate and drank on horseback, held meetings and played games there; it was their central point of reference, the measure of all wealth.

Their stage was vast and pitiless, marked by temperature variations as great as any on earth and a prevailing aridity relieved only intermittently by storms of awesome power. Here pastoral prosperity could be wiped out by a single visitation of the dreaded *jud,* a freeze so intense that it turned miles of rich pasture into wasteland virtually overnight. Since ruminants and horses succumbed far more quickly than humans when deprived of nourishment, herds caught in such a frost frequently died in place before undamaged pasturage could be found. It is easy to see why the speculative nature of shepherding yielded a pronounced insensitivity toward death and a penchant for organized theft.

The quickest and most direct means of rebuilding a herd was by raiding

Bad-Weather Ponies

They were already out there waiting, superbly adapted to the harsh endless grasslands. Built close to the ground, the steppe pony rarely exceeded thirteen hands. Chunky and well insulated, they were heavy in the head and shoulders with full manes and tails. Like the felt tents, or yurts, of their masters, their shaggy coat was a shield against most weather; warm in the winter, cool in summer, and greasy enough to keep out everything but driving wet snow. They were even camouflaged, mustard-colored with black trim; standing still in certain lights they faded into the background at 200 yards. They lived on grass alone, so food was everywhere. In winter they had the ability to nuzzle through deep snow to find forage. No breed of horse could live better off the land.

As mounts, steppe ponies were short-legged but surprisingly swift. Yet their true virtues lay in ruggedness and agility. With a low center of gravity and plenty of muscle, they could dart and turn like a jackrabbit, and recover from a fall with the resilience of a bouncing rubber ball. They were the original cow (and sheep and goat) ponies ideally suited to the professional needs of their riders. They were also the original war horses, for many of the same reasons. Size and strength mattered little to a mounted archer intent only on finding the enemy through all kinds of terrain and weather and then staying out of range.

The steppe pony's sole drawback was endurance, not the capacity to persevere, but simply the proverbial lack of stamina that came with eating only grass. They needed time resting and grazing to recover. Hence, the more remounts the better. This was why wealth on the steppe would always be measured in horses, and why a pastoral army on the move so resembled a roundup.

the four-legged assets of a neighbor—an act guaranteed to stimulate further raiding along with a unique set of military skills. The swirling tactics—the feints, the ruses, and reliance on surprise—all were mainstays of rustling. Steppe aggression spurred a second great pastoralist theme, the urge to move on. The low energy levels of grasslands along with the nature of nomadism meant that human populations here would always be thinly distributed. It also must have stimulated rapid migration into the seemingly endless lands to the east—especially if someone else was giving chase. The presence of the pastoralist Afanasievo culture on the grassy plateaus of the Altai range—virtually at the gates of China—by the middle of the third millennium B.C. indicates that infiltration of the deep steppe was nearly complete.

Soon after pastoralists came to form a thin but contiguous mass, the dynamic of their aggression was transformed. So long as nomadic groups remained tiny disjointed entities, their raids were likely to have been random and self-initiated. But once Inner Asia had been populated, however sparsely, pastoral assaults generated great ripples, as one hard-pressed group attacked and displaced its neighbors, until those at the edge crashed into the agricultural societies of both East and West.

Because these outer pastoralists had been pushed off the steppe and lacked a ready avenue back, their emphasis shifted from raiding to occupation and even conquest. Beginning in the early second millennium B.C., groups like the Kassites and Hittites descended upon various Middle Eastern polities to become their rulers. Their footprint remains faint, however. The very act of leaving the steppe robbed them in short order of their central advantage—the military skills automatically instilled by a life on horseback chasing sheep and stealing them. Meanwhile new rulers, illiterate and devoid of administrative skills, were forced to turn to the same class of bureaucrats whom they had so recently overturned, or become just like them. In either case, within the space of a few generations they found themselves irrelevant. The culture of the steppe was simply too austere to direct the sophisticated agricultural tyrannies that grew up around it.

Nevertheless, the process of raiding and conquest persisted, based on the twin pillars of necessity and opportunity. Despite their fundamental self-sufficiency—pastoral nomads were proverbially covetous of material goods, especially luxury items, and these would act as a magnet drawing them irresistibly into the world of agriculture. Frequently they came to trade, bringing their carpets, and pelts, and horses in numbers unattainable elsewhere. But the temptation simply to take was always present, since they understood their own military superiority, even if the paladins of imperial agriculture had a dangerous tendency to forget.

In combat the dynamic mobility of the steppe trooper enabled nomad commanders to practice battle tactics fundamentally different from anything familiar to agriculture-based infantry. Speed was the basis of all—swiftness of attack, alacrity in retreat, and a lightning volley of arrows cast atop a hapless foe. "Feint was a favorite device," explains steppe historian Stuart Legg. A reckless charge followed almost immediately by a headlong withdrawal over the horizon left the adversary believing the onslaught was finished. An hour or a day might pass. Then, without a hint of warning, another charge of redoubled violence was mounted, followed by a third and a fourth, until the victims were exhausted and ready for annihilation. Whole armies simply disappeared.

From the few survivors, a shroud of myth embraced these ghostly

Portrait of a Horse Archer

The power of the steppe was based on the individual pastoral unit, the man on horseback. By all accounts, he was a unique creation, singular in his abilities and outlandish and terrifying in the eyes of victims, so much so he frequently defied description. Aesthetically, he left much to be desired. Clad shabbily in boots and trousers—both inventions of the steppe—kept supple through liberal portions of left-over butter and grease, he was likely a pungent warrior, especially since he himself never bathed. Upper garments were composed of crudely stitched pelts, valued only for warmth and protection. Strapped to his back was a quiver full of carefully crafted arrows and his formidable bow, both encased against the elements due to their extreme vulnerability to moisture. A well-cast bronze dagger would have completed his personal arsenal, since the steppe's rich copper and tin deposits were exploited almost from the beginning of penetration.

It was horsemanship that set the pastoral trooper apart. Under ordinary circumstances control was exerted by reins attached to a bit—sometimes copper or bronze, but also bone or hemp. Saddles were blankets and hides. There were no stirrups, not before A.D. 500 at the earliest, so balance was based on experience and skill. Over time a horseman's thighs and knees grew so sensitive to his mount's movements that it became possible to maintain a firm seat at full speed using legs alone. The net effect was a union that left some wondering where the man left off and the horse began—the Greeks, for instance, imagined a race of centaurs, wild and unpredictable, humans and equines joined at the hip. Others were less fanciful, but nearly all who crossed his path were amazed by the steppe horseman's ability to let go the reins and launch a rapid-fire barrage of arrows at full gallop through an arc of 270 degrees or more. He was as dangerous in retreat as moving forward—his fabled rearward Parthian shot brought an end to a legion of pursuers. No one was more lethal in the ancient world.

legions. Though often outnumbered, they were frequently described as "hordes" and demons beyond counting. Fantastic rumors proliferated until a climate of disbelief arose and commanders were lured once again into combat and defeat. The sacrifice was unnecessary. Until Genghis Khan's assault on China, nomads lacked siege engines and farmers were safe behind their walls. Yet defeat is often a matter of misunderstanding.

Pastoral nomads constituted military power of a different order, an instrument of violence that transcended (or simply ignored) the most basic assumptions of armies representing settled societies. Agriculture dictated that war among the settled was about territory, but for steppe warriors war was a matter of theft and predation. In a famous passage from Herodotus, when Idanthyrus the Scythian nomad is asked by Darius the invading Persian king why he kept running away, he replied: "If you want to know why I will not fight, I will tell you: in our country there are no towns and no cultivated land; fear of losing which, or seeing it ravaged, might indeed provoke us into hasty battle."[1]

The record of pastoral aggression is fragmented, reflecting the unpredictable nature of militant pastoralism itself. One scenario of aggression from the steppe combines the ripple effect of local disturbances with the emergence of a charismatic leader—both virtually random events—to generate a military explosion. A second milder version has outer steppe dwellers infiltrating surrounding areas and exerting a more gradual military influence, even as they were assimilated and enervated.

China it seems experienced both versions. For several thousand years traditional Chinese culture, including agriculture, writing, pottery, and the production of bronze, coalesced gradually and without evidence of organized violence. Then, sometime after 3000 B.C., the process accelerated, and with it came unmistakable signs of war and with war the beginnings of the dynastic tradition. Most tangibly, towns began building walls around themselves using the laborious hang-t'u, or "stamped earth," technique. Occupants of these sites must have felt seriously threatened. But by whom?

"Each other" was the answer traditionally provided by historians and archaeologists. But why had it taken so long? And why did fortifications appear in the north but not the south? An alternative is provided by the redating of the horse's domestication and the early arrival of the Afanasievo culture in the eastern Altai range; it was pressure from steppe warriors engaged in a pattern of intermittent raids. Not only does the chronology work, but it serves to explain certain peculiarities of Chinese military history. If it is true that a distinctive Chinese culture was already well on the way to formation when the first nomads could have arrived, it follows that their impact would have been limited. In fact, there is little evidence that the development of language, religion, or the familial basis for politics in China was significantly altered. What did change had to do almost exclusively with what nomads did best—fighting, along with its implements and customs.

These effects are encapsulated in the rise of the Shang, the most famous and best documented of the early dynasties, founded according to tradition in 1766 B.C. Echoes of Shang rule are captured by thousands of inscribed oracle bones recovered from the capital An-yang. The record was one of leaders constantly on the move and on the make—inspecting the realm, collecting tribute, rewarding or punishing—a perpetual odyssey of arm twisting and instructive violence.

Yet the shadow of the steppe nomad hung over the whole enterprise. While the Shang military was clearly representative of an agricultural people, their armaments were not. They relied heavily on composite bows similar to those carried by steppe nomads. Even more suggestive is the Shang preoccupation with horses and the chariot as the preferred platform for elite combatants. Like the composite bow, these sophisticated vehicles emerged full-blown; rather than carts or wagons, they were the first Chinese wheeled conveyance.

That steppe nomads might have brought the chariot to China seems implausible. As a weapons platform the agile steppe pony was vastly superior and a great deal easier to acquire. Yet in this case the evidence speaks for itself—rock drawings clearly depicting chariots at six principal sites across the vast expanses of Central Asia. Add to this chariot remains recovered from the steppe along the Russia-Kazakhstan border dated as early as any record of them in the Middle East.

And why not chariots on the steppe? Who would be more likely to appreciate a conveyance that survived on its ability to apotheosize everything equine? Chariots were light and easily towed by a people already dependent on wheeled transport. Nomads would have found them an affordable luxury for leadership transport, racing, and ceremony—not, though, as a serious weapon. This they would leave to those who had either lost their skills on horseback or never had them . . . like the Chinese.

Rock drawings of chariots from sites in Central Asia.

The Shang had many enemies. But their greatest foe was the Ch'iang. Representative of rim land pastoralists to the north and west who had spilled off the steppe and become more or less settled, the Ch'iang were distinguished by the importance of horses to their way of life. Although far removed from the Shang heartland, the Ch'iang employed chariots and were considered extremely threatening. Repeated Shang expeditions were mounted against them—one numbering 13,000. The frequency and duration of these forays indicate less than total military success, but they did generate war prisoners. They were then recruited, apparently to raise horses and drive chariots, or they were ritually killed.

War for the Shang and later dynasties was never heroic in the Western sense. Instead of literary epics and stalwart equivalents of Achilles and Roland, China produced strategy—a body of prescriptive thought that has come down to us in a series of seven military classics, by far the most famous of which is Sun Tzu's *Art of War.* Although the military classics (written for the most part in the fourth century B.C.) crystallized military thinking for the future, they also seemed to reflect an earlier time, when peaceful farmers were first subject to attack. Thus strategy combined the nomad's preference for ruthlessness, deception, and surprise with a distaste for violence exemplified by Sun Tzu's statement "subjugating the enemy's army without fighting is the true pinnacle of excellence."[2]

China was marooned on a treadmill. The contradictions in its military thought were never resolved, and the nomads never disappeared. Around 800 B.C. the threat grew worse. Great ripples from the steppe propelled true horsemen through the outer shell of buffer pastoralists and reintroduced them into Chinese history, this time in sufficient numbers not just to terrorize but to threaten and keep on threatening the political order. Henceforth, Imperial China suffered unnumbered assaults. Periodically whole dynasties were swept away. Yet there was no adequate response, the military advantage of the horse archer was simply too great. So a symbolic defense was mounted.

There is a substantial body of myth associated with Chinese barrier defenses. Today's Great Wall, for instance, cannot be seen from the moon. Nor were earlier linear fortifications its direct ancestors. But it is true that Imperial China expended huge amounts of energy building them—Ch'in Shih-huang-ti's wall was said to have required a force of 300,000 laborers. Nor is there any doubt that they were primarily devoted to keeping steppe

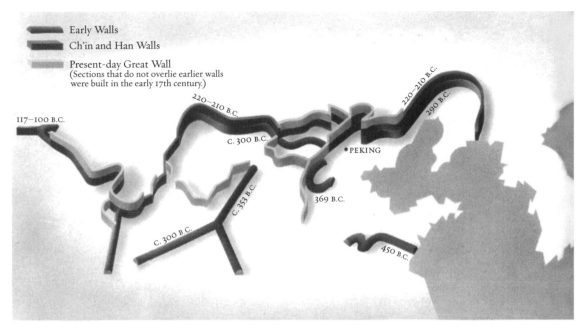

117–100 B.C.

220–210 B.C.

C. 300 B.C.

220–210 B.C.

290 B.C.

•PEKING

C. 353 B.C.

369 B.C.

C. 300 B.C.

450 B.C.

Today's Great Wall and its predecessors from antiquity.

nomads out. They simply failed. To be effective, barrier fortifications had to be effectively manned, and China could never man a line that stretched hundreds and even thousands of miles. So the walls became strategic sieves.

So why waste the effort? Because walls stood for something. If circumvallation brought a measure of safety from rampaging pastoralists to agricultural villages, then symbolically these walls could do the same for all of China. They constituted tangible evidence of China's will to resist and draw a line between itself and those who roamed outside. Unfortunately horse archers remained unimpressed.

The alternative to macro-circumvallation—creating an equivalent force of horse archers—met an equally dismal, but even more ironic, fate. On several occasions such forces were trained, equipped, and even had some success fighting the real thing. Yet the resource drain proved ruinous, and the Chinese version of horse archers had a persistent tendency to desert and become steppe nomads themselves!

In the end China could do little but persevere. Incursions from the steppe were turned away when possible and outlasted if necessary. Even if the worst came to pass and conquest ensued, the Chinese learned that soon enough the new boss would become just like the old boss. Much the same happened far to the west.

After the initial incursions by Kassites and Hittites early in the second millennium B.C., infiltration into the Middle East by former steppe dwellers settled into a pattern. They mixed with locals, losing their riding skills though not their ability to raise horses and employ them with chariots. Two such groups were probably the Mittanni and in particular the Aramaeans, who moved into lands controlled by the Assyrians and began causing trouble during the reign of Tiglath Pileser I (1115–1077 B.C.).

Approximately four centuries later, just as it was happening to China, the Inner Asian ripple effect became much stronger and began thrusting whole groups suddenly and intact off the steppe. Once loose in the agricultural Middle East they proved impossible to eliminate. The Cimmerians were first. They quickly fanned out into the interior of Asia Minor, where they wreaked havoc for at least the next half century, overrunning first Phrygia in 676 B.C. and then Lydia in 651 B.C., killing its king, Gyges. Ever aggressive Assyrian overlords proved unable to do much. Armies such as theirs were simply too plodding for an enemy that could outrange and outrun them in virtually every circumstance. The best the Assyrians could do was a combination of diplomacy and slow reconquest of areas the Cimmerians had already swept through. It must have been like boxing shadows. To make matters worse, the next people to arrive, the Scythians, were even fiercer . . . terminally so. In 612 B.C. they likely joined in sacking Nineveh, the Assyrian capital.

Nearly six centuries later, Crassus came to Syria hungry for a victory. He was the richest man in Rome, but the other two members of the First Triumvirate, Caesar and Pompey, were great soldiers with fresh triumphs. Crassus had only defeated an army of slaves led by the gladiator Spartacus. It would not do. So, without permission from Rome and without provocation, he marched on the Parthians with seven legions and 4,000 horses.

The Parthians were steppe dwellers who had come to rule substantial portions of what had once been the Persian Empire and later the Macedonian-led Seleucid kingdom. But they remained aloof and avoided enervation. They employed Greek functionaries to administer their possessions; but they continued to speak their own language (Pahlavi), and left no written literature, or art, or architecture. In doing so they retained the essence of their military skills: they were still lethal horse archers.

As he drew his army out of winter quarters, Crassus was warned that the Parthians "had a new and strange sort of darts, as swift as sight, for they

Herodotus and the Scythians

Archaeological finds are quickly adding to our knowledge of ancient horse nomads. But the words of Herodotus of Halicarnassus bring them alive. As a young man he traveled to the Black Sea and far up its rivers, where he could observe them directly. By the time Herodotus got to them, the Scythians had been in and around the littoral for upward of two centuries. Some had settled down and become agricultural, but those inland remained nomadic and lethal. His view of their arrival from the deep steppe was clear: they were chasing the Cimmerians and being chased in turn by the Massagetae . . . the ripple effect in microcosm. Herodotus's observations on Scythian living conditions were acute. His comments on their metalworking skill, their penchant for horse and human sacrifice, the use of skulls as drinking vessels, the nature of their tombs, the relative equality between men and women, the smoking of hemp, and a host of other details have been confirmed by modern scholarship.

He also understood their military power: "The Scythians . . . have managed one thing, and that the most important in human affairs, better than anyone on the face of the earth: I mean their own preservation. For such is their manner of life that no one who invades their country can escape destruction. . . . A people without fortified towns, living, as the Scythians do, in wagons, which they take with them wherever they go, accustomed, one and all, to fight on horseback with bows and arrows, and dependent for their food not upon agriculture but upon their cattle: how can such a people fail to defeat the attempt of an invader not only to subdue them, but even to make contact with them?"[3]

pierced whatever they met, before you could see who threw them."[4] The King of Armenia suggested that he proceed through the mountains to reduce the advantage of cavalry. Crassus paid no attention. Instead he was drawn deep into the arid plains by his retreating foe, until he was near the town of Carrhae. Here the Parthians took a stand . . . of sorts. They approached to within fifty yards of the Romans and began letting loose with their composite bows. At this range their "darts" were devastating, pinning the Romans' stout shields to their forearms, even nailing their feet to the ground. They continued charging and retreating, shooting continually, never letting the Romans near them. The disciplined legionaries formed a hollow square and waited for the Parthians to run out of arrows. They never did. The Parthian general, known to us simply by his clan name, the Suren, had arranged for a caravan of camels to resupply his men with ammunition. Finally, in desperation Crassus sent out his son Publius with eight cohorts—500 archers, and 1,300 cavalry sent from Gaul. Within hours Publius returned—or at least his head did, impaled on a pike by the mocking Parthians. Out of an army of 43,000, only 10,000 escaped. Crassus was killed during a parlay, his head cut off as a present to the King of Parthia.

The Death of Publius and Birth of the Cataphract

Nearing the end of his death ride, the younger Crassus must have noticed that a number of Parthians looked different than their waspish archer compatriots. Mounted on huge stallions, both these riders and their horses were covered with coats of chain mail or iron scales. Rather than a bow, their primary weapon was a lance so long the Greeks called it a *kontos*, or barge pole. Charging directly at the Romans, these kontos-wielding armored tuskers took advantage of their height and at least part of their momentum to impale multiple victims. The desperate Romans had some success by swarming them with swordsmen. But the combination of horse archers plus these armored rams was simply too much to survive.

The appearance of these cataphracts (from the Greek "covered over") marked the emergence of a fundamentally new and durable weapon system. Like most things having to do with horses, their origins were on the steppe, where certain groups began experimenting with coats of mail for protection. Sarmatians first used thinly sliced horse hooves, but metal was better. Because horses were big targets, it also made sense to extend the coverage to them. The result, however, was one weighed-down pony. The Parthians solved the problem by taking over the Nesaean stud, the huge imperial herds of Persia's rulers. Kept in Media, horses here were carefully bred to combine speed with size and strength—enormous chests, barrels, and haunches but fine lower quarters and feet . . . Clydesdales in track shoes.

The Parthians made good use of them, organizing their nobles into a force built around the great steeds. Their size, speed, and resulting momentum naturally suggested a role as an instrument of collision and the lance as its essential agent. It proved an enduring combination. One key problem remained: keeping horse and rider attached at the moment of impact. Like nomads everywhere, Parthian cataphracts rode virtually bareback. To be fully effective this would have to change.

Rome, like China, was subjected to wave after wave of barbarian attacks, all of them ultimately driven by pressure from Inner Asia. As elsewhere, the earliest of these migratory elements probably had no direct contact with actual steppe culture; but as time passed the threat became increasingly more mobile, until actual Central Asian horse archers, the Huns, burst into the empire.

To deal with these successive invaders, after A.D. 250 Rome placed increasing emphasis on a cavalry-heavy mobile strike force. Given the Western animus against the bow and the fact that the initial cavalry threat was spear-based, Romans naturally turned to the kontos-cataphract combination. With time, protection grew heavier until the Emperor Julian (circa A.D. 360) described cataphracts as looking like "glittering statues." The lance was gripped with one hand and steadied by the other, which also held the reins. A definite saddle had evolved, though not built up at front or rear, and most importantly not including stirrups. Without a firmer seat the mounted lancer's true potential as shock cavalry could never be realized. Instead, he was forced to temporize, carefully thrusting, ever mindful of being thrown to the ground where the weight of armor would render him nearly helpless. Rome may have fallen anyway, but at this point the cataphract was in no position to save her.

Rome had been crushed into a thousand pieces, and the political edifice that grew up in her place, the Carolingian Empire, was little more than a facade of rubble powerless to defend against a string of horrific and unpredictable raids by seaborne Vikings. In an attempt to generate a measure of responsiveness and mobility, Charles the Bald in A.D. 864 ordered every Frank who owned a horse to serve mounted, thereby institutionalizing heavy cavalry.

But as more and more of the initiative for defense slipped into local hands, fortifications sprang up, at first throughout rural France and later practically everywhere in Europe. As the sole refuge for people terrorized by the latest wave of marauders, these nascent castles were a good investment, enabling their masters to exact a growing share of the increasing agricultural yield. This formalized the manorial system and it supported a new class of equestrian warriors. The power of castles was defined by the range of the heavy cavalry within—about ten miles if you wanted to be home by dark.

Thus cavalry and castles froze the diffusion of power across the countryside. If charging knights ruled the field, they could not reduce a castle. A defeated, outnumbered, or otherwise hard-pressed vassal could usually take refuge behind his walls and wait out external authority. Politically it spelled the collapse of central government.

For Want of a Stirrup...

Sometime around A.D. 500, horse nomads in Inner Asia began experimenting with foot loops, contrivances that quickly developed into true stirrups. Not only did they improve control of their mounts, but stirrups allowed archers to stand while shooting, further absorbing the motion of the horse and rendering them still more accurate and lethal. Yet the ultimate significance of stirrups had more to do with the lance than the bow.

In the early 1960s, a period when military technology was the focus of strategic thinking, historian Lynn White built a considerable reputation maintaining that the use of stirrups by Charles Martel's Frankish warriors—who traditionally fought on foot—enabled them to defeat the more lightly armed Islamic cavalry at the supposedly critical battle of Tours in A.D. 733. He further theorized that this victory marked the beginnings of feudalism in Europe. More recently it has been shown conclusively that the Franks did not yet have the stirrup at the time of Tours, nor did feudalism take its final form until fifty years after the death of Charles Martel's grandson, Charlemagne, in A.D. 814. This has led some to dismiss entirely the validity of White's thesis.

This is a mistake. The cataphract of the late Roman Empire, just a few centuries before, incorporated all the essentials of medieval shock cavalry save one—the stirrup. It was this device that provided lateral stability. Prior to the stirrup's introduction, any blow to the right or left was likely to unhorse a lancer. Of equal importance, stirrups allowed a rider to brace himself at the moment of impact and to couch the lance, relying on the horse's full momentum to deliver the blow. Further improvements—the deep-seated saddle with built-up cantles, even means of locking the lance in place—virtually welded the rider to his horse. But it was the stirrup that transformed cataphract into knight, and thereby changed the course of military history in Europe.

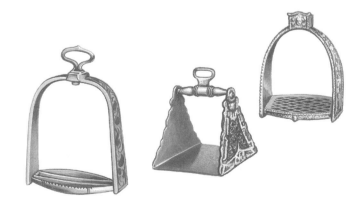

The crucial stirrup:
some examples.

Prior to modern times the relationship among weapons, styles of warfare, and politics was particularly intimate, and feudalism provided almost a textbook example. War, in the absence of centralized and ambitious states, cast aside much of its predatorial flavor and took on more of the characteristics of intraspecific aggression.

Fighting became a matter of class. The aristocracy monopolized the right to bear arms, and henceforth elite soldiers were knights. Membership, though increasingly hereditary at the upper reaches of the nobility, was primarily defined by the ability to fight as shock cavalry.

From the greatest lords to minor henchmen, all shared the socialization process surrounding this most demanding form of combat. Beginning at puberty and often entailing long residence and training among peers in the households of the great lords, the experience generated a corporate self-consciousness confident in its solidarity and reveling in its traditions—a cosmopolitan fraternity of arms destined to dominate European military history up through the early stages of the Industrial Revolution. Underlying it all was the importance of individual prowess.

There were numerous instances of large-scale battles and disciplined maneuvers by squadrons of knights, but the emotional focus of battle drew down to the level of the single combatant. As a motive for violence political issues never entirely faded, but the fragmentation of power encouraged the usurpation by other values. Gradually fighting became a formalized ceremony for deciding dominance, remarkably similar to analogous rituals among other animals, particularly mammals.

Tournaments appeared—elaborate and deadly games of war to which the knightly class became addicted beginning in the early twelfth century. Vassalage and later primogeniture yielded a surplus of younger sons, who were left with the choice of the priesthood or a life of adventure among wandering bands of peers, the so-called knights errant. Given the nature of genetic strategies, many chose the latter path. "These adventures," historian Georges Duby argues, "were also revealed as quests for wives, perhaps first and foremost. . . . All *jeuvenes* were on the lookout for an heiress."[5] The risks of the knightly life in general insured a steady supply of young widows, and marriage to one was for a knight errant the surest path to support and status. But there was bound to be competition, and given the importance of prowess at arms, the development of tournaments where fighting skills could be tested and observed was entirely logical.

The redoubtable William Marshal was a testimony to the ideal of the knight errant. He was literally a product of the tourney. The younger son of a minor vassal but a formidable fighter, Marshal spent fifteen years wander-

Training devices and knights jousting in a tournament.

ing from event to event, building a near legendary reputation that earned him an heiress, great estates, and eventually the regency of England—the medieval dream.

War provided similar opportunities, but they were less observable. Besides, even participants might have had trouble telling the two forms of combat apart, so laden were they with intraspecific ritual. The very nature of jousting or war making—high-speed, head-on collision with protruding tusklike lances—invites comparison with the spectacular ceremonial matches of a number of hoofed animals—all nonpredators. Combat was individualized and hemmed in by an elaborate web of submission-related rules, the essence of which has come to be known as chivalry. When Richard *Coeur de Lion,* without hauberk or mail, was assailed by William Marshal, a vassal of his enemy Henry Plantagenet, he called out: "By the legs of God Marshal do not kill me. That would not be right for I am unarmed."[6] Marshal, grumbling, lowered his lance and merely ran through Richard's horse.

Fighting could still be deadly but killing had become a secondary objective. And the state of arms reinforced this situation. Although the mounted lancer possessed extraordinary offensive power, he was also extremely well protected and his kite-shaped shield deflected most opposing lances. Knights occasionally turned on the lightly armed infantry of the day with deadly effect. Yet real combat—the combat that mattered in the eyes of the participants—was fought between equals in arms as well as class. Marshal did not so much spare Richard's life as refuse to fight him without the proper equipment.

*Examples of medieval
mail tunics.*

The armaments of shock cavalry proved highly durable, dominating Europe until well into the fourteenth century. Deadly to outsiders and those below, they were eminently usable among the ruling classes—ideal instruments of the status quo. Like weapons in nature optimized to enforce dominance, the appliances of heavy cavalry were impressive visually, relying on a heavy measure of implied threat. The Middle Ages was a time laden with symbolism, and the charging armored juggernaut and crenelated battlements of the medieval castle were images so strong that they would endure to condition weapons choice right into the industrial era.

Nevertheless, the medieval system was too dependent on shock cavalry. Like the Greek phalanx, it was certainly a potent instrument of coercion, but one specialized to a degree that it became self-referential. Well before the emergence of the gun, there were indications of fragility. One was the reappearance of the crossbow in the West. Unlike the Greeks who suppressed it, Europeans found it extremely useful in the Middle East. This was especially evident during the Third Crusade when it became a key element of the sophisticated infantry-cavalry teams crafted by Richard *Coeur de Lion* to repel the hit-and-run tactics of Islamic forces. Yet the implications of the crossbow were threatening to the knights themselves. Unlike the reflex bow, it could be shot accu-

*Medieval helmets
for heavy cavalry.*

*The crossbow —
so lethal it
provoked the first
formal attempt
at arms control.*

rately with very little training. Even worse, after 1100, when the square headed, armor-piercing quarrel became available, crossbows were being used in Europe by common foot soldiers to shoot knights out of the saddle at ranges of up to a hundred yards.

Appropriate to an age dominated by religion, the Church stepped into the breach. In 1139 the second Ecumenical Lateran Council outlawed the use of the crossbow among Christians. As humankind's first formal attempt at arms control, the edict deserves more attention than it usually receives. Its effort to enforce weapons symmetry and the technological status quo proved typical of arms control in the future. There was also an element of cynicism, or at least self-interest, in that the council's edict in no way discouraged the use of crossbows against Muslims.

This also appears to have been a key to the feudal nobility's open-mindedness to weapons-related innovations while fighting in the Holy Lands. Given the apparent bipolarity of aggression, it was not surprising that the decorum of war and the aversion to weapons deemed less than honorific were sharply reduced when fighting against infidels deemed less than human. Unfortunately for the flower of knighthood, this proposition was reversed when true horse archers from the steppe injected themselves onto the battlefields of Europe one last time.

In 1222 the Queen of Georgia dispatched an urgent letter to the Pope: "A savage people hellish of aspect, as voracious as wolves and as brave as lions, have invaded my country. The brave knighthood of Georgia has hunted them out."[7] As she wrote, her knights were being shot to pieces by the mysterious invaders, who then simply vanished into the Caucasus. Gradually it was learned they had been on their way to Russia, where they would destroy an army led by the Prince of Kiev, only to disappear again for fifteen years. "Who were they?" wondered settled folk from Constantinople to Rome.

They were in fact Mongols, a reconnaissance party led by Subotei, marshal and chief advisor to Genghis Khan, the apocalyptic leader who had succeeded in uniting "all who dwelt in tents of felt," and then turning them on China and the Khwarizmian kingdom of Persia, to amass an unbroken string of victories and corpses numbering in the millions. "They are the soldiers of the Uruds and Manguds," warned one nomad who had seen them fight. "They pursue men like game."[8] As a killing machine and instrument of conquest the Mongols and their allies remained unsurpassed until the maturation of the gun. They were the final iteration of steppe warfare; built around horse archers, they were supplemented by a cadre of mobile cataphracts, and had picked up a siege train in China, enabling them finally to overwhelm fortified sanctuaries. Their wild swirling tactics were no longer even improvised; all were meticulously disciplined and minutely organized. Nothing could have stopped them.

Yet medieval Europe had only the slightest conception of its true peril. Rumors persisted that the Mongol expansion was really Prester John, the legendary Christian imperator of the Far East, on the march to rescue the Frankish kingdoms of the Holy Land from the resurgent forces of Islam. When the Mongols materialized once more, moving across the Volga in December 1237, they were again led by Subotei and it quickly became apparent they were anything but redeemers.

In short order they sacked Moscow, Susdal, and Vladimir, followed by Yaroslav and Tver, at which point they moved into the open plains of Ukraine and smashed Kiev. Poland was next. On Palm Sunday 1241, Kaidu, a young Mongol prince, destroyed the lovely city of Kraców, and then moved to Liegnitz, where Duke Henry of Silesia and the Teutonic Order had collected a heterogeneous force of around 25,000 Polish and German knights to bar the way into the Holy Roman Empire.

Coming upon what appeared to be a ragged band of nomads bent on a hasty retreat, Henry and his heavy cavalry gave chase in open formation. Suddenly the rabble on ponyback turned in unison and let loose a cloud of armor-piercing arrows, followed by a devastating coordinated lancer charge. The Europeans broke almost immediately. Before nightfall Henry's head graced a pike and the Mongols had collected nine sackfuls of right ears from dead Templars, Hospitallers, and Teutonic knights.

Meanwhile Subotei had invaded Hungary with three widely separated columns, which then converged on Budapest, where King Bela IV had concentrated nearly 100,000 defenders. But instead of investing the city, Subotei began a slow retreat, luring Bela and his forces after him. Six days later, at Mohi on the Sajo River, the Mongols turned on their pursuers, pinning Bela down, then surrounding him, and finally leaving a small corridor for survivors to escape. These were run down and killed at a total cost of 65,000 men.

On Christmas Day 1241, Mongols rode across the frozen Danube and seemed to be headed toward Vienna. Soon after another column reached the Dalmatian coast, and still another was within sixty miles of Venice. There was nothing to stop them, the rape of Italy and Germany seemed about to commence. Then word of dynastic quarrels far to the east brought relief; as enigmatic as ever, in the spring of 1242, the nomad thrust melted away with the snow, never to return. Chivalry had put up a defense of sorts, but in the end the forces of agriculture were still no match for militant pastoralism and were reduced to the simple prayer: "From the wrath of the Tartars, Lord deliver us."

Chapter 5

IMPERIAL TREADMILL

City-states metastasized into empires. Earlier we left the process of governance still in an adolescent state of tension between the demands of the ruler and those of the ruled, so as to examine the possibilities of the phalanx. At this stage war between independent city-states like those of Sumer was pursued essentially as a means of balancing their relative power. Yet neither domestically nor internationally was the system anything more than metastable.

On the home front, the vast amount of drudgery required to keep the agricultural system going, combined with periodic bouts of famine and epidemic disease, favored an increasingly more centralized mechanism of control—bureaucratic hierarchies, state religions, despots, and, propping up the whole tottering structure, armies. When famine struck, the stores of others could be taken. When pestilence reigned, fresh laborers and breeders could be captured. Alternately, when populations rose, new territory could be conquered, or, in the case of defeat, excess soldiers eliminated. Warfare was never a precise equilibrator, nor was it necessarily pursued with these ends in mind. They simply set the underlying conditions that made aggression appear attractive to those in charge.

While balance-of-power systems typically produced limited results, this was seldom the objective of the respective players. Desired solutions characteristically took on hegemonic overtones, thwarted only by the rapid formation of countervailing alliances. Yet the net effect was little praised or apparently even understood. It worked in spite of the participants, playing on their foibles, but ever vulnerable to a master of the game.

Historically, this was first recorded in the rise and triumph of Sargon the Akkadian. He was born humbly, the illegitimate son of a Semite girl who abandoned him like Moses in a cradle of reeds, only to be fished out of the Euphrates and raised in Kish, where he rose quickly to the post of royal cupbearer. Sumerians might have been more wary of the foundling.

Upon gaining control of Kish, Sargon marched on Uruk, where he defeated its ambitious king, and brought him back in a dog collar. This was insulting, but likely of far greater consequence to the citizens of Gilgamesh's ancient home, he tore down their walls. This he did repeatedly as he picked off city after city—Ur, E-Ninmar, Lagash, Umma—until the entire alluvial plain was his. Walls were the bedrock of the balance, a last refuge in defeat. By removing them Sargon signaled that the rules had changed and also the game.

It was the middle of the twenty-fourth-century B.C., the dawn of a new day in Mesopotamia. Sargon moved to implement a blueprint for tyranny that would provide the basic floor plan for empires across the ages. His agents fanned out across Sumer, framing the structure with tax lists, garrisons, trustworthy locals, and royal governors. Yet all beams and joists pointed inward toward his capital, Agade, where he stayed, surrounded by royal officials.

More critical was the standing army that he kept at his side—inscriptions report that 5,400 soldiers ate daily at the huge palace complex. Gone was the concept of warriors-by-consensus, replaced by an army of retainers beholden to no one but Sargon. Having stripped Mesopotamia of its local defenses, keeping his empire intact required that he pacify the hinterlands, basically a never-ending task. There was more to it than simply securing borders and ridding the area of potential marauders; Sargon waged an extended campaign of extortion that stood as a model for future tyrants. For a time the uplands were safe enough to float barges heavy with silver, cedar, and other goods down to Agade. Yet appearances can be illusory, and few more so than the stability of empire.

In old age, after decades of consolidation, Sargon discovered his vassals suddenly in revolt and Agade surrounded. He crushed them. Still there would be other rebellions to put down. He died counting thirty-four victorious campaigns to his credit, but never having found sustained peace. Things went little better for his heirs. Rimush, his own son, was greeted with general revolt, and spent the next nine years reconstituting his father's empire, before falling prey to a palace conspiracy. The grandson, Naram-Sin (he liked to call himself "King of the Universe"), filled sixty-four years on the throne with almost nothing but military campaigns, all on the

Naram-Sin and the Composite Bow

There is a famous stele, now in the Louvre, showing Naram-Sin exactly as he would have wanted to be remembered. Sporting a horned helmet, he stood atop a pile of his fallen foes, an arrow protruding grotesquely from the neck of one of them. Leaving little doubt as to how it got there, Naram-Sin was armed with a composite bow, the first known depiction of such a weapon in the Middle East.

His choice of arms represented a significant departure from Sumerian warriors, all of whom were shown equipped for close combat. Ancient weapons authority Yigael Yadin speculated that the Akkadians relied heavily on such bows. Unlike the phalangites of Sumer, rank-and-file infantry of despotic empires dependent on irrigated agriculture were proverbially unwilling to fight at close quarters. Their stake in society being marginal, they would fight only on the margins. Such troops had to be provided with an appropriate long-range weapon—in better armies a powerful composite bow, but frequently a loosely strung device more appropriate against small game than well-armed and armored aristocratic warriors. Battle among the forces of despotism was anything but a level playing field.

The Stele of Naram-Sin.

periphery. His son and successor would hang on for a few years before disappearing along with Agade and the political structure Sargon meant to last forever. Yet the world's first emperor had created a lasting if evanescent monument to his life's work, a despotic apparatus so logical and compelling that it would reinvent itself with only moderate variation across the broad face of time and geography.

The legions of tyranny were models of the societies they represented—portable pyramids of power with the minority at the top cornering every advantage. This had substantial implications for an ordinary combatant's chances of survival. Skeletal evidence from both the Old and the New Worlds demonstrates that high-ranking individuals were markedly taller (perhaps six inches) and more robust than members of the lower tiers. Bringing to mind the larger and better-armed "soldier" class of certain ant species, these human analogues—almost certainly the product of better nourishment, particularly access to meat—would have had a similar advantage in battle. Superior strength and vigor was likely complemented by a significant edge psychologically. Epics all used imagery emphasizing the size, strength, aggressiveness, and terrifying demeanor of elite warriors to make the point that lower-class combatants stood little chance against them. Further compounding their advantage, elite fighters in most imperial armies had much better weapons and protection than common soldiers, since among the leadership's key prerogatives was access to metal. In close combat this was critical. Unlike stone or wood it was unlikely to shatter or break at a critical juncture.

Copper was the first metal regularly exploited by humans, smelted far back into the Neolithic. For use in large tools and weapons, it was characteristically cast as a substitute for stone in ax and mace heads along with dagger blades. Yet its softness precluded much more in the way of new types of arms.

This changed dramatically with the discovery that copper could be combined effectively with arsenic or tin to produce a far harder but still ductile alloy, bronze. Not only could it be cast into the most complex shapes, but after cold-working, yielded weapons of a hardness and tensile strength rivaling those of iron, until Roman times, when tempering came to be understood. The toughness and ductility of bronze made it possible for the dagger form to be stretched to generate a true sword by the middle of the third millennium B.C. Such an instrument, by virtue of its superior reach, maneuverability, and capacity to inflict both slashing and puncture wounds, was ideal for the kind of close combat that was the specialty of the elite warrior class. Bronze also substantially increased the penetration of holdovers like the spear, arrow, and battle-ax, which, in combination with the sword, rather quickly brought forth defensive reciprocals in bronze and bronze-reinforced helmets, shields, cuirasses, and greaves, to produce a metal-clad combatant largely immune to any but similarly accessorized adversaries.

War's appetite for bronze fed on itself, further encouraging political centralization and the dominance of military elites intent upon controlling the sources of supply. Deposits of tin, in particular, were scattered and relatively difficult to extract. Literary allusions and other records from the Bronze Age make it clear that the metal and its constituents remained valued, rare, and monopolized by those in control.

Iron changed things somewhat. Anatolian armorers had experimented with the metal, probably derived from meteoric deposits, to produce blades as far back as 2500 B.C. Iron weapons were tough and held an edge, but were subject to rapid and continuous deterioration through rust. Rather than superiority, its large-scale use was driven more by the relative abundance of ferrous deposits. Once the higher heats required for extracting terrestrial iron were mastered, it could be produced in quantities necessary to begin to provide whole armies with at least some metallic implements.

But bronze had staying power. Because it was only minimally affected by corrosion, it could be used repeatedly and for different purposes simply by melting and recasting. Iron emerged as a red-hot pasty ball that had to be worked by hammering, rather than as a liquid that could be poured into molds. Shaping remained difficult and labor-consumptive. Gradually, a better understanding of metallurgy—basically tempering and the development of steel—led to an increased reliance on ferrous-based weapons, particularly during the Middle Ages.[1]

Yet in the age named for bronze there was no question who and what ruled the battlefield. Here there were two classes of opponents: a relative few wielding and protected by bronze, and the nonmetallic masses, essentially designated victims. Although armies ranged from the low thousands up to around 20,000, battles could be and were decided by several hundred elite fighters. A combat environment in the Bronze Age typically consisted of opposing lines of archers (often supplemented by slingers and javelin throwers) exchanging desultory fire, while champions on each side sought each other out to wage what amounted to individual combat. Yet prior to and particularly after the leadership fought each other, common soldiers were subject to promiscuous killing.

While the symmetrical matching of archers and of champions took place under fairly strict rules of engagement, this was hardly the case when members of the elite confronted social inferiors. Common soldiers were treated virtually as prey—killed in the most cold-blooded fashion. In epics this often took the form of a martial warm-up, a prelude to a meeting of major combatants. Of more historical significance were the bloodlettings that transpired after elite confrontations. There are numerous recorded examples of such slaughters, much of which occurred during the "pursuit" phase of battle. Unlike defeated aristocrats who retained ransom value, members of the losing rank and file had few bargaining chips in the eyes of the victors. If captured and it was convenient, they could be recruited or enslaved. But it frequently made more sense to just catch and kill them, especially since their aristocratic tormentors were equipped with the ideal means of running them down. Chariots may have been invented on the steppe, but they epitomized war in the sphere of imperial agriculture and also its tactical limitations.

Arrogant Chariot

Picture a crowd of agricultural laborers recently turned soldiers, facing the charge of a thunderous line of chariots. Consider the visceral tug-of-war between fight and flight as the chariots drew close and the horses' pitiless hooves threatened. Was it any wonder that so many turned and ran? And once they did they were helpless. A relentless killer of men, this was the chariot of mythic memory. Among Achaean Greeks, Assyrians, Egyptians, Vedic Indians, and Chinese of the Shang and Chou dynasties—virtually across the gamut of hydraulic tyrannies—the horse-drawn chariot marked the focal point of aristocratic warfare. For a weapon's ability to inspire fear will always be at the core of its existence. Yet throughout military history, the actual effectiveness of major weapons systems was often at odds with the impression they left.

Detailed iconography along with actual examples recovered from tombs have made it possible to construct accurate replicas of ancient chariots. They proved rickety vehicles of suspect stability, barely controllable on all but the smoothest ground. In confrontations between aristocratic fighters, a close reading of epic literature reveals that chariots were used less as battle platforms than as battle taxis ferrying champions to the scene of combat. For those subsequently indisposed they were also a handy means of escape . . . and, when it came to commoners, for preventing escape.

When they ran, the real work began. Egyptian and Assyrian reliefs typically show charioteers racing ahead, furiously bearing down on intended victims with bows resolutely drawn. But a more leisurely rendition seems probable. How was a warrior expected to let go with both hands, while standing on an erratically bouncing platform, and shoot with any degree of accuracy? The chase was prosecuted with a series of starts and stops—drawing up to a victim and either shooting or simply stabbing him. In this manner the slaughter must have proceeded until the field was strewn with an impressive number of bodies. Win or lose, the social order was preserved.

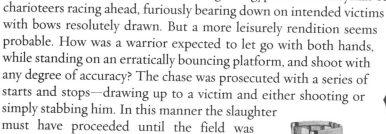

An Egyptian chariot—better at inducing terror than casualties.

Beyond considerations of how the odds were stacked, there was another dimension of lethality that helped explain why ancient warfare was such an engine of mortality. From an actuarial perspective a number of things could happen to a soldier caught up in a military campaign: he could get sick; he could be injured; or he could be wounded in battle. In all cases there was little that could be done for him. The elite were better treated, but they took more chances.

An ancient army was a medical disaster in progress. Until the introduction of modern military health care in the early twentieth century, far more soldiers died of contagion then were ever felled by weapons. A military force was almost an ideal incubator of pestilence—a large compact mass of hosts under nearly continual stress, steadily infecting and reinfecting itself and also the enemy. Repeatedly, the mere introduction of armies appears to have had a catastrophic effect on previously unexposed populations. And the process could be reversed, especially given the run-down condition of the aggressors. Marches were not conducted at the leisurely pace of lightly burdened hunter-gatherers, but at a forced pace of around ten miles a day by troops loaded down with weapons, equipment, and supplies. The result was a campaigner virtually programmed to get sick.

Immobility was even worse. Outside of the Romans and Egyptians, it was a rare ancient army that paid any attention to even rudimentary sanitation while camped. At least camps were transient nests of pestilence. Infection would grow into a certainty with the spread of siege warfare, which often devolved into a contest to see which side got sicker faster.

Of course, the point of an army was not to wait for opponents to die. Ancient armies were made up of sharp or otherwise harmful implements. In modern times there have been numerous experiments with replica weapons, aimed at determining their actual lethality. Most passed the test easily; even the hapless chariot was capable of killing under the right circumstances. This is fully supported by the ancient evidence, scattered though it is. One Egyptian mass grave from around 2000 B.C. yielded sixty bodies sufficiently preserved to reveal gaping wounds, arrows still lodged, and severe injuries from maces. Martial epics also provide a statistical glimpse at wound profiles. Of the 147 wounds described in the *Iliad*, all of those to the head and 77.6 percent of the total were lethal, swords and spears being the prime killers.

For those who survived the immediate effects of an injury, the chances of subsequent infection and death remained substantial, even with a minor wound. There is good evidence that two of the major causes of such mortality, gas gangrene and tetanus, were present as far back as the Bronze Age. Although the ancients did have some effective techniques for dressing wounds—honey and lint, for instance—other practices such as packing them in dung and binding them tightly enough to cut off the blood supply were more dangerous than the original injuries. Finally, it is apparent that in most ancient armies combat surgeons were so few that only elite warriors received any care at all. While this may have been a blessing for many of the untreated, the general profile of health and trauma care, wounds, and differential armaments leaves the indelible impression of vast human wastage.

So why were masses of weakly motivated and marginally effective combatants included in such armies? The psychological impact of large numbers must have been viewed as a significant intimidator. But there were deeper motives at work, ones that strike at the heart of war's role in societies living at the outer edges of possibility. Armed aggression could gain more land in times of overpopulation, or, if defeat resulted, eliminate excess mouths to feed. These ends were logically addressed through large numbers of basically expendable soldiers. They became the mediums of exchange and battles were key mechanisms for transferring energy from one political entity to another.

There were other factors. The greed and ambition of rulers plainly had an impact, particularly in the timing and frequency of wars. The simple fact of invasion left little in the way of options to the masses, except to fight the invaders any way they could.

Yet below it all the basic demographic equation continued to grind away, though with considerable variability. At one end of the scale were the forces of classic Oriental despotisms, whose tolerance for casualties and general military fecklessness provided the purest examples of the phenomenon. The attitude was reflected in armaments. Egypt, shielded by desert and long free of invasion until the incursions of the Hyksos around 1730 B.C., fielded troops with notably crude weaponry. As late as the Middle Kingdom, armament consisted of little more than flint-tipped spears, axes, lightly strung bows, and, discounting an occasional shield or helmet, no protection at all. Even after becoming caught up in the Darwinian environment of Middle East power politics and belatedly developing a chariot force, little was done for the infantry.

In a similar vein the Persian Empire's second invasion of Greece under Xerxes provides another example of military wastage and the logic behind it. While the Persian force clearly numbered far below Herodotus's estimate of 1.7 million, it was by all accounts huge and largely made up of troops armed far below the standards of the Greek hoplites they would have to face. At Thermopylae, even after the force of just 300 Spartans had been hopelessly outflanked, Xerxes' troops had to be whipped into battle. Then, after the decisive naval defeat at Salamis, Xerxes simply withdrew, abandoning a force of at least 100,000, which would be annihilated at Plataea the next year.

It had been a march to nowhere. Militarily, Xerxes would have done just as well by limiting his land forces to the Ionian Greek hoplites who were under his control, along with elite elements of the Persian cavalry. The rest were spear fodder. Worse still, Xerxes and his kin provoked the undying enmity of the Greeks, a hatred that would help to propel Alexander's invasion and conquest of Persia a century and a half later. Once again, all the Great King provided in the way of defense was a huge force structure and the blood of innocents.

Fate was kinder to ancient India, but not to its foot soldiers. Very little is known about ancient Harappan civilization along the Indus River, but the apparent lack of city walls indicates that it was not highly militarized. Its successor, based on Indo-European-speaking Aryan culture, left a rich literary legacy in religious texts known as the *Vedas* and the *Mahabharata* epic. In these and later sources warfare was portrayed as aristocratic to the extreme. Charioteers armed for close combat but fighting primarily with bows settled all issues, leaving the infantry not only ill-equipped but stigmatized by racial inferiority. To compound matters Indian elite warriors would be the first to move beyond chariots, to a still more lofty and casual killer of foot soldiers.

Wastage could be overdone. Winning still mattered, and one obvious route was increasing the number of effective fighters through better armament and training. States hard pressed by circumstance and geography, or those preoccupied by ambition or militarism, tended to move in this direction. Assyria was all of those things, and created an army like none before.

Assyria came to live by war "like a wolf on the fold," in the words of Lord Byron. It was not always so. Early Assyrian history down to at least 1800 B.C. was dominated by trade, but this was followed by a period of weakness and subordination to a Hurrian aristocracy known as the Mitanni. Around 1350 Hittite pressure weakened the Mitanni sufficiently for Assyria to reassert itself, this time through conquests by a string of warrior kings, the most successful being Shalmaneser I (1274–1245), Tukulti-Ninurta I (1244–1208), and Tiglath-Pileser I (1115–1077). Assyria was well on the way to becoming one of history's great empires, a dominion that would eventually swallow all of Mesopotamia, Syria, Palestine, and finally Egypt. Its history was not without reverses, and more than once the empire shrank practically to the gates of the capital, Nineveh. Yet Assyria kept coming back. It was wed to war and its borders pulsed to the beating heart of its military.

The Assyrian army described in the Old Testament and historical accounts was essentially the product of the last 300 years of imperial

Panzer Pachyderm

In theory, it was virtually an ideal weapons system—lumbering but nearly as fast as a horse, strong enough to carry on its back a "castle" or cupola accommodating up to six archers, and equipped with a natural arsenal of trunk, tusks, and feet that could batter and squash virtually anything alive. If a horse could terrorize, imagine the impact of an elephant up to thirteen feet tall and weighing in excess of 10,000 pounds, running down a human combatant. Elephants fit the profile of favored weapons almost exactly. They were loud, trumpeting their presence with piercing screams; they were naturally thick-skinned (and frequently fitted with armor); they were fast, and above all they were huge.

Yet, like chariots, what they symbolized and what they could actually accomplish in battle diverged markedly. The reaction of Alexander the Great and his Macedonian army, the first Europeans to encounter elephants in battle, captured both sides of the great beasts as war machines. The Macedonians' major test with elephants came in 336 B.C., when they invaded the Indus valley and encountered the forces of the Pauvara raja (Porus in Greek) commanding a large force, including 200 war elephants. Tactically, the invaders adapted almost immediately, creating panic among the animals and gutting them from beneath using *sarissas*. Despite having neutralized the elephants so quickly, the Macedonians were deeply impressed. Alexander and especially his successors went to great effort and expense to import war elephants, fueling a pachyderm arms race between the Hellenistic monarchies. Among the feats for which the great Carthaginian general Hannibal is most remembered has to do with transporting just a few elephants over the Alps in his invasion of Italy. Yet all of this produced minimal results, and the elephant disappeared from European armies shortly after the Punic Wars.

Pachyderm armor.

In the East, war elephants endured, and their presence on the battlefield says a good deal about armies in this sphere. Never truly domesticated and frequently unable to distinguish friend from foe, the beasts were dangerously unpredictable in combat. The squad of infantry assigned to protect the animals' vulnerable underbelly frequently became its first victims in case of wounds. Elephants were also relatively easy to panic with flaming javelins or other incendiaries, and once this happened their rage was murderous for troops near them. For despots this was part of the natural order of things. Panzer pachyderms suited their purposes admirably.

history. As a force structure it marked the culmination of Bronze Age military trends allied with something entirely new: combined arms. The Assyrians used with ruthless efficiency all known weapons and added new ones when opportunity arose. It marked the subordination of tradition over experimentation, and also demonstrated the degree to which military power had overcome competing claims on resources.

As is was for other contemporary armies of the region, the bow was at the center of the Assyrian arsenal. But theirs was a carefully crafted composite weapon of extraordinary power, ends characteristically curled forward to resemble the bill of a duck. These bows were primarily in the hands of foot archers, either deployed as skirmishers or in massed formations. But the Assyrians departed from their rivals in their protection and the care which was taken to integrate them with other functions. After the reign of Ashurnasir-pal (885–860), ranks of archers were depicted as not only screened by shield bearers and heavily armed spearmen, but themselves dressed in long coats of mail and conical helmets. Such measures consumed metal on a grand scale, especially iron, which the Assyrians pioneered using in quantity. Yet ironcladding achieved an important result—sufficient stability in these formations to exploit other tactical possibilities.

This resort to marginally effective cavalry said a good deal about the Assyrian outlook toward war and weapons. They stopped at nothing and spared no expense to feed an appetite that called for deliveries of as many as a hundred horses a day to Nineveh, and yielded archaeologists 160 tons of iron from the palace arsenal of Sargon II (721–705) alone. This was a state obsessed with war, led by rulers who filled their chronicles with endless

*Assyrian armored archers
and slingers. Relief from
the southwest palace of
Sennacherib, Nineveh.*

Uneasy Rider

The Assyrians were dedicated to relentless pursuit no matter where it led them, and chariots were virtually useless in hilly and mountainous country. So they became the first agricultural people to employ mounted soldiers. Progress with cavalry, beginning around 875 B.C., was painfully slow, as chronicled by a series of revealing reliefs. Assyrian horsemen were first depicted riding bareback, uncertainly and in pairs, with one rider grasping the reins of both horses, as the other attempted to discharge his bow. Next they were found walking their mounts through rugged territory. Finally, around 750 B.C., we see them poised more comfortably on saddlecloths above the horse's withers, operating with some degree of confidence in hilly terrain. Assyrian cavalry almost never was shown in true battle scenes, only in pursuit on ground inhospitable to chariots. They were never natural horsemen, or anything close. Pastoral nomads might manage the acrobatics necessary to shoot a bow accurately at full gallop; but the Assyrians were simply interested in obliterating retreating adversaries.

recitations of "I destroyed . . . I devastated . . . I conquered . . . I impaled . . . I burned with fire . . ." For Assyria's enemies there would be no haven, no zone of safety. Prior to 880 B.C., walled towns sometimes fell to starvation, treachery, undermining, or even direct assault with scaling ladders and crude rams—really only poles with sharp tips. But, other than that, attackers lacked reliable means of reducing stone and masonry fortifications, and consequently avoided them. Assyria, marching to a different cadence, turned to innovation.

Assyria paid a heavy and increasing price for its warlike ways. Although military expeditions were still undertaken for strategic reasons, after 800 B.C. they began to resemble gangsters collecting protection money. At times, through lax management, they even seemed to encourage revolt, enabling them to squeeze the perpetrators all the harder—Metenna of Tyre was forced to pay Tiglath-Pileser III 150 talents of gold, while Sargon II relieved the city of Musasir of in excess of five tons of silver and more than a ton of gold. Unquestionably Assyria's lords of extortion profited handsomely, but the evidence also indicates that most of the take was consumed by the army and relentless combat—180 of the 250 years between 890 and 640 B.C. were devoted to war. Combined arms proved prodigally expensive—plainly exceeding the resources of an economy ultimately based on near-subsistence farming. So the army had to keep on fighting and sacking, and in the process war became an end in itself.

"I Besieged..."

The Assyrians were the first masters of siege craft and the perfecters of its most reliable implement prior to the invention of firearms: the battering ram. It would prove to be one of the most devastating weapons in history. Battering rams first appeared in Assyrian reliefs around 900 B.C., already in an elaborately developed form. Built on massive six-wheeled frames up to eighteen feet long and covered with wicker protective plates, the machine's heavy projecting beam was internally suspended to operate like a pendulum, though it could be adjusted to strike horizontally or upward as the situation demanded. Different bits could be fitted to the ram for brick or conglomerate, and there was even a cupola for defensive fire.

Visually the Assyrian battering ram was a startling prefiguration of a twentieth-century armored vehicle. But whereas tanks would evolve to fight other tanks, battering rams were aimed asymmetrically against noncombatants. Siege engines implied total war. "The city I destroyed, devastated. I turned it into mounds and ruin heaps, the young men and maidens in the fire I burned," raged Ashurnasir-pal after sacking one town.[2] Not every power would lay sieges with such voracity, and some, like the Hellenic Greeks, would abstain for substantial periods. But with the onslaughts of the Assyrians the notion of sanctuary had been breached by a member of the same agricultural clan who had once huddled behind ramparts to escape war. In this regard the battering ram opened the way to a day when urban centers filled with civilians would be bombed "strategically" and targeted with nuclear-tipped rockets.

Assyrian battering ram, one of the most fearsome and effective weapons in history, in relief from the northwest palace of Ashurnasir-pal Kalakh (Nimrud).

Assyria spun into oblivion. As early as 1000 B.C. tax records reflect serious problems maintaining population levels. Soon after, transfers of defeated peoples began on a massive scale—a dragnet that would ensnare an estimated total of between four and five million souls during the final three centuries of Assyrian history. Yet they do not appear to have worked as ballast for the top-heavy ship of state; instead Assyrian freeholders found themselves reduced to serf status.

Meanwhile, the army fought on until it inadvisedly broke into Urartu, a buffer kingdom to the north, and created a breach filled by a flood of Cimmerians from the deep steppe. The empire's days were numbered, and in 612 B.C. Nineveh fell surrounded by enemies—many of them former victims who had learned her ways. Doubtless, the Assyrian capital's walls were breached by the same types of siege engines her rulers had so confidently introduced. Military innovation has a way of backfiring.

All-consuming militarism was a potential end-state for any ancient tyranny based on irrigated agriculture. Organized coercion was necessary to generate the drudgery required to keep the society running, while a competitive external environment naturally raised the levels of armaments and military expertise. It was a feedback loop for which there was no obvious stopping point.

Therefore, what proved true for Assyria had similar though temporary results in China, a far less militaristic entity. After the fall of the Shang dynasty (1122 B.C.) and the disintegration of its successor, the Western Chou (771 B.C.), China came to be composed of around 170 competing feudal statelets. During the Spring and Autumn Period (722–481 B.C.), the number was winnowed down to seven powerful survivor states, who then engaged in a grinding free-for-all culminating in 221 B.C. with unification under Ch'in Shih-huang-ti (Qin Shi Hang in the modern pinyin system), the first emperor of the new China. Throughout this whirlpool of political amalgamation the frequency and intensity of combat increased; until the Chinese had true professional armies.

Ch'in Shih-huang-ti did not have a good reputation. Within a year of his death rebellion broke out, ending only in 206 B.C. with the establishment of a new and far different dynasty, the Han. Unification triumphed, but the violence and militarism leading up to it was decisively rejected. Effective government might ultimately depend upon compulsion in such a society, but it was to be justified in moral and intellectual, not military terms. Civil administrators, not warlords, would rule, and territorial aggression and the capture of peoples as social equilibrators remained largely in abeyance. There was a price to be paid in terms of nomadic invasions and armies

Underground Army

If you really can't take it with you, Ch'in Shih-huang-ti must have been terribly disappointed. He is remembered today as proof of the strange extremes to which militarism can drive men and societies. His most ambitious project in a reign of ambitious projects was the construction of what amounted to a mortuary city near the Ch'in capital of Hsien-yang (modern Xian). To protect his complex the Emperor buried an entire army, fashioned from over 8,000 life-sized terra-cotta figures of soldiers and horses. They were equipped with real weapons and accoutrements from his armories. For over 800,000 days his subterranean retainers stood guard, silently and invisibly, until archaeologists began digging them up in 1974.

Together these terra-cottas constitute the best evidence of how the Chinese military of the third century B.C. looked and fought, and how they were armed. More than 10,000 weapons of the period have been excavated. Long-range offensive arms, bows and the crossbow, were distributed to the vanguard, flanks, and cavalry. (Of particular interest was a block of 334 crossbowmen, anticipating the concentrations of arquebusiers in early firepower armies.) Heavy infantry carried long-shafted offensive weapons such as the spear, halberd, and *yue* (Chinese battle-ax). Elite troops were provided with swords and the *jin gou,* a curved knife used for hooking.

Only lances were fashioned from iron. All other bladed weapons were made of bronze. Recent laboratory tests show that the bronze of the Ch'in was coated with a ten- to fifteen-micron layer of chrome, which not only kept them sharper longer but testified to some very sophisticated and costly technology.

Care was taken to protect key elements of the rank and file. While skirmishers were given little in the way of defensive equipment, heavy infantry, halberdiers and crossbowmen, all wore plate armor over long heavy coats. They were also known to have employed shields and helmets, although these were omitted here, presumably to better display the facial features of the soldiers. The very fact that artisans were charged with crafting each soldier into an individual is a clear indication that the infantry was considered more than faceless rabble.

Combined arms and close coordination were critical to the desired end of great mobility and flexibility. This buried force structure points to what Sun Tzu and the other strategists called the "ordinary" (*cheng*) and "extraordinary" (*ch'i*). The former was used to fix the enemy in place, and the latter to attack unexpectedly and decide the issue through shock and surprise. But of course Ch'in's army also was a towering act of conspicuous consumption, not only representing the excesses of war in the realm of the real, but extending them to the afterlife. This, rather than military preeminence, proved historically the stronger message.

slaughtered through incompetence. But China survived, and the formula of the Han endured virtually intact for two millennia.

Things were different in Rome. She was built on agriculture, but not irrigation, and as a result escaped from the confines of pure tyranny. Dry farming was hard work, but it never included the drudgery of constructing and maintaining vast waterworks. So its inhabitants did not have to be chained to the food machine, but could be treated more like citizens. Rome was never democratic, but she was inclusive; her citizens returned the favor with fierce and enduring loyalty. In military terms this meant a consistent readiness to fight at close quarters. When fully exploited, it turned the Roman infantrymen into among the most destructive and effective foot soldiers in history.

The experience of fighting nearly every military power in the Mediterranean basin prodded Rome to experiment until a formula was found that best utilized the aggressiveness of her soldiers. It was never perfect—with a few notable exceptions her generals were plodders and could be fooled, and her cavalry was frequently victimized. Yet when it came down to the individual infantryman the system never failed. In part this was due to the recognition that bravery, even among Romans, was capricious and short-lived. The Roman aim was to increase the chances that bravery would appear at the right time by insuring that troops, up to the moment of battle, were as well armed, trained, fed, and rested as planning and logistics could make them. More than anything, maintenance of her manpower distinguished Rome from her many adversaries.

Rome began its military history dependent upon an aristocratic heavy cavalry. Contact and possibly conquest by nearby Etruscans, a mysterious and thoroughly Hellenized people, exposed the Romans to the military institutions of the Greeks, and led them to adopt the phalanx as their own. Membership remained limited to those who could afford a Homeric panoply until the ten-year siege of the Etruscan town of Veii, a campaign so long that regular pay had to be introduced and participation broadened. This was only the beginning.

Ten years later in 390 B.C. disaster struck in the form of 30,000 Gauls, who crossed the Apennines looking for plunder. A wild people given to drunkenness and oratory, they rushed naked into battle screaming and wielding long iron swords. Physically much larger than the Romans and outnumbering them by as much as two to one, the Gauls engulfed their

phalanx at the river Allia and swept on to thoroughly sack the city. The
historian Livy portrays the Romans watching glumly from a nearby hill
and vowing to look in the future "solely to their shields and swords in their
right hands."[3]

As good as their word, the Romans engaged in a series of military
reforms over the next century that transformed their army from an under-
sized porcupine into the human version of a school of piranhas. The days of
being outnumbered were made part of the past. Payment for military
service was continued, and participation further widened until it included
the entire body politic. Not stopping here, Rome expanded across the Italic
peninsula ready and willing to share her citizenship with those incorporated. Thinking themselves Romans they fought as Romans. Revolts of
"allies" were not unknown, but by and large they remained remarkably
loyal. In contravention to the Assyrian approach, Rome used fair treatment
to gain unprecedented access to a manpower pool of vast proportions.

She would learn to wield it with devastating effect. The brittle Roman
phalanx was abandoned, and the thrusting spear largely cast aside in favor
of the sword. Two thirds of the dominant heavy infantry—the first two
lines of a three-line formation—were equipped with the pilum (weighted
throwing javelin) to momentarily break the enemy's cohesion; then the
Romans would weigh in with swords—at this point about two feet long and
ideal of stabbing. Only the third line—older men used as a screen behind
whom the others might retreat—retained the spear.

Regular pay enabled all infantry to be well equipped defensively, with
helmet, breastplate, a single greave, and the best shield in ancient warfare, a
stout rectilinear affair with metal-bound edges. Rome had literally staked
her future on the shields and swords of her fighting men.

The verified existence as far back as 435 B.C. of the *spolia optima*—granted
only when a Roman commander killed his opposite number in single
combat—indicated that individual fighting had long been a Roman preoccupation. The conditions encountered during the wars of the fourth
century, particularly the Samnite Wars (343–290 B.C.), accentuated this
emphasis. Samnium was rugged, hilly country. To maneuver effectively the
Romans had to open their formations, leaving each soldier a designated
area of thirty-six square feet to protect.

Beyond his personal territory, the Roman soldier remained a team
player. As part of the tactical reforms, the basic three-line formation was
divided vertically to generate rectilinear tactical units called maniples of
120 men, each one capable of individual maneuver. Normally a Roman
legion of thirty maniples lined up in a checkerboard pattern; but if the

*Roman pilum
and shield.*

opportunity arose, maniples could shift position or even change form, providing unparalleled tactical flexibility. Over and over, enemy formations were shattered by a few spare maniples coming from an unexpected direction.

Unlike the phalanx, the manipular order was not friendly to amateurs. Complex and opportunistic maneuver combined with the most lethal sort of hand-to-hand combat meant incessant practice. Soldiers had to be thoroughly conditioned to react automatically.

Flavius Vegetius Renatus, chief recorder of Roman training methods, noted that particular attention was paid to swordplay. Recruits were first given wooden staves (double the weight of real swords), which they used to pummel human-sized posts twice a day. Gradually more advanced techniques were introduced—vital points on the body, the means of delivering a blow without opening up for a counter, emphasis on more lethal puncture wounds—until the legionary emerged a professional sword fighter.

Rome's commitment to the new fighting model can also be gauged by the degree to which it foreclosed on other options. Romans epitomized the

---

Roman gladiator.

Blood Sport

The Roman addiction to gladiatorial combat is remembered with disapproval bordering on revulsion. But the murderous games appear to have originated with a serious purpose in mind. Gladiatorial combats were first staged by private parties to commemorate the lives of famous relatives as part of their funerals. The first record of them in Rome dates back to 264 B.C., shortly after the military reforms had transformed the mode of fighting. The timing was unlikely to have been coincidental.

Skill was critical in sword fighting, but so was cold-bloodedness. Killing took getting used to. Matches between gladiators performed a function training could not—they showed men mutilation and death, hardened them to its sight. The casual attitude toward killing epitomized by the games may still be decried. But for a state that tied its future to the sword and the shield, the act of combat was bound to have been a focus of attention. Rome needed men who killed readily and killed individually. The games served both ends.

Western way of war—their objective was to close with the enemy and chew through his formations. They scorned the bow for the same reasons Homeric heroes did: it did not connote sufficient aggressiveness. Their determination to fight on foot was similarly unbending, and the weakness of Roman cavalry reflected this pedestrian outlook. Hannibal took masterful advantage of this opening, repeatedly peeling away their screen of horse and leaving the infantry blind to his presence. Romans even did it to themselves. Forgetting that cavalry should act as the eyes of the army, Roman horsemen had a tendency to dismount and join the melee as infantry.

If Romans lacked an appreciation of horses as fighting platforms, they cannot be accused of ignoring the virtues of mobility, especially strategic mobility. They were memorable marchers, capable of a steady twenty miles a day weighted down by as much as sixty pounds of equipment. This was more than a matter of sheer endurance, it stemmed from a highway system that surpassed in quality and extent anything prior to the great American interstate network. As Rome extended her empire, meticulously crafted roadways inevitably followed, eventually knitting the entire Mediterranean basin together with a web of paving stones that enabled the realm to be held and defended by an army that often did not exceed 125,000.

Measures taken to hasten troops on their way found a reciprocal in the attention lavished on the act of camping out. Castramentation—the habitual construction of elaborate fortified camps at the end of each day's march—was at the heart of the Roman military doctrine. Precisely laid out in an unvarying rectilinear pattern, the camps offered a familiar retreat in hostile territory. Customarily Roman generals only accepted battle if they were a short distance from their camps in case of disaster.

But they never refrained from assaulting the strong points of others. Among the greatest military builders in history, they were also demolition experts par excellence. Siege warfare was practiced from the dawn of the republic. After the desperate struggle with Hannibal—and later during the campaigns of Julius Caesar—the Romans refined their sieges into an ordered series of steps that drew a noose of men and machines around a town until it submitted or fell. This systematic approach was their primary contribution. The actual engines of destruction they borrowed from others, primarily the Hellenistic kingdoms, whose mechanical artillery constituted some of the most remarkable devices in the ancient world.

This sort of esoterica was lost on the Romans. Through the late republic and early empire they remained intent simply on adopting better killing machines, whether they were massive contrivances or simply contributed to the primal act of sword fighting. During his campaigns on the Iberian

Artillery and the Birth of R&D

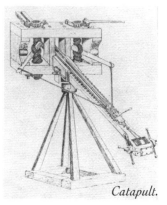

When the Hellenic monarchies centralized weapons manufacture into state-owned arsenals, it had a catalytic effect on ingenuity and images of what might be possible in the way of war machinery.

The most significant product was the torsion catapult. The first arrow-shooting catapult was said to have been developed at Syracuse for the tyrant Dionysius around 400 B.C. The power of this weapon—likely derived from the suppressed Greek crossbow, gastrophetes—was limited by the strength of its flexing self bow. This problem was directly addressed in the arsenals. Shortly after 300 B.C., the bow was replaced by two vertical cylinders of twisted fiber (women's hair was best) that attached to the bowstring by means of inserted wooden arms. Not only was this catapult vastly more powerful (it could pierce armor at a quarter mile); it was the first weapon consciously developed to capitalize on a physical principle—torsion, or the reactive torque that an elastic solid exerts as a result of being twisted. Stone throwers soon followed, first double-armed and later with just a single arm propelled by a large horizontally mounted fiber cylinder.

Catapult.

Practical application reached a crescendo with the war machines of Demetrius, son of Alexander's general, Antigonus the One-Eyed, employed during the monumental siege of Rhodes in 305–304 B.C. Demetrius's centerpiece was the Helepolis, a gigantic tower with a base seventy-five feet square, tapering upward to a height of 150 feet— nine stories packed with catapults and sheathed in iron plates, all of it mounted on eight huge swiveling casters. Supporting the Helepolis were two gigantic battering rams, each 180 feet long and employing crews of 1,000 men each, sheltered beneath massive wheeled sheds. None of it worked. Rhodians eventually immobilized the great machines by pouring quantities of water at their base, sinking them in a sea of mud.

Stone thrower.

Demetrius soon withdrew, carrying with him the title Poliorcetes (Sacker of Cities), an ironic sobriquet if there ever was one.

Innovation continued, nonetheless. Greeks at Alexandria experimented with repeating catapults, along with engines powered by metal springs and compressed air. None proved practical; but they remain significant since they embodied the deliberate exploration of physical and mechanical principles to improve armaments …true military research and development.

peninsula Scipio Africanus, Hannibal's nemesis, became acquainted with
the gladius, or Spanish short sword, and issued it to his own troops. Soon it
became emblematic of legionaries everywhere. Rather than bronze or iron,
these blades were forged of tough Toledo steel, making them virtually
unbreakable. Relatively heavy and sharpened on both edges, the gladius was
equally suited for slashing and stabbing, basically a meat cleaver with a
point. The effects were commensurate, shocking even the tough Macedonians when first exposed to the results. Livy wrote: "When they had seen
bodies chopped to pieces by the Spanish sword, arms torn away, shoulder
and all . . . they realized in a general panic with what weapons and what men
they had to fight."[4]

The continuing evolution of the pilum, or throwing javelin, exemplified
homegrown improvement. The long-pointed iron shank fitted to a wooden
shaft was made gradually thinner, so as to bend on impact and prevent pila
from being thrown back. Around 100 B.C. Gaius Marius went one better by
attaching the head to the shaft with a wooden dowel designed to shatter on
impact. The passage of 200 years found the pilum considerably shorter and
now with a lead-weighted ball just below the shank to facilitate armor piercing.

Roman gladius.

Defensive equipment followed the same restless path. Marius's troopers
had discarded their greaves, but now wore an expensive iron-ring mail shirt
of Celtic origins. Helmets, however, were still fashioned from bronze, and
shields retained their traditional form. By A.D. 20 legionaries continued to
wear mail armor, but now had iron helmets and carried a modified shield
shortened to save weight. Eighty years later legionaries reached their defensive peak, fitted with plate armor carefully segmented to preserve freedom
of movement, and helmets with elaborate baffles to deflect blows. The net
effect was one of the best-protected infantryman in all history . . . yet he
was still an infantryman.

Through sheer political will and methodical determination, Rome
turned him into the terror of the agricultural world. Yet the saga of unend-
ing frustration with Parthia exposed a fundamental weakness. Horse
archers remained superior, virtually unbeatable in open terrain by even the
best infantry. The equation never changed: what could not be caught could
not be killed.

Imperial infantry had one final bastion: the New World, in particular what
is now Latin America. The civilizations of the New World evolved independently and developed their own versions of intensified agriculture,

hierarchically organized populations, institutionalized religion, and monumental architecture.[5] Seen in this light they act as experimental controls
for the study of complex societies, which shared versions of these traits. In
war, too, the New World was separate yet similar.

People came to the Americas across the Siberia–Alaska land bridge in
waves, beginning perhaps as early as 40,000 B.C. (the date is extremely
controversial) and continuing through around 12,000 B.C. At this point the
only domestic animal available to join them was the dog, and the groups
themselves were too small to sustain the bacterial and viral infections
responsible for recurrent epidemics in Eurasia. Nor did they encounter
indigenous cattle, sheep, goats, or horses that might have promoted
pastoralism and acted as vectors for plagues. Hence complex societies grew
up in the New World largely without the pressures exerted by mass contagion and independent herdsmen. As a result the transition from hunting
and gathering to agriculture, from wandering to fixed settlement, from
peace to war all transpired at a much more leisurely pace.

Warfare thus became more ritualized and ceremonial than in Eurasia.
Without independent pastoralists, there was no sustained raiding from this
quarter; nor could there be horse archers without horses. Lacking a fundamental urge to build and huddle behind walls, population centers in
Central and South America grew up basically unfortified. War clearly
played a role in societal consolidation, but it was more exclusively political.
Looser settlement and the absence of persistent epidemics made it likely
that the populations of complex New World societies were more stable
demographically, and less in need of warfare as a balancing mechanism. The
result was a different martial tradition—one that largely excluded noncombatants and remained a matter for elites.

Still it was no game. The civilizations of the Americas had a number of
functional shortcomings and vulnerabilities—protein deficiency was a
chronic problem particularly in Mesoamerica, and on the whole their
ecologies were fragile and climates prone to disastrous fluctuation—all were
reflected in the nature and frequency of war. On the Andean plateau and in
the Basin of Mexico improvements in agriculture and distribution culminated in marked population increases . . . paralleled by an intensification of
warfare and more frankly expansionistic empires shortly before the arrival
of the Spanish.

The people we know as Aztecs provide a telling example of war in the
New World, being at once extraordinarily militaristic—virtually the Assyrians of the Americas—but still fighting in a way that could only be described
as quixotic. The Aztecs were newcomers, having wandered unwelcome into

the Basin of Mexico sometime during the twelfth century A.D. to begin painfully carving out a place for themselves. They built upon a long tradition of hegemonic societies, but one without written record, since the peoples of the basin were never truly literate. There remained only the faint voices of tradition and silent ruins.

The most impressive could be found at Teotihuacán, twenty-five miles northeast of the Aztec capital, Tenochtitlán (now Mexico City), a complex of public structures including the huge Pyramids of the Sun and Moon. The surrounding city may have held as many as 200,000 inhabitants, since the rulers at Teotihuacán moved just about everybody in the basin to the capital. As an imperial economic mechanism it worked for a long time, ensuring political control and providing the labor to hydraulically intensify agriculture, while facilitating effective food distribution in a world without draft animals or even wheels. Ultimately, however, the combination of an overworked environment, systemic inefficiencies, and trade interruptions dragged Teotihuacán into decline and then collapse. After A.D. 600 there was a decided increase in martial imagery, including a tradition of military-related human sacrifice and possibly cannibalism. Yet in the end the city suffered little physical damage; some monuments were burned but not much else was touched.

The Toltecs, the dominant people in central Mexico between the tenth and twelfth centuries, marched along a similar path. Their capital at Tula contained as many as 125,000 occupants, but was also unfortified. Yet there was plenty of evidence of military activity. Relics indicated that the Toltecs fought in formations numbering in the thousands, divided into projectile specialists and a small number of elite combatants. Archaeologists at Tula also discovered a rack resembling those used by the Aztecs to display enemy skulls, along with what are thought to be receptacles for the hearts of sacrificial victims. Meanwhile, large numbers of scattered human bone fragments provided ample food for thought regarding cannibalism. Despite this, when Tula collapsed suddenly in the late twelfth century only the central ceremonial precinct was destroyed; the rest was simply abandoned.

This was what the Aztecs first encountered—the tradition of Teotihuacán and Tula and a political landscape dominated by groups living out in the open on the valley floor engaged in a kaleidoscopic competition for hegemony. At first the Aztecs fared poorly, driven from place to place until finding sanctuary on an island in Lake Texcoco, a settlement that grew into Tenochtitlán. From here they participated in the local power balance as junior members of the Tepanec alliance, earning a reputation for desperate ferocity as warriors, along with a growing appetite for land and tribute. By degrees the ruling hierarchy and its religious ideology became increasingly militarized. Critical was the emergence of the imperialist cult of Huitzilopochtli, a god with an insatiable lust for the hearts and blood of war prisoners, a hunger that would one day leave Aztec society marooned on an accelerating treadmill of war, capture, and human sacrifice.

A crisis came when the leadership decided to challenge Tepanec. The commoners did not want the fight, and some even suggested surrendering the cult image of Huitzilopochtli. The war party's proposal was still more drastic. "If we do not achieve what we intend . . . You can eat us in the dirtiest of cracked dishes so that we and our flesh are totally degraded."[6] Judging by other evidence, the offer was not entirely metaphorical. The gamble worked, however. Tepanec fell in 1428, leaving the militarists permanently in charge. The die was cast.

From here on the Aztecs turned Tenochtitlán into what one historian called "a beautiful parasite," the center of a tribute empire resembling in its gross functions that of Assyria. There was the same lust for payment administered through an equivalently jerry-built structure of control, suspiciously configured to encourage revolt, reconquest, and still higher exactions. Like the Assyrians, as military obligations grew, so did imperial levies of tributary manpower—predictably divided into massed archers and projectile troops, fortified by a cadre of elite fighters armed for close combat.

Yet a good look at an actual battle would have revealed some startling differences. The link between the political and military was rendered entirely ambiguous by the behavior of the elite combatants. Plainly, their aim was not to destroy the enemy by killing as many as possible; it was capture through individual combat. Battles did include other types of fighting. Barrages of projectiles dominated the initial stages, though the penetrating power of Aztec arrows was so weak that invading Spaniards stopped wearing armor. There were also regular attempts to break enemy formations. But at the core was still capture; a man's prisoner was his future, the key to rank and privilege at home. This was so true that the Aztecs came to wage a special kind of warfare designed to produce captives.

The ritual killing of war prisoners accelerated steadily among the Aztecs, reaching monumental proportions in the late fifteenth century. Most notorious was the dedication of the temple of Huitzilopochtli in 1487, a 20,000-victim extravaganza: four lines of prisoners each two miles long shuffling toward the central pyramid where teams of priestly executioners zealously ripped out their hearts night and day until sufficient blood was spilled. Even more stark was the destination of the resulting corpses . . . the stewpots of their captors. Discounting sensational and

Flowery War and the Thorns of Battle

When asked by one of Cortés's conquistadors why the Aztecs had not bothered to subdue the local Tlaxcaltecs, the captured emperor, Motecuzoma, replied, "We could easily. But then we would have nowhere to train our youths except far places. Also, we wish to have people at home to sacrifice to our gods."

Instead, the Aztecs engaged them in Flowery Wars, carefully arranged contests for the elite, the sole result being prisoners—some of the purest examples of the ritualization of human warfare ever recorded.

Yet there was nothing aberrant about the Flowery Wars for Aztecs. The fighting was simply a distillation of what normally took

Mid-sixteenth-century indigenous rendering of the conquest, the Lienzo de Tlaxcala. Note the arms of the Aztec warriors.

place on the battlefield. Nor were the armaments any different in form or function; in both instances their chief purpose was to facilitate capture, not to kill. An elite fighter was hard to miss, but easy to wound. He was marked for identification by a towering feather "headdress," a decorative structure built on a light wicker frame strapped to the back so as not to impede movement. Each combatant also carried a cane and leather shield made recognizable through a personal insignia, but providing little in the way of protection. He had no helmet, only a decorative cap. Armor consisted of a feather suit worn over cotton padding. The preferred offensive arm was a long oak sword, its shallow edges lined with razor-sharp obsidian shards fashioned to slash rather than to cut deeply.

In a matched duel the aim was to bleed an opponent, gradually weakening him, so he could be brought down by cutting a hamstring or crippling a knee, leaving him set up for hog-tieing and capture . . . a flowery victim.

---

unlikely claims that human flesh constituted a significant element of the Aztec diet, there is overwhelming evidence that they did in fact eat those they sacrificed. The obvious gusto surrounding the custom, the care with which corpses were divided, the recitations of recipes, the pirating of bodies, all argue that this was a significant part of the warriors' motivational package, in essence, prestige protein.

Cannibalism made better sense in light of a population explosion that had driven the number of residents around the shores of Lake Texcoco into the millions, a disastrous famine in the year 1 Rabbit, and numerous other examples of dietary stress. These phenomena were hardly unusual among other societies dependent on hydraulic agriculture; but in a place where people customarily ate other people, it must have taken on an entirely different cast. Both myth and reality pointed to cannibalism as not just a reward but a threat. By this time not only captive warriors were sacrificed; a growing class of slave merchants were supplying victims from the ranks of malefactors and the improvident. Even sympathetic historian Inga Clendinnen calls Tenochtitlán "a startlingly violent place," a place where thousands upon thousands of skulls were racked and proudly displayed, where men ran through the streets wearing the flayed skins of others, where even children regularly observed the clergy ripping out the hearts of fellow humans.[7] It is not hard to reach the conclusion that this was a society spinning out of control—not only overpopulated, but hated by outsiders, increasingly feared by its own members and malnourished to the point of consuming itself.

Onto this stage strode Hernán Cortés and a tiny band of 500 Spanish soldiers. Within the year they brought the Aztecs to their knees. Although these conquistadors possessed weapons representing an order of magnitude increase in military power, the Aztecs fell because they were already tottering. Imperial agriculture was always an unbalanced entity, which fought as much for equilibrium as for anything else. The Aztecs, exemplified by their capture, not kill, weaponry, were a particularly exaggerated example, and collapsed accordingly. Empire was a Faustian bargain from the beginning, and war remained only its treacherous servant.

Chapter 6

AT SEA

Wood floats. This simple proposition was at the root of everything nautical from the moment the first damp soul—perhaps a member of the clan *Homo erectus*—clambered aboard a drifting tree trunk and discovered it would bear weight. The swimmer was suddenly a sailor, and set about putting his stamp on nature's ready-made vessels. Rather than simply straddling the log, hand axes were applied to dig out a place within. Not only did this result in a more stable perch, it substantially increased the log's buoyancy and pointed to a second key discovery—loads could be transported on boats, far heavier loads than anything available on land.

Propulsion became an issue. Currents provided a free ride downstream, but the return trip required some form of traction. Punting poles were useful in the shallows; but the minute water of any depth was encountered, paddling became necessary—hands first, then, rather quickly, large wooden hands in the form of paddles and eventually oars, when the principle of leverage was applied. Experience in large rivers and coastal waters must have led to a third great nautical insight—propulsion without effort awaited those who could harness the wind. Providing an air trap in the form of a sail was reasonably intuitive, and probably exploited fairly early through animal skins sewn together and jury-rigged from a temporary mast. Surfaces large enough to propel substantial craft would have to await the production of textiles in quantity, the first tangible evidence being found in Egyptian sailboat drawings dated around 3200 B.C. The rigs depicted show a mast stepped amidship with a single broad square sail hung from it, an arrangement that was the basis for the subsequent wind-powered vessels in the ancient Mediterranean world.

Funerary vessel of Fourth Dynasty Egyptian king Cheops (c. 2680 B.C.) exemplifies early built-up hull.

Cheops' rebuilt vessel, 142 feet long and 19 feet wide, is seen bow on. The framework enclosing the cabin may have been a form of air conditioning by supporting dampened mats.

Below: A diagramatic overview pinpoints the canopy at the bow (*right*) and a stubby gangplank at the waist. Cutaway sections reveal the inner structure, and an overturned portion of deck planking (*left*) shows its ribbed construction.

Size was an issue extending beyond sails. Advantages in carrying capacity, seaworthiness, and even speed could be realized by making hulls larger. The first barrier to overcome was that of the trees themselves—trunks were long but inherently too narrow. At some point in the fourth millennium B.C., probably in Egypt, the concept of a built-up hull took hold. Rather than use the trunk as the main body of the vessel, it became an anchor point, a proto-keel to which planks could be attached to broaden and deepen the entire structure. The planks themselves were joined to each other to build the hull to a desired size and shape.

In the Western world since the late Middle Ages wooden ships had been built on internal frames, with planks attached to these structural members. Strength and ease of construction predisposed historians, in the absence of tangible evidence, to assume something similar or at least analogous in the case of ancient shipwrights. The advent of scuba-based marine archaeology changed this fundamentally. Key finds, particularly one in 1982, definitively established

Egyptian coastal trader.

that the ancient means of construction was to join planks at the edges using mortise and tenon joints (a series of notches driven into both edges and then connected by inserts locked into place by dowels), and then into this tight-knit structure was inserted a frame—essentially the reverse of the familiar Western technique. It was highly labor-intensive and demanded great skill, but it worked.

The emergence of the ancient ship was an amazing achievement. It was a virtually magic vehicle, capable of smashing heretofore impenetrable barriers. Archaeologists now have evidence of Upper Paleolithic and Neolithic terrestrial trade networks reaching out across hundreds of miles. But they were limited to compact luxury items by the difficulties of land transport, a condition that would not change fundamentally even with animal traction, the wheel, and road networks. Until the exploitation of the steam engine, long haul bulk traffic over land was impossible.

Canals and inland rivers filled the gap in China, Mesopotamia, and Egypt; but it was the ancient seagoing ship that enabled the exchange of products in quantity across the Mediterranean basin. From the beginning of recorded history vessels plied the coasts of the Arabian Sea and the eastern waters of the Mediterranean carrying cargoes measured by the ton rather than the pound. The spectrum of products suitable for trade broadened accordingly, and their growing transit revealed the outlines of supply and demand—that goods plentiful in one area could bring dramatic profits in places where they were scarce. From here it was but a short step to value added—the notion that materials might be processed or created in ways that made them uniquely valuable, generating even broader demand and still larger profits.

This had revolutionary potential. Unlike prior means of subsistence, it was an open-ended proposition, theoretically without limits on the wealth it might generate. Free exchange, manufacture, and innovation were self-reinforcing processes, which played upon the uniquely human capacities for reciprocity, complex communications, and invention to raise the prospect of a future vastly exceeding anything nature had as yet rendered possible. The full flowering would have to await modern times; but very early, societies emerged that exploited the fundamentals to create something new in human history, true mercantile communities.

This first occurred on the island of Crete. Around 2000 B.C., Minoan culture emerged, built on a trading network that extended throughout the Eastern Mediterranean to generate an array of manufactured products that brought in even larger quantities of wealth and still further development . . . a true takeoff verified by excavations at a series of "palaces" (probably as much factories as royal residences) that revealed remarkable levels of economic activity.

They also revealed nearly a total absence of walled fortifications. Thucydides, the great Greek historian of the fifth century B.C., provided the traditional explanation. "Minos was the first to whom tradition ascribes the possession of a navy." Crete's cities did not need stone walls, because their ships provided wooden equivalents, protecting the island and their trade routes, which stretched toward Asia Minor, Egypt, and mainland Greece. The Greeks had a name for it, "thalassocracy" (rule from the sea).

Now, thanks to excavations on the island of Thera, we even have representations from a fresco of Cretan naval cruisers. They were slender and looked fast, but also appeared quite primitive—paddled rather than rowed in some cases, and with just a single steering oar. But these were plainly fighting vessels. In one image they were shown forging through waters filled with drowning men and scattered weapons, a marine armed with a long lance manning a low parapet in the bow.

While the Thera frescoes are fragmentary, they nevertheless tell us quite a lot about the naval state of around 1600 B.C. Distinctive warship designs had already split off from the basic transport form. These ships were far too elongated to have emphasized carrying capacity. Although they mounted sails, Minoan vessels relied on human propulsion. This would remain so for much of Mediterranean history. The wind was capricious. Fighting at sea required superior speed on demand, and this meant rowing. Ancient warships were by definition galleys, differentiated from each other by the number and arrangement of their oars. While bigger crews did not always mean more speed, oared warships were from the beginning labor-intensive.

Seaborne commerce—large vessels carrying valuable cargoes and small crews—had brought forth a reciprocal: faster, better-manned vessels dedicated to preying on them—pirates. The origins and existence of the

Minoan navy provide the best sort of circumstantial grounds for believing the dangers to overseas commerce were significant.

Thera blew up. Around 1470 B.C. an explosive volcanic eruption not only shattered the island, but created a tsunami that crashed without warning into the north coast of Crete, seventy miles away, very probably shattering its ports and swamping any ships they held. The Minoan thalassocracy was finished. Replacing it for a time were their former allies and subordinates, the Mycenaean Greeks, who took over the largest Minoan palace complex at Knossos.

Things changed under new management. Minoan society was basically peaceful, learning to use force late and pragmatically to protect their commerce. The Mycenaeans were a different breed. They knew war long before they knew trade, and they never forgot it. As beneficiaries of the Minoan emporium they were troublemakers, trading but also raiding and pirating and meddling in local politics throughout the Eastern Mediterranean. Quite probably they bit off more than they could chew when they attacked the rich city of Troy at the entrance to the Black Sea. Troy fell around 1240 B.C., but the exhausted Mycenaeans never recovered. This had been no way to run a thalassocracy, and the Mycenaeans' demise coincided with an era of maritime chaos. Raiders were loose all over the high seas, and in 1190 B.C. a confederation of "sea people" (Philistines and possibly Etruscans along with some Achaean Greeks) descended on the Nile. To this day the results speak for themselves.

Around 900 B.C. the nautical spotlight shifted north, focusing on two emergent societies, the Hellenic Greeks and the Phoenicians. While the artistic and intellectual triumphs of the Hellenes continue to dazzle, from a developmental perspective the Phoenicians may have been even more interesting. Located on the edge of the Levant, the key centers—Sidon, Tyre, Byblos, and Berytus (Beirut)—evolved into almost purely urban commercial societies, utterly dependent on trade and manufacture. Basically without landward possessions and walled against invaders, they withstood the threats posed by terrestrial empires like that of the Assyrians by using access to the sea as an escape hatch. Phoenician cities were the first to survive essentially through long-distance trade, generating and transporting goods and receiving wealth and sustenance in return. In doing so they broke the heretofore inescapable shackles of food production, setting up a far-flung network of trading bases, which by 800 B.C. dotted the shores of the Mediterranean as far west as Gibraltar. Their aim was clearly profit, not conquest. But the Phoenicians realized sea lanes were vital, and out here were a match for anyone.

The Phoenicians were renowned throughout antiquity for seamanship;

Decision in the Delta

To celebrate his great victory over the sea people, Pharaoh Ramses III erected a temple at Medinet Habu and ordered his artists to commemorate the event with a carefully carved relief on its walls. The resulting details form an invaluable basis for comparison between the two forces and their fighting styles. The ships of the sea people were clearly less specialized and more oriented toward navigation at sea, with high stem and stern posts decorated to look like duck heads, dual steering oars (henceforth a standard feature on ancient vessels), and depending on sails for propulsion. The warriors aboard were plainly equipped for close combat with swords, shields, and breastplates, their aim to close and board, thereby generating the nautical equivalent of a land battle—a technique seen throughout ancient naval history.

The Egyptian ships were more specialized—galleys propelled by a single line of approximately ten rowers on each side, though steered with only one oar. Judging by the armament of the onboard combatants—composite bows and javelins—they should have used their superior mobility to avoid contact and subject their enemies to withering long-range fire. But the evidence in the mural poses several contradictions. First, the Egyptians appear quite aggressive, fighting in close rather than staying out of range. Further, one of the sea people's ships was shown as having capsized. How was a mishap of this magnitude possible as the result of arrows and javelins? On close examination the bows of the Egyptian ships are plainly lower than the sterns, and tipped with protruding feline heads. The purpose could have been purely decorative; but these appendages also might have served a destructive function—a proto-ram to inflict structural damage. Without further evidence this must remain only a suggestion, but as the subsequent course of naval architecture showed, things were certainly heading in this direction.

Reconstruction of the Medinet Habu relief.

but they were also excellent shipwrights, exploiting the most advanced construction techniques. And this was true of warships as well as merchant vessels. An orientation toward trade and the sheer distances traversed had an impact on naval designs, leaving them with fuller lines and more freeboard than rival warships. Yet they were far from tubs. The Phoenicians were very probably responsible for a number of key naval innovations, and not far behind in others. It is likely, for instance, that they introduced two rowing levels, and possibly three, but because they never depicted their ships, it is impossible to say with certainty.

Future events were not kind to the Phoenicians. The Greeks challenged their entrepôts with larger, land-hungry colonies of their own. One by one the cities of the Levant were overcome. Eventually even Carthage, the sole Phoenician foundation with imperial ambitions, would be ground down by Rome. In the end all that remained of their literature were funeral dirges and the alphabet. Meanwhile, history was recorded by the winners.

Aristotle called the Greeks "frogs." Like amphibians they were at home on water as well as land, and learned to fight on both. Unlike the Phoenicians, they depicted their weapons on anything that would accept an image—frescoes, statuary, temple pediments, ceramic vases and urns — founding a tradition that is responsible for much of what is now known about ancient warships. These stone and clay snapshots definitively establish that a new type of naval combatant had come into being by around 850 B.C.

Powerful rams now jutted indisputably from the prows of Greek galleys. The naval vessel had been transformed into a projectile, capable of inflicting lethal damage with a single blow. Whether its actual inventors were Egyptians, Phoenicians, or Greeks, the ram was the single most important innovation in weapons at sea prior to the introduction of the gun, an instrument of destruction so long lasting that the developers of the early steam battleship resurrected the device and attached it to the noses of their smoking leviathans. Oars or steam, propulsion and deliberate collision were joined in the naval imagination. Whether ramming or boarding, this was a confrontational style of combat that neatly fit the precepts of heroic warfare. It was as if Homer had been a naval architect.

Sometime around 700 B.C. shipwrights escaped the design cul-de-sac. Creative seating engendered a new type of hull. The standard twenty-four oarsmen on each side were divided into two superimposed banks, twelve on top rowing from frames set above the gunwales, and twelve below, their oars extending through ports in the hull. To allow oars to move freely, they

Propitious Penteconter

The first Greek rams were propelled by a line of ten oarsmen on each side, for a total of twenty. Yet speed and power were now paramount if the enemy was to be successfully impaled, so this meant more rowers. Triaconters, thirty-oared craft, were introduced, followed by boats with fifty oars, or penteconters.

Here the process stopped. Adding to the files of oarsmen had stretched the galley's length from about fifty feet to over double that. Since width had barely increased, the result was a pencil-thin hull form. Wood (generally an oak keel with pine planking) was only so strong, and the penteconter had reached that limit. Any longer and two waves might have hit the bow and stern simultaneously, causing the boat to bend, or "hog," in the middle. Also boats of exaggerated length were slow to turn, potentially a fatal flaw in naval battles.

Warship design hit a plateau during the eighth century, leaving the penteconter to rule the waves with only minor changes. The ram—layered in bronze and tapered to a sharp point—had the nasty habit of becoming stuck. Later models grew blunt, with the aim of inflicting concussive structural damage rather than simple and dangerous punctures. Meanwhile, the possibility of boarding—both offensive and defensive—was reflected by decking on the stern, bow, and along a narrow walkway down the centerline, creating a fighting platform for marines.

Despite the decking, war galleys remained low in the water and not very seaworthy. This factor, along with the absence of space for provisions, made it necessary to beach the ships at nightfall and usually to hug the coast. When traveling, the ships were sailed, not rowed, employing removable masts and a single square sheet—not particularly efficient, but it saved labor. On either side of the penteconter's stern were the steering oars, first mounted in parallel and later asymmetrically to increase responsiveness. A true rudder was never developed.

The net result was a sort of lethal racing shell. Penteconters and their descendants were useful instruments of naval power, but also fragile and expensive ones. Operating a penteconter required approximately ten times the crew of a sailing transport of equivalent displacement. Just as the preindustrial world's other major weapons systems were restricted to certain types of societies, navies demanded concentrations of wealth and seafarers—perfect for urban trading entities, but difficult if not impossible for others. And the subsequent course of warship design would only raise the ante.

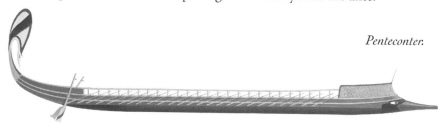

Penteconter.

were staggered so those above were positioned over the space between the lower ports. The resulting vessel was not only inherently sturdier and more seaworthy—increased in both freeboard and width—but it was also shortened by a third, making it a smaller and more maneuverable target. The evidence points to the Phoenicians as the source of the innovative hulls, but the superiority of these so-called biremes was such that they quickly spread across the Mediterranean where all the major naval powers—the Carthaginians, Etruscans, and Greeks—took them up. As good as they were, the biremes' reign as capital ships was short-lived.

Literary tradition points to the bireme's successor emerging once again from the shipyards of the inventive Phoenicians, possibly as early as the middle of the seventh century B.C. Referred to as "three-fitted," or triremes, they could be assumed to have been an evolutionary development of the two-level warship, basically three-deckers. Yet the evidence, such as it is, leaves many questions.

The introduction of the trireme coincided with the appearance of naval warfare on a massive scale. The ship's superiority caused navies to become basically standardized in terms of equipment, thereby placing heavy emphasis on quantities. It was not unusual for opposing fleets to number 100 triremes each, and on occasion 150 or more. Hence the great era of Hellenic naval warfare—encompassing the Persian Wars, the Peloponnesian War, and the campaigns leading up to Alexander's invasion of the East—was dominated by large sea battles, and without question favored wealthy states. Financially and operationally it was the antipode of the amateur-centric, come-as-you-are phalanx warfare so favored by agricultural Greeks.

Triremes were not just expensive to build and accumulate. Their light construction and the stresses of rowing and ramming insured almost continuous repairs and a short life span, necessitating steady replacement. Meanwhile, the human capital involved was exceedingly expensive and even more perishable. Adding officers and marines to the trireme's gang of rowers resulted in a crew of 200, which, when multiplied over 100 ships, added up to at least 20,000 men. Unlike later times, citizens, not slaves, manned the oars, demanding not just food but pay. Since naval campaigns lasted weeks and even months, the financial and logistical burdens were massive. Nor did the challenge end here.

Extensive training was also a requirement. Handling an oar effectively and working together could not be taken for granted. The two basic ramming maneuvers, the *diekplus*—a breakthrough of the enemy's line followed by a sudden turn to strike from the rear—and the *periplus*—an end run around the adversary formation followed by a similar stern attack—

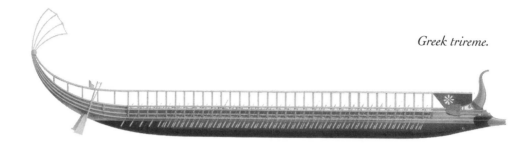

Greek trireme.

The Trireme Controversy and Its Solution

The ancient trireme dominated naval battles for 200 years, and continued to serve as useful members of fleets down to almost the end of the Roman Empire. By all accounts they were expensive and hard to build, but made up for it with an extraordinary combination of speed, power, responsiveness, and maneuverability that was only barely surpassed in centuries of naval experimentation. What was the trireme's secret? The quest for an answer generated a major scholarly controversy and in the end led to the ship's rebirth.

Phoenician triremes were known to be spacious and high out of the water, indicating that they had been created by simply adding another deck to the bireme. Besides raising the center of gravity, elevating the third level of rowers this high must have robbed them of much of their power, since the top oars would have struck the water at an uncomfortably sharp angle—an implausible model for a ship known for swiftness and agility. To approximate these qualities, the trireme had to have been lower to the water.

This led W. W. Tarn, the famed classicist of the 1920s and 1930s, to the radical conclusion that the Greeks solved the problem by putting groups of three oarsmen on one level in the herringbone pattern similar to that employed by Venetian galleys during the Renaissance. Tarn's solution ran afoul of the evidence. Athenian naval inscriptions showed that practically all rowers used oars of the same length—impossible in such an

demanded speed, coordination, and clean execution. There were other variables, but basically the boats with the best crews won.

Yet keeping crews together, both physically and in terms of capabilities, was a task, when multiplied over hundreds of boats, that stretched the capacity of even the richest and most determined naval powers. Whole fleets could disappear with frightening suddenness. Twice during the Peloponnesian War, at Syracuse and Aegospotami, Athenian squadrons, each numbering almost 200 triremes, were trapped on the beach and completely lost, along with their crews.

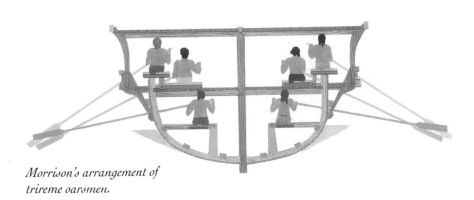

Morrison's arrangement of trireme oarsmen.

echelon configuration—and a stone carving, the Lenormant Relief, plainly depicted a trireme with several levels.

A decade later Sinclair Morrison came up with a promising solution; his model was a bireme with an outrigger, which enabled a third bank of oarsmen to sit only slightly higher than the second, but well outboard of them. The resulting design retained a low profile, but still allowed for oars of the same length. Subsequent experiments to determine mechanical efficiency added weight to the argument.

Yet the controversy had gone on so long, and the arguments had become so tangled, that only a full-scale demonstration would settle the issue. In the summer of 1987 the Trireme Trust launched a full-scale replica, carefully constructed under the supervision of J. S. Morrison and J. F. Coats to test the outrigger theory. One hundred and seventy young oarsmen climbed aboard the narrow 120-foot-long craft, shoehorned themselves into the superimposed benches, and began training. Within weeks they achieved spectacular results, demonstrating a top speed of over ten miles per hour, and managing a 180 degree turn in under a minute while carving an arc of only two and one half boat lengths. Dramatically more powerful and maneuverable than a penteconter, it was still light enough to be dragged up on the beach at night. Here at last was a craft truly reflective of the history it made over 2,000 years ago.

Storms were even more dangerous. Since oars could not be worked in high seas and triremes necessarily rode low in the water, they could be quickly swamped. Whole fleets were wiped out in sudden gales, with staggering death tolls.

It now appears that chronic shortages of experienced oarsmen were key in charting the subsequent course of military naval architecture. At the very beginning of the fourth century B.C., with Athens and the other major Greek naval powers exhausted, references to ships beyond triremes ("fours" and "fives") began to appear.

*Greek naval
ramming
maneuvers.*

In the wake of the Peloponnesian War the Greek city of Syracuse on Sicily became enmeshed in a naval rivalry with Carthage, founded on the coast of North Africa by Phoenicians, but unique in its size and success. Always traders and mariners, Carthaginians had grown gradually more aggressive and determined to foster their own colonies on Sicily. Not only had they honed and expanded their navy, but Carthaginian shipwrights, building on the Phoenician tradition of innovation, apparently improved the basic naval unit. The chief weakness of the trireme always had been the mechanical difficulties posed by rowing from the top level. Now Carthaginian designers provided help by adding a second seaman working the uppermost oar. The resulting arrangement—a four—was a major success, the fastest and handiest of all ancient war galleys.

What had been a matter of choice for Carthage was likely a necessity for Dionysius, the inventive tyrant of Syracuse, who was determined to build a massive fleet of his own. He simply did not have enough trained oarsmen. Yet the Carthaginian four pointed toward a solution. With more than one man to a sweep, only the rower at the end required skill; the others supplied only muscle. Ship designs could even be adjusted to the magnitude of the shortage, since more than two rowers could be fitted to a sweep. Dionysius's counter to the Carthaginian four was a five; yet this could have taken several forms. Depending on how far he wanted to stretch his skilled rowers, it might have been a single-decker with five men at each oar, a bireme with three working the top-level oars and two operating those below, or a three-decker (the ancients never went any higher) with two, two, and one.

Multiple rowers added power; but they subtracted speed and responsiveness, since they necessarily reduced the number of oars per crew member and also the number of strokes per unit of time. All of the subsequent numbered galleys except the four were slower and clumsier than the trireme, and with respect to each other, those with the fewest levels were the least mobile. On the other hand, they were wider, more stable, and because of their added power could carry more men and catapults on broad fighting platforms.

These factors came into full play after Alexander's death, when his successors began building navies of their own. Lacking skilled rowers but rich in Asian and Egyptian manpower, they indulged in what became the greatest arms race in ancient history. Not inappropriately, Demetrius Sacker of Cities touched it off in 315 B.C., when in addition to fours and fives he had his father's Phoenician shipyards turn out some sixes and sevens. By 301 he had accumulated eights, nines, tens, one eleven, and a single monster thirteen, a fleet of floating dinosaurs to which he ultimately added a fifteen and a sixteen.

His colleagues were not deterred. Demetrius's rival, Lysimachus, after inspecting the City Sacker's bloated flotilla, responded with the *Leontophoros*, a ship so large that its fighting deck could accommodate 1,200 marines. The size of this throng pointed to the character and intent of these megagalleys. The biggest had grown into lumbering barges for unleashing catapult barrages, boarding, and, ultimately, intimidation. Size, at sea as well as on land, assumed a quality all its own; these were nautical analogues to Homer's champions, grown gigantic in pursuit of an intraspecific image of dominance that would condition weapons design right down to the present. This Hellenistic Greek naval arms competition resembled nothing so much as the great international race to bigger dreadnought battleships that occupied the first four decades of the twentieth century.

Antigonus Gonatus, a true son of his father, Demetrius, made a splash by launching a galley even bigger than the *Leontophoros,* this in the face of chronic instability and poverty at home. His gesture, however, was dwarfed by Ptolemy II of Egypt, who, shortly before his death in 246 B.C., had managed to accumulate a fleet made up of 17 fives, 5 sixes, 37 sevens, 30 nines, 14 elevens, 2 twelves, 4 thirteens, 1 twenty, and 5 thirties. Not to be outdone, his grandson Ptolemy IV constructed a forty, a ship so huge that it inspired detailed descriptions by the ancients, but still flummoxed contemporary historians, who struggled to explain how it might have been configured.

Romans, practical to their core, proved far more serious about turning sea battles into land battles. Archetypical landlubbers, in 264 B.C. they

Galley Cat

According to Plutarch, the forty scarcely "differed from buildings which are rooted in the ground, and had great difficulty putting to sea." No wonder. The ancient writer Athenaeus described it as fifty feet wide, festooned with three banks of oars, the uppermost reaching fifty-seven feet. It was manned by 4,000 rowers, 400 other crew members, and fully 2,850 marines . . . a total of over 7,000, the population of a medium-sized ancient city. Given the strength limits set by wood, no hull could have been made long enough to accommodate such a crowd. Yet Athenaeus's and Plutarch's accounts could not be dismissed as simple exaggerations, since other sources had traced a steady progression of numbered galleys from fours up through thirties.

A puzzling portion of the description points to a solution. The forty was said to have been "double-prowed" and "double-sterned," along with having four steering oars. It follows that the forty as well as the thirty were catamarans, with twin hulls supporting an extraordinarily broad fighting deck. These ships constituted a sort of reductio ad absurdum of bringing land tactics to sea—an entire oar-powered battlefield. Plutarch noted that the forty was "only for show," as was likely with the thirty. In ancient history as in modern times, size in weaponry was almost always more a matter of psychology than function.

Ptolemy IV's forty.

found themselves locked in war with Carthage, owners of the best navy on earth. As had been the case with Dionysius and the Carthaginians, the casus belli was control over Sicily. Hostilities began well enough for the Romans. In the first two years they successfully landed an army on the island, took control of the countryside, and set about laying siege to the important towns. A few fell, but with the rest along the coast, Rome ran into a literal and figurative stone wall. Unable to break through their landward fortifications, the besiegers were thwarted by Carthaginians in fives who kept starvation at bay by supplying from the sea.

In the early spring of 260 B.C. the Roman Senate bowed to reality and decided to build a navy aimed at directly challenging Carthage. It was an audacious choice, backed by characteristic resolution. Within sixty days a hundred fives were nailed together. Although certain features of a beached Punic five were incorporated, naval historian Lionel Casson reasons that the Roman version was probably not a sophisticated multidecked copy, but simply a single-tiered galley propelled by five men to an oar, most hastily taught to row on great stages. Both the timber and the men in these ships must have been very green. Nonetheless, in June the new fleet left Rome's harbor at Ostia, and sailed down the boot of Italy to rendezvous at the friendly city of Syracuse. It was likely a scary trip, and a prelude to a scary battle. The Romans didn't seem to have a chance against the swift rams and experienced sailors of Carthage. The fledgling navy seemed doomed. But then, just in time, a miracle arrived—borne not in the talons of a Roman eagle but in the beak of a crow (see Corvus, page 108).

Rome was far from giving up, but neither was she finished with catastrophe. The fleet was rebuilt and sent back to Sicily in 250 B.C. to blockade Carthage's bases. But a year later it too was shattered, half of it in a sea battle, and the rest in a storm at the exact point where the first gale had struck. Meanwhile, the Carthaginians were constructing warships as blockade runners, making them even lighter and faster. The clumsy Roman vessels didn't have a chance of catching, much less dropping a corvus on them.

The crow had to go and a new fleet had to be built to Carthaginian standards. Roman oarsmen, grizzled by long years of service, were as good as any Punic counterparts; they just needed equivalent ships. Using a crack trireme manned by a picked crew, the Romans managed to run down the fastest five in the Carthaginian navy. It took until 242 B.C. for the weary Romans to reproduce it 200 times, but when they were finished they had finally turned the oar blades on Carthage. It was their fleet that was bigger, newer, and better trained, and when they met the next spring off the Aegates islands they crushed the Carthaginians, sinking fifty and capturing seventy of their fives.

At last the war was over, but at a terrible price. Strictly a land power, Rome had only reluctantly gone to sea, and suffered the consequences. The Greek historian Polybius put Roman losses at 700 fives to the Carthaginians' 500. A race of nonswimmers, it is entirely possible that Roman and allied war dead approached 400,000, a figure unprecedented in military history.

Nevertheless, the stage was now set for Roman naval hegemony. Carthage fought Rome twice more, but never again posed a serious challenge at sea. Meanwhile, the Romans made short work of the Hellenic naval powers of the east, and, just seventy-five years after the Senate's fateful

Corvus

Technology saved the day. Syracuse had a long tradition of naval innovation, and it is likely that some local inventor, possibly Archimedes, came up with a device that suited their ally's needs.

Roman infantry were unbeatable if they could somehow grapple the elusive Carthaginians. Their fives were too slow to initiate contact, but they did make excellent targets. The solution came in the form of a thirty-six-foot-long boarding bridge tipped with a long spike extending downward. This device was mounted in a vertical position near the bow. When a Carthaginian galley approached to ram, the securing tackle was cut, the bridge pivoted down, and the spike was driven into the opposing deck, followed within seconds by eighty Roman legionaries with murderous intent. The motion of the device reminded someone of a pecking crow, who dubbed it corvus.

The corvus had a wildly successful introduction. In August a superb Carthaginian fleet of around 130 warships was ravaging the northern coast of Sicily near Mylae, when a slightly larger number of Romans crept out to challenge it. Utterly confident, the Punic galleys charged into the naval novices, only to be caught in the claws of their crows. Before the day was over the Carthaginian admiral saw forty-four of his ships captured and 10,000 of his men hacked to pieces. For nearly four years the Romans trained with the new device and added ships, all the while avoiding contact with the frustrated Carthaginians. Then in the summer of 256 B.C., Rome took the offensive, sending a huge invasion force to North Africa protected by the fleet, now swollen to 350 warships. While the land portion of the campaign proved indecisive, it provoked two great sea battles, one off the promontory of Economus and one at Cape Hermaeum. In both instances the Romans used the corvus to good effect, virtually destroying the Carthaginian navy.

Then disaster struck. Headed home, its ranks swelled by 114 prizes, the Roman battle fleet ran into a gale near the southeastern tip of Sicily. The storm made short work of the low Roman galleys, rendered even less seaworthy by the weight of their boarding bridges. When the skies cleared, nearly 200 of 364 ships had sunk and over 100,000 men drowned. What the crow had saved, it now took away.

decision to go to sea, ruled the waves from the Strait of Gibraltar to the coast of Syria. Dominance would last nearly 425 years, until Rome's decline opened the door for new naval developments.

By A.D. 678, the Eastern Empire had been pushed to the edge of extinction. An Arab army was camped along the six miles of walls that separated the imperial capital, Constantinople, from the Balkan landmass, while poised in the Sea of Marmara was the largest Islamic fleet yet assembled, threatening to cut off the city altogether and starve it into submission. Against this force the Byzantines sent out their own recently rebuilt fleet; but it was smaller and the best they could achieve was a stalemate. Unless they could find a way to break the siege, the end loomed.

Since A.D. 513 the Byzantines had managed to dominate the Mediterranean with dromons, or "runners." These ships were new designs, probably biremes, generally with one man to an oar. They ranged in size from 120 oars down to much smaller vessels but were all called runners, and all carried two masts, a main and a foremast, and the biggest may have mounted a mizzen also. Later models were fitted with lateen rigs, and sails were carried into battle. All of this implies a more efficient use of wind power, a true synergy of sail and oar, which at the least must have substantially improved high-speed endurance. Early dromons were fitted with rams, but gradually the offensive emphasis shifted to incendiary weapons.

These were not entirely new. As far back as 190 B.C. several trapped Rhodian men-of-war staged a spectacular escape with experimental fire pots extending from the bows and then dropped on the enemy—a sort of corvus au flambeau. The Byzantines went several steps further. They installed catapults designed to launch flaming projectiles, and they developed a siphon, based on preheating and pressurization, that spewed burning liquid over short ranges. The problem was that the naphtha (crude oil) laced with sulfur incendiary had to be lit before it was delivered, making it inherently unreliable. When it worked it must have been devastating; when it didn't it was harmless. This flaw must have brought about the initial stalemate of 678. But things were about to change.

Callinicus, an engineer from Heliopolis driven from his home by the Muslim occupation, slipped into Constantinople bearing a secret weapon. Just what is still uncertain. The best candidate remains saltpeter (potassium nitrate) . . . that and the discovery that its addition to the heated incendiary mix rendered it capable of spontaneous combustion. The results spoke for themselves. In late 678 a Byzantine squadron sailed out and challenged the Islamic fleet, leaving its ships, according to the chronicler Theophanes, "engulfed in flames" by a "marine fire," or "Greek fire" as it was subsequently called. Like napalm it stuck to everything, devouring all manner of organic substances—ship's timbers, oars, sails, clothing, and flesh. Even jumping into the water would not douse the all-consuming flames. With shocking suddenness the Muslim naval force was reduced to ashes, the siege lifted, and the city saved.

Two generations later the drama resumed, providing what historian Alex Roland calls "another stolen glimpse of this mysterious weapon." This time the besieging Islamic force was much larger—reputedly 200,000 troops and 7,500 ships—so large, in fact, that it ran out of food and had to be resupplied by a flotilla numbering in the hundreds of Arab transports. Once again, the Byzantine dromons sprang into action, this time equipped solely with siphons disgorging Greek fire. Any relief vessel that did not go up in flames was captured by the Christians, leaving Constantinople flush and the Arabs to starve and withdraw within months.

Greek fire constituted one of history's purest examples of technological surprise—at once qualitatively different from anything before and a harbinger of a new day in war and weapons. Culturally, it pointed to a time in the West when the primary instruments of destruction would shoot from afar, negating the Homeric tradition of confrontation, and necessitating a redefinition of courage itself. Even more fundamental, Greek fire marked a resort to chemical energy, the basis for the firepower revolution. The probable use of saltpeter along with sulfur foreshadowed by centuries the discovery of gunpowder. Certain accounts even reported a loud boom accompanied by a cloud of smoke when the liquid fire left the tube.

Greek fire also presaged an era when science and technology would be deliberately applied to "infernal mechanisms." Greek fire was so secret, and the rulers of Byzantium so obsessed with keeping it, that they invented compartmentalization, the same veil that shielded the Manhattan Project. People worked on different aspects of the system—the specially modified dromons, the siphons, and the liquid itself—but only the ruling family and, legend has it, the descendants of Callinicus knew the exact recipe and how everything meshed. The veil worked too well. When it was removed, the secret was gone, forgotten somehow, perhaps as early as A.D. 750 but surely by the time Crusaders from the West sacked Constantinople in 1204.

A better parable of "ultimate weapons" would be hard to find—if shared with too few they would be rendered useless, if employed promiscuously they would be copied. This would be the legacy of the coming of the gun.

TUBES OF FIRE

Europeans, seeking to explain how "Gonnes" had been invented, conjured up the shadowy figure of Bertholdus Schwartz, a Faustian tinkerer who "devised by the Suggestion of the Devyll himself, a Pipe of iron and loaded it with powder, and so finished this deadlye and horryble Engine."[1] It seemed logical. Who but Satan would wish such an affliction on humankind. Who but a Westerner possessed the skill to bring it to fruition?

Until recently the sudden and unprecedented appearance of a vase-shaped, arrow-shooting metal artillery piece, depicted in a 1327 manuscript by Englishman Walter de Milamete, was believed to be the product of indigenous development. The absence of evidence for previous European experiments with gunpowder-based weaponry was attributed to these clues having been lost. It was recognized that the Chinese had experimented with gunpowder somewhat earlier; but the actual appearance of firearms was believed to be roughly simultaneous and a product of independent paths of development both East and West. This was wrong. Schwartz or his equivalent never existed. It is now known that firearms were brought to Europe from the outside fully formed.

The invention of firearms begins and ends in China. Historian Joseph Needham and his colleagues, through close study of early manuscripts, have put together a convincing sequence of events that chronicles the gun's Eastern evolution. The initial step was gunpowder. It first appeared around A.D. 850, as the result of long experimentation by Taoist alchemists looking for elixirs of life and material immortality. Instead of ying they arrived at yang. Randomly mixing and burning ingredients, until they discovered that the

*Gun depicted in Walter de
Milamete's manuscript.*

combination of sulfur, charcoal, and the key energizer, saltpeter (potassium nitrate), burned and sparked intensely. If more nitrates were added, it became truly explosive.

The material was soon applied to warlike devices. By around 1000 simple bombs were being lobbed by trebuchets (lever- and weight-based throwing machines), their development progressing from thin-cased "thunderclap" devices to thicker "thunder-crash" models. Their effectiveness was attested to by a twelfth-century print showing bombs being cast over the Great Wall onto a body of Inner Asian horse archers, whose dismembered corpses were left strewn over the steppe. The projection of these devices was depicted as purely mechanical, but experiments were underway that used gunpowder as a propellant.

A cardinal feature of Chinese science and technology was a belief in action at a distance—a key element in the discovery of the magnet. This was paralleled by martial traditions in China, where the crossbow had driven the chariot from the battlefield and the heroic image of close combat did not exist. Guns not only made sense from a cultural perspective; but China was blessed with a ubiquitous natural material that prefigured the form they eventually would take . . . bamboo. Shortly before A.D. 1000 these natural tubes were cleared of internal matter and filled with gunpowder to create a sort of flamethrower, which was soon being employed by the Sung Dynasty against invading horse archers.

From here development forked along two distinct paths. Around 1150 some user of the flamethrower, familiar with its recoil effect, drew the brilliant conclusion that it could be fitted backward onto an arrow to create a self-propelled projectile, which, when lit, would whiz away in the general direction of the target. The rocket was born. With the addition of balance weights and the technique of hollowing out the charge to produce a smooth simultaneous burn, it evolved into a useful weapon, even reaching the point of two separating stages during the fourteenth century. Yet until recently the rocket was never a primary weapon. This status was reserved for the second path of development, that of the projectile launcher or gun.

With increasing use of flamethrowers, the strength limitations and flammability of bamboo led to its replacement by tubes of cast iron and bronze. The resulting leap in durability allowed for the use of nitrate-enriched powder, generating explosive forces and a massive increase in the velocity of the burning gases emitted from the muzzle. So great were these forces that their usefulness as a propellant for a projectile must have been obvious. At first this took the form of bits of metal and ceramic shards packed in front of the charge; but soon arrows were being cast from the muzzles of these improved, now vase-shaped fire tubes. The problem was that a great deal of the force flowed out and around the arrow. Nevertheless, this was an important stage from a historical perspective, and a 1985 discovery inside a cave temple documented just such a vase-shaped device carved in high relief in the year 1128. Needham clearly shows that the final stage of the gun's basic evolution, a ball sized to approximate the diameter of the barrel—the first real bullet—also took place in China, but not before 1280. This was suggestive, since it is a virtual duplicate of the arrow shooter, not the ball thrower, that first appeared in Milamete's 1327 manuscript. The logical conclusion was that the transmission to Europe took place sometime before 1280.

The Chinese marginalization of firearms remains one of the major puzzles in world history. Gunpowder weapons in the hands of China's rulers offered an extremely promising means of thwarting the threat from the steppe, along with a ready-made instrument for maintaining public order at home—always a major dynastic concern. Certainly guns posed fewer psychological and cultural challenges to a military traditionally devoted to long-range engagement than they did to the Homeric warrior caste of the Occident. Yet they failed to thrive.

This had to do with the company they kept. The development of firearms in China took place amidst a tumultuous economic takeoff high-

Relief of Chinese vase-shaped gun carved in 1128.

Last Gift from the Steppe?

Prior to heading west, Genghis Khan's Mongols attacked China in 1205. From the beginning, the campaign was slowed by the horse nomad's ancient nemesis, walls. Unlike earlier steppe invaders, the Khan's forces refused to be thwarted, picking up a siege train consisting of anything that would help them break through fortifications quickly. This included gunpowder, which the Mongols routinely used to blow open city gates. They also must have known about many of the firearms the Chinese had available, including the vase-shaped arrow shooter, already in existence for at least seventy-five years.

When the Mongols did turn west to crush the Khwarizmian kingdom of Persia and then Georgia, Muskovy, Poland, and Hungary, they brought with them not only siege engines but also Chinese technicians adept in their use. It is logical that the train included individuals familiar with, or actual precursors of, the vase-shaped weapon that later appeared in Milamete's work. Just how this monumental transfer of technology took place—recovery of an abandoned weapon, the defection of a Chinese operator, or indirectly through Middle Eastern sources—remains unknown. But it now seems entirely possible that the Mongol invasion was the source of the West's introduction to firearms.

If this was the case, then the Mongols were the deliverers of their own destruction. For there was no more fertile ground than Europe for the seeds of gunpowder. It was the West that would take up firepower with a vengeance, and generate the armaments and procedures destined to render the horse archer defunct. Given the march of technology and the unchanging nature of steppe society, this fate was almost certainly inescapable at some point. But it might have been postponed, perhaps for centuries, had only the Mongols left firearms in China. For in the East, authority would circumscribe the gun until all revolutionary potential was suppressed.

lighted by a nascent industrial revolution. Improvements in transport, particularly canal building, led to the rapid commercialization and specialization of agriculture, generating volumes of cash and credit, which was poured into other segments of the economy. Most remarkable was the growth of metallurgy, particularly the iron and steel industry. Here, the substitution of coke for charcoal in the smelting process enabled production to surge dramatically to 125,000 tons per year in 1078, a figure 66 percent higher than the entire ferrous metal production of England and Wales 700 years later. It was the attendant metalworking skills that allowed Chinese technologists to transform the discoveries of alchemists into tangible and even mass-produced cannon. In outline form, all of this seemed to draw a picture of unstoppable technological progress analogous to the path eventually followed by the West. But it was nothing of the sort.

The governmental ethos of China was ultimately a hierarchical monolith, a mandate from heaven that had little place for the swirling forces of the market, technology, and war. Confucian mandarins were devoted to a strict ethical code, and in their eyes it was vital that entrepreneurs and soldiers remain subordinate. Consequently, when the economic takeoff of the first millennium created not only wealth but a growing chasm between rich and poor, it was suppressed. The rights of the officials to intervene was never in doubt, and the economic miracle withered just as surely.

The gun's demise was no less certain. War had always been viewed as a dangerous intruder and the excesses of Ch'in Shih-huang-ti had formalized this judgment in dynastic policy. True military competence was not only dangerous, but potentially all-consuming. In this context the gun was as much of a threat as a solution, and its continued development in the hands of the empire's soldiers was even more frightening than the threat posed by China's enemies. Better to temporize and bribe the outsiders than to let these fire-breathing mechanisms loose on the body politic.

Europeans were ready for the gun. The warrior class continued to cherish Homeric combat values; but by the late Middle Ages there were already signs of their passing. Thus in 1346 at Crécy-en-Ponthieu, England's Edward III achieved the first in a series of stunning victories over the flower of French knighthood by using highly skilled longbowmen firing at extreme ranges. The key English stratagem was the establishment of a narrow-fronted killing zone into which several thousand arrows could be launched in sheets at ten-second intervals. It took utter advantage of the charging French cavalry's confrontational urge to close, and also provided a glimpse of the randomized killing on the gun-based battlefield. When Edward approached Crécy-en-Ponthieu, he also brought a small fire tube fabricated by his armorers in the Tower of London. It had no effect on the battle, but does seem to have been shot.

A century later the gun was no longer a toy. It now fired spherical projectiles out of a cylindrical, not vase-shaped, barrel at much increased muzzle velocities. Key to these strides was the discovery, around 1420, that fine gunpowder could be mixed with water and then dried to produce a granular or "corned" equivalent that was significantly faster burning and more powerful. Corned powder combined with improvements in metallurgy allowed gunsmiths to fabricate so-called bombards—siege guns put together like beer barrels out of forged iron staves reinforced by hoops. They fired stone projectiles up to thirty inches in diameter and weighing in excess of 1,500 pounds. Bombards had a nasty proclivity to blow up in the faces of their users; but they were also crushingly potent against conventional curtain-walled fortifications.

European bombard.

Just how potent was demonstrated by Charles VII of France, when he effectively ended the interminable Hundred Years War by liberating over seventy English-held castles between May 1449 and August 1450, sometimes beating them down at the rate of five a day. The lesson of the war, officially over in 1453, received a crushing confirmation that same year when Turkish Sultan Mohammed the Conqueror captured the ancient city of Constantinople with a bombard so big it had to be put together on the spot by Christian technicians from Transylvania.[2]

By shattering masonry, the gun made a major impact on warfare in the West within a century and a quarter of its introduction. A number of historians argue that the true military revolution took place considerably later and was due as much to changes in organization and the sheer growth of armies as it was to the arrival of firearms. Yet it was the gun that forced these changes, the engine of destruction that sent reverberations across the West so profound that the year 1453 is often called the end of the Middle Ages.

The challenge posed by guns was not necessarily welcomed with enthusiasm in the West. Unlike China, however, there was no centralized authority capable of suppressing them; quite the opposite, guns were generally employed by rulers to broaden their authority. Similarly, the rising class of European financiers, who would soon play such an important role in the transcendence of firepower warfare, were viewed at best as a necessary evil. But gun-based militaries were massively expensive, as were the industries upon which they rested, and there was no place else to go for the money. So an alliance was forged among kings, capitalists, and cannon.

The West may have been behind China in overall technical infrastructure, but it did contain several areas of excellence that proved particularly important in the production of firearms. New mining techniques—among them the use of gunpowder for blasting hard rock—would open the rich argentiferous copper deposits of Central Europe to systematic exploitation. Also, the rise in church construction during twelfth and thirteenth centuries led to parallel advances in the bell caster's art, creating a reservoir of skills unsurpassed elsewhere. These two fields dovetailed to support major improvements in firearms.

After 1465, fabricators in France, Burgundy, and the Low Countries realized that artillery of greatly diminished size could inflict the same damage as huge bombards if tubes could be made strong enough to fire smaller but denser iron projectiles. The solution lay in a retreat to Homeric bronze, which was not only tougher than the iron used in bombards, but could now be cast with the necessary precision to thicken the critical area around the combustion chamber, while progressively tapering the barrel toward the muzzle where the pressures were reduced. Bronze remained expensive, but it was rust-free, could be recycled, and—thanks to new mining techniques—was available in sufficient quantity for fairly large-scale production. The resulting tubes were far safer and a great deal lighter . . . so light they could be accommodated on special two-wheeled gun carriages. Whereas ponderous fifteen-foot-long bombards had to be either built on site or moved by water, the new cannon, mounted on trunnions near its center of gravity and equipped with long "trails" extending to the rear, could now be transported cross-country by horse teams no more numerous than those that pulled large wagons. Going into combat,

Firepower on the Move

Charles VIII, the clueless king of France, had no idea he was a revolutionary, standing at the head of a new kind of army about to inaugurate a new era of war and politics. As he moved into Italy in 1494, he wouldn't have even called it an invasion. He was on his way to a Crusade, hoping to embark from Naples to reconquer Jerusalem. Instead, he lit a firestorm of war, sixty-four years of almost continuous fighting in Italy, followed by a leap northward to ignite the Low Countries, France, and eventually Germany. It was guns that generated the sparks and fanned the flames, and it was Charles who played the role of Prometheus.

The key to his army was the new kind of cannon—shiny, bronze, horse-drawn pieces, assembled for the first time in such numbers and drawn over the Alpine passes on a march that constituted one of the more notable strategic surprises in history. Italy, particularly its cities, had no answer to such a combination of power and mobility. Overawed, Naples and the Pope surrendered almost without a shot fired. The sole resister that year, the Neapolitan border fortress of Monte San Giovanni, made famous by withstanding a prior investment of seven years, was reduced to rubble within eight hours. This debacle set off a frantic but successful search for better defenses, enlisting Italy's best minds, including Leonardo da Vinci and Michelangelo. But before the solution could take hold, the scales of war remained unbalanced.

*Earliest known
handgun.*

artillerists simply unlimbered the trails, rolled the carriage into position, adjusted the barrel's elevation using the trunnions, and the gun was ready to fire . . . all within a matter of minutes. This constituted truly mobile firepower and meant that no fortified site in Europe was safe, especially in Italy.

While better siege artillery was knocking down walls, better guns were knocking down soldiers. Relatively little is known about the evolution of the earliest handguns, and only three examples dated before the year 1400 are known to survive. Physically, they were nothing more than small metallic tubes fired by means of a touch hole. Soon after, some were joined to a pike handle, marking not only the origins of the stock but the beginning of a relationship between pike and gun that would last hundreds of years.

The first European general to make effective mass use of small arms was Jan Zizka, who, around 1430, equipped his beleaguered Hussite Bohemians with primitive handguns, fired from behind carts with the specific intent of breaking through the double layer of mail and plate worn by men-at-arms as protection against armor-piercing crossbow quarrels. By the time the fighting ended it was said that more than a third of the Hussites were armed with firearms, many of whom were later employed by other armies as mercenaries.

During the last half of the fifteenth century several technical improvements were made that turned the fire tube into something recognizable as a small arm today. The addition of a true gunstock allowed the user to aim with two hands while absorbing the recoil with his shoulder. The first simple matchlock mechanically delivered the glowing tip of a slow-burning fuse to the touch hole, thereby removing a good deal of the randomness from the act of discharge. The result was the first widely used small arm, the arquebus. Although slow firing (no more than one round every several minutes), it packed a lethal punch. By the 1530s, according to Biringuccio, a master weapons craftsman of the time, arquebusiers were doing "in battle what the archers used to do."

Politically disastrous, the Italian Wars constituted a fertile test bed for exploring the new dimensions of firepower. This experimentation took the form of a series of set-piece battles, stretching from Cerignola in 1503 to Pavia in 1525, when the new defenses took hold and brought a temporary, generation-long halt to major European field engagements. The most obvious contemporary impression, the dramatically higher level of violence, was directly attributed to the introduction of firearms on a large scale. Cannon capable of killing 700 men in three minutes (Novara, 1513), or cutting down thirty-three men-at-arms with one ball (Ravenna, 1512), shocked Italian professional soldiers used to far fewer casualties. In the heat of anger they reacted by denying quarter and, in the case of one Paolo Vitelli, by plucking out the eyes and cutting off the hands of those caught with arquebuses.

In the longer run, though, the major response was symmetrical, with participants acquiring similar arms for themselves. As this occurred, casualties mounted still further, until men-at-arms such as Blaise de Montluc came to hate the gun: "Would to God that this unhappy weapon had never been devised, and that so many brave and valiant men had never died by the hands of those . . . that would not dare to look in the face of those whom they lay low with their wretched bullets."[3]

There was no going back. The gun constituted military reality. Ways had to be found to fight with it more safely and effectively. Not surprisingly the participants clung to the past, referring almost compulsively to the body of military writing preserved from the Greeks and Romans. Yet answers were not easily derived from the ancients. Firearms, by virtue of their operation, raised a host of use-related issues that had profound battlefield implications.

The vulnerability of cannoneers and arquebusiers while loading injected an element of tactical instability exploitable through maneuver, the judicious use of reserves, and close cooperation of separate arms—all of which took on increased importance and transformed command into an activity a good deal more cerebral than it once had been. The military leader was increasingly a firepower-savvy battle manager. Thus the cannonball's ability to fell body after body, given terrible vent at Ravenna, was taken carefully into account by commanders, who henceforth avoided massing deep infantry formations within easy range of artillery. Gonsalvo de Córdoba's extraordinarily effective use of arquebusiers packed in field fortifications, such as those at Cerignola, likewise demonstrated a clear understanding of the basic advantages of the defense in engagements featuring infantry armed with muzzle-loading small arms. Cavalry, on the other hand, was primarily victimized by firearms in the Italian Wars, and would not be positively affected until the wholesale adoption of the newly invented wheel lock pistol.

Hand- and shoulder-fired weapons had not driven, nor would soon drive, all other weapons from the field of battle. The sword and shield combination and the bow went fairly quickly. But several traditional members of the previous era's arsenal found secure futures. The success of the arquebus and later the musket in the field would be predicated on a symbiotic relationship with the pike phalanx, which kept off attackers and broke opposing formations destabilized by firepower. Cavalry eventually took a step backward, casting aside their pistols and taking up once again the saber and the lance. Such retrogression raises a fundamental issue.

Why did guns proliferate on battlefields after the year 1500? It could certainly be argued that a good bow had considerably greater range, accuracy, and rapidity of fire than contemporary small arms. Compounding matters, arquebuses and pistols were exceedingly prone to misfires and virtually useless in the rain. It added up to a substantial indictment, one that left a number of skeptics wondering if the adoption of small arms wasn't some sort of historical accident.

Wheel lock spanner.

It was not. Smoothbore firearms may have been slow-firing, unreliable, and inaccurate; but they were killers. A slug of some weight hitting a human torso at near supersonic speeds not only pierced any body armor available, but shattered bone, induced trauma, and inflicted a large jagged wound. Unlike an arrow or quarrel, which, barring a lucky shot, required some time to induce sufficient internal bleeding to debilitate, a single hit by a gun generally put a combatant down to stay. And, although operating a muzzle-loader was far from simple, it was a skill that could be taught over a far shorter period than was demanded by the bow or the traditional arms used by the aristocracy. The gun continued the precedent set by the crossbow; it opened the use of sophisticated weaponry to relatively vast pools of agricultural laborers, and offered the prospect of turning them quickly into lethal combatants. Finally there was the sheer visceral impact of firearms. Humans are instinctively frightened by loud noises, and there were numerous instances of soldiers, both European and non-European, breaking in terror when first exposed to the reports of guns. Soldiers did become conditioned to such sounds. But eyewitness accounts indicated that the sheer noise of firearms continued to be impressive, and the cacophony they

Killer Clockworks

The wheel lock provides an excellent example of the serendipity of technologies that was so much a part of the gun's triumph in the West. The evidence indicates that its invention took place around 1510 in the workshops of southern Germany famous not only for their arms and armor, but also their clocks. In fact, the wheel lock was essentially a clockwork mechanism. The heart of the device was a small serrated wheel attached to a powerful steel spring. Employing a spanner, the pistoleer wound the wheel, tightening the spring, until it engaged a catch linked to the trigger. When pulled, the trigger released the catch, causing the wheel to spin against a lump of iron pyrite, which generated sparks, igniting the powder in a priming pan, the lid of which was also retracted during the firing sequence. Complex and ingenious in the manner of a cuckoo clock, the wheel lock was self-igniting and therefore addressed a key problem of guns on horseback: the rush of air that usually extinguished the matchlock's slow-burning fuse.

Capable of reliable one-handed fire, the short-barreled guns were an immediate success with horsemen. Lances discarded, they went into battle literally packed with pistols most carried a brace in saddle holsters plus one stuffed in a boot, but some managed four or even five. Once adopted, the weapon caused mounted tactics to shift dramatically from the headlong charge to the ponderous caracole, a maneuver aimed at blowing holes in infantry formations through the repeated discharges of ranks of horsemen, who then attempted to reload in the saddle. It didn't work very well, but still demonstrated the power of guns to bend the conduct of battle.

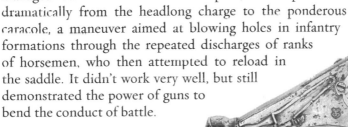

Wheel lock pistol.

produced remained very much a part of the confusion and terror of battle. To the Western mind steeped in Christianity there was also something very appropriate about these smoke-belching infernal engines. If battle constituted hell on earth, then why not fire and brimstone?

Nonetheless, the tactical implications of handguns were at odds with prior military experience. In order to maximize firepower and avoid being slaughtered in clumps by cannon, optimal deployment of small arms infantry maneuvering out in the open demanded that they be stretched as far as possible. On the other hand, stringing men all over the battlefield led to dangerous instability, especially if these lines were taken in the flank.

Despite this fundamental contradiction, infantry formations generally followed the dictates of the gun. As early as 1516, the Spaniards began experimenting with the countermarch, an internal maneuver that saw successive ranks of arquebusiers, or later musketeers, each firing a volley and then retreating between the files to reload. When first introduced, the countermarch required a minimum of ten ranks to keep up a steady rolling fire. Yet as infantry trained, rates of fire increased, allowing more men to be removed from the files to further elongate the lines, until by the end of the seventeenth century it reached a standard three ranks. Although it invariably worked better in training than in battle, the impact of the countermarch was pervasive, since it established the original conditions for the loading drill, which remained at the heart of infantry tactics.

It is said that the Italian Wars marked the reemergence of infantry as the centerpiece of field warfare. The issue is more complex. Each of the three basic components—infantry, artillery, and cavalry—had strengths and weaknesses, which were consistently exploited by the others. Without field fortifications, which mercenaries were proverbially unwilling to dig, infantry in elongated formations were at risk. So too was the relatively slow-firing artillery. On the other hand, cavalry, once the caracole was shed, possessed the mobility to consistently endanger those on foot—though it was terribly vulnerable to firepower. Tactics, then, rested on a balance of the three elements. Most battles eventually turned on a symmetrical match between rival infantry lines, but seldom without regard for the dangers and possibilities posed by the other two elements.

Yet the lessons of firearms were mostly processed through minds conditioned by the military beliefs and expectations of the previous era. This was the framework on which new weapons were hung if at all possible. It was symptomatic that until the seventeenth century artillerists were grafted onto armed forces as independent commercial entities rather than integrated into the force structure. Similarly, ranks of arquebusiers were attached initially to blocky field formations of pikemen like tactical cummerbunds.

Nowhere was this atavism more evident than among the aristocracy. Thus before Ravenna, the first major field action to be decided by an artillery barrage, the rival commanders exchanged formal defiances by trumpet. Efforts were also made to perpetuate individualized combat in the face of firepower's tendencies toward capricious killing. As late as 1536 Emperor Charles V saw nothing odd in challenging the King of France to single combat, a prospect that was only stopped by the intercession of the Pope.

The warrior code was also behind the persistent urge to close. Not just mounted knights, but infantry commanders continued to hold the hand-to-hand melee in high esteem and seldom missed a chance to join one. In 1525 Francis I, finding his army surprised and misaligned at Pavia, lead charge after charge into the Imperialist lines in a desperate attempt to hold off inevitable defeat, leading to his capture and the greatest slaughter of noblesse since Agincourt. Fate was still tougher on young Gaston de Foix, commander of the French army at Ravenna. Having already won the day, he and twenty other gentlemen sportively charged a retreating column of desperate Spanish arquebusiers, who, being in no mood for chivalry, gunned them down to a man.

The implications of such incidents were ratified by available casualty figures for men-at-arms. At Ravenna eleven of the twelve colonels commanding units for the losing Spaniards were killed; while on the winning side virtually all the officers of the German mercenaries died fighting. Similarly, at Marignano (1515) and particularly La Bicocca (1522), where officers were coaxed into fighting in the front ranks, the Swiss command was virtually wiped out. For a time men-at-arms took refuge behind ever thicker and more complete plate suits; but by 1530 the introduction of the heavier musket, capable of piercing any armor at 200 yards, left them once again very vulnerable. The rapid adoption of the pistol can in some measure be attributed to the fact that as a short-range weapon, it afforded horsemen the firepower equivalent of hand-to-hand combat. But it was altogether more deadly and less discriminating than traditional close-in weapons, and in a melee anyone—noble or commoner, skilled or unskilled—could bring down the most celebrated of warriors.

The course of arms resembled the evolutionary concept of punctuated equilibrium, with long periods of stability interrupted by short bursts of very rapid change, generally centering around one key breakthrough. This was certainly true of the gun's arrival in Europe. Just as occurred at the very beginning of the atomic age, the period of the Italian Wars marked one of the rare points at which persons of towering intellect became involved with weapons development. Thus Einstein, Bohr, and Kapitsa had Renaissance equivalents in Leonardo, Michelangelo, and Dürer. None were warlike men. All were simply swept along by the events of the day, convinced, despite reservations, that the importance of arms advances warranted their participation. The phenomenon amounted to a trans-European eruption of death-dealing ingenuity. They looked out hundreds of years into the future and delineated the weapons types made possible by chemical energy. Doubtless an absence of appropriate power sources and materials fore-

The Duke of Parma's Lunch

The gun was ravaging the soul of the warrior. To many among them, virtually the whole purpose of battle was to demonstrate courage. Custom dictated that an international corps of heralds hung like scavengers about the battlefield, ascertaining brave deeds to be recorded by chroniclers. Bullets were making the whole process ridiculous; the standards of courage were becoming the standards of idiocy. Insistence on close-in fighting, elaborate rituals of identification, and pairing off were not just inappropriate on a battlefield full of guns; they helped reveal the impotence of the ruling classes.

The bullet-riddled environment of the sixteenth century demanded a basic redefinition of what constituted courage. This would take time; but an incident near Brussels in 1582, during the Dutch rebellion, foreshadowed the direction it took. It was an early-spring afternoon and Alexandro Farnese, the Duke of Parma, Philip II of Spain's most famous general, decided to dine with his staff outdoors, near the trench works. No sooner had they sat down when a cannonball took off the head of a young Walloon officer, and a skull fragment also struck out the eye of another gentleman. The table was cleared only to have a second ball kill two more of the guests. Their blood and brains strewn over the previously festive board, the remaining diners lost all appetite and got up to leave. Yet Parma calmly insisted his guests resume their places, ordering his servants to take away the bodies and bring a clean tablecloth.

A traditional hero might have charged the cannon . . . not Parma. His response was passive disdain. If flesh and bone were unequal to flying lead and iron, the spirit was. Parma's defiant hospitality was a prototype. One day men of courage would be inclined to stand fast and take it. Rather than ferocious aggressiveness, not flinching became the sine qua non of the warrior class.

Jan Zizka's war carts.

closed on the exploitation of some; still, the vision remained penetrating and clear from the beginning.

As far back as the 1430s, the blind visionary Jan Zizka not only played an important role in the introduction of small arms; but he and his Bohemian Hussites anticipated amazingly the armored column with their *Wagenburgen,* caravans of horse-drawn, armor-plated carts equipped with firing ports and portable bombards. Screw-based breech-loading and exploding shells, two of the prime factors in the dramatic artillery developments of the nineteenth century, similarly were known before 1500. Considering that the characteristics of smoothbore small arms largely defined the nature of firefighting through the Mexican-American war of 1846, it is also notable that rifling was used for hunting pieces by 1525. By this date there had also been numerous experiments with multiple barrels and at least one documented attempt at a Gatling-type machine gun. Hand grenades, land mines, and a variety of incendiaries also appeared. By 1530 many of the lethal instruments of the early twentieth century had actually been born. But still more impressive was what existed in the mind's eye of the era's greatest genius.

The immediate future, however, belonged to protection. As far back as the 1440s, Italian architect Leon Battista Alberti speculated that fortifications would be more effective against gunfire if they were "like the teeth of a saw," even suggesting an overall star shape. Alberti's thoughts were not

Leonardo's Notebook

He hated war, calling it the *"bestialissima pazzia"* (the most bestial madness). Nor did his weapons fantasies occupy more than a small portion of his voluminous manuscripts. Yet Leonardo da Vinci, perhaps more than any other major scientific thinker, was gifted with what he described as "the imagining of things to be."

Almost unique among military thinkers he was unconcerned with operational issues such as doctrine and tactics, concentrating instead on the mechanisms of Mars. Characteristically, his visions of future weapons were precise and mechanically credible,[4] destined to serve as blueprints for the Western technical imagination. Leonardo described at one point or another cartridges for small arms, explosive projectile firing mortars, air guns, a Gatling-type gun, steam catapults, rocket launchers, chemical weapons, armored vehicles with dramatically sloping ballistic protection, submarines, and flying machines. Certainly key supporting technologies for many were missing; but it is still accurate to say this one man caught with a sweeping glance the future of weaponry virtually to the end of World War II.

published until 1485, and the vulnerability of high, thin curtain walls still remained. Rather quickly, though, the possibilities of shock absorption became apparent. In 1500 the Pisans, under siege and desperate, threw up a rampart of earth and found that the dirt swallowed cannonballs like quicksand. Serendipitously, the ditch formed in front of the barrier from digging created an additional obstacle, a kind of negative wall. These two concepts, Alberti's and the idea of ramparts, combined to open the way to a fundamentally new kind of defense.

As walls were built lower and thicker, the ground directly below was no longer visible to defenders, raising the possibility of unobserved assaults. The counter, effective flanking fire, could only come from rectilinear gun towers, or bastions, projecting at an angle from the ramparts. Cannon mounted on these bastions did not just sweep the ground below, but harassed opposing siege guns. Plane geometry along with accommodations for terrain variations enabled designers to cover every angle of approach. In the process they generated polygonal-shaped forts that were actually variations of Alberti's basic star. With time, refinements were added, including mutually supporting outworks—ravelins, crownworks, hornworks, and redoubts—along with stone facing on the slope of the ramparts to prevent soil erosion.

The system came to be known as the *trace italienne* and it was extraordinarily effective in protecting strongholds and their outlying territory. It spread quickly through Italy after 1515. But the *trace* was extraordinarily, even ruinously, expensive. Both Rome and Palmanova had to significantly scale back their construction programs, while the Republic of Siena spent so much that it had no money to hire defenders when an attack came, thereby losing its independence.

Nonetheless, the proliferation of new fortifications so clogged the landscape that after 1525 field battles ceased. The Italian Wars became a matter of drawn-out sieges. As historian Geoffrey Parker explains, a heavily defended fortress that sheltered multiple thousands of soldiers became too dangerous to an advancing army's supply lines to bypass. It had to be taken down. Yet the new fortifications were so effective that their defeat usually required time and unprecedented quantities of munitions and numbers of men. Normally siege tactics relied on picks and shovels, creating vast trench works (hired laborers, not mercenaries, did the digging), which first surrounded the stronghold and then closed in with zigzag slits and concealed batteries. Once they reached the ditch or moat, sappers drove deep mines to the ramparts, which they filled with gunpowder in hopes of creating a substantial breach. Often defenders dug equally

Probing for Weakness

Every defended place had to have a way in and out, and if it was a city or a significant fortress this meant a door wide enough to admit supply wagons . . . or a body of attacking troops. Into this potential breach was thrust the petard, a demolition device designed to blow open a defended entrance. Shaped like a bell with handles, it was filled with explosive and sealed at the mouth. The idea was to attach it to a door, light a fuse, and then blast it off its hinges.

There were problems. While a simple door required only a small, very portable petard, a formidable gate fronted by iron grating might demand a bomb so large it could only be transported by cart. Nor could it simply be leaned against its target; directing the blast through the portal meant that it be firmly spiked or chained into place. All of this was noisy and took place precisely at the point most likely to be closely guarded. Therefore, the life span of petardiers tended to be short. If not killed outright by defenders firing triple-shotted flanking cannon, captured demolitionists frequently met with what amounted to poetic justice. "For 'tis the sport to have the engineer/Hoist [blown apart] with his owne petar," Shakespeare reminded us in *Hamlet*. Accordingly the heyday of the petard did not extend much beyond the middle of the seventeenth century.

More lasting was the mortar. In 1520, when artist/arms aficionado Albrecht Dürer visited the Netherlands, he did not fail to sketch an impressive piece of artillery called a mortar, one of several "wonderful things" he discovered in a visit to a gun foundry. If direct-fire artillery had great trouble firing through the *trace italienne,* the mortar was to fire over it with a distinctively high trajectory. It was short, squat, and thick-walled, its downward recoil demanding that it be firmly anchored to the ground, not a carriage—a natural siege piece. Mortars during Dürer's time were limited to solid projectiles, the largest able to cast stones of more than 300 pounds. Yet Leonardo's notebooks included a drawing of a mortar firing bursting projectiles, and within a few years they had become a reality. Launching them was ticklish. The bombardier had first to charge his mortar and slide the projectile down the short muzzle, fuse facing upward. Then in rapid succession, he lit the projectile, the charge, and took cover. For if his timing was off, the fuse burned too quickly, or the charge was a dud, the results could be extremely unpleasant.[5]

Mortars were, in the end, weapons of attrition, designed less to break down fortifications than destroy the property and persons within. They might allow attackers to hurt what the new defenses protected, but not break in.

tedious and dangerous countermines aimed at thwarting the advance. Thus siege warfare settled down to a nasty laborious equilibrium.

The *trace italienne* thrived, crossing the Alps and finding new homes in

Bronze siege mortar.

places that could afford it. Soon over a hundred Italian military engineers were employed in France bolstering the kingdom's northern borders, which by 1544 were defended by fifteen of the new type of forts studded with over a thousand cannon. At the same time, other Italians were busy in the Habsburg Netherlands, building more than twenty-five miles worth of modern defenses between 1529 and 1572, at the staggering cost of 10 million florins. The same process took place wherever there was money.

The strategic map came to be a kind of chessboard. Those areas left unpenetrated by the new defenses might still feature more decisive battles in the field. But where the *trace italienne* had taken hold, sieges could not be avoided. Cannon and cannon-proof defenses thus insured that European warfare in areas of the greatest wealth and most dense populations grew more protracted and less decisive—virtually the polar opposite of the Greek phalanx's effect—and so expensive that only princes could afford them.

The gun had a rapid and profound impact on international affairs in the West. By expanding the scale and economics of war, firepower pushed princes to the center stage of high politics. Hence, European statecraft became even further personalized, cast in a great melodrama of family feuds and personalized disputes. Yet changes in the power of players and the implements of power meant changes in the way the game was played and the rules that were followed. Contemporaries recognized a new ruthlessness that accompanied the spread of firepower, and seldom failed to attribute it to a disgraced Florentine diplomat and the author of the greatest political "how to" manual in Western thought.

Every age has its master metaphor, a concept so powerful that it comes to pervade not just its own area of influence, but numerous others—even when the fit is bad. Recently it has been the computer. A century ago it was evolution. In the time of the gun's adolescence it was commerce and capital formation, fiduciary mechanisms that structured a host of endeavors, including war. It was no accident that *condottiere* literally meant contractor, or that mercenary bands were termed "companies." They were exactly that, corporate entities dedicated to making a profit. If the medieval nobility's monopolization of war left few opportunities for mass military participa-

Mentor of Princes

Though a republican at heart, Niccolò Machiavelli was no man to turn a blind eye to the facts. His book was about princes, for princes, dedicated to a prince, and, so no one missed his point, entitled *The Prince*.

Considered the quintessential guide to the balance of power, *The Prince* recommended transcendence whenever possible. Nor was Machiavelli less than clear on the central means of achieving it. "A Prince should therefore have no other aim or thought, or take up any other thing for his study, but war and its organization and discipline."[6] Naive as to the tactical impact of firepower, Machiavelli was nonetheless acutely aware that warfare had changed fundamentally, and sought to develop appropriate guidelines. Just as Homer had codified etiquette for the age of swords, Machiavelli pared it down in a way that matched the new tools at hand. The randomness and impersonal killing capacity of the gun was paralleled by the ruthless pragmatism and deceptiveness he recommended for princes. The only rules to be followed were those absolutely necessary to retain loyalty and credibility. Nonetheless, Machiavelli remained suspicious of the international band of mercenaries firearms had fostered. So too did he accurately forecast that no prince could fully control and utilize his military until he had a national army at his back. Yet this also exposed a key contradiction of the time. If princes had no earthly superior, they had at least a ruthless taskmaster, and that was money.

tion, paid combatants and armies for hire quickly filled the vacuum. Yet as metaphor-stretch was prone to do, this one had some very unfortunate consequences.

The treatment of artillery provided an example. Virtually from the arrival of cannon in Europe, their military integration rested on the girders of capitalism. Artillerists formed independent commercial entities, which not only cast tubes and procured powder and shot, but also tended them in battle. Throughout the sixteenth century they remained autonomous, retaining an intimate knowledge of all phases of manufacture, testing, and use. They were not soldiers, but more akin to alchemists, members of a guild, which, disdaining military discipline, passed its secrets under oath to apprentices. Until the time of Gustav Adolph in the seventeenth century, when they were finally integrated, about the only thing a commander could count on from his artillerists was that they would try to leave if not paid.

Nor did the buck stop here. If the easy-to-learn crossbow brought forth the mercenary infantry band, the arquebus and later the musket only rein-

forced the trend. Rather suddenly very large numbers of agricultural laborers became potential soldiers. Many joined, intent upon becoming professional soldiers. Training and especially combat accomplished just that,
creating a class of veterans, whose usefulness in battle often proved well out
of proportion to their numbers. Companies of veterans became valuable
commodities, very much worth importing, especially to heretofore peaceful
areas, where their fighting advantage was maximized. This process first
occurred in Italy, but as the wars on the scorched peninsula finally burned
out, companies of veterans were observed trudging north.

History had seen mercenaries before, but none so armed or so systematically avaricious. It was this combination that brought an entire new and
vicious cast to the conflicts that followed. War hung over France, the
Netherlands, and finally Germany like a festering contagion, and armies
assumed the role of parasites, living off and reinfecting the host. Spanish
troops stationed in the Low Countries during the Dutch rebellion mutinied
incessantly when left unpaid, and then took it out on the inhabitants. In
Germany and what is now the Czech Republic armies of the Thirty Years
War crept back and forth over the countryside, cutting swaths of destruction, burning and looting and eating the land into chronic desolation and
poverty. Frequently missions, especially among garrison troops, were undertaken for no other reason than profit—to steal some food, to raise some
money, or to take a valuable hostage for ransom. Despite the religious
nature of the conflicts, the utilitarian ends of capitalism were frequently
served by filling out depleted force structures with the ranks of the
defeated, whatever their faith. Veterans were too valuable to waste, or to let
free when not needed, in which case they were shot. Yet these excesses
proved one thing: certainly business would continue to be associated with
war, but war was not a business.

In the midst of excess there was also moderation, or at least its
promise. Techniques were developed by the Dutch that would one day
bring the gun under control. Maurice of Nassau, son and successor to
the martyred William the Silent, was faced with the same problem as
his father—how to get rid of the Spaniards. After a generation of fighting, the prospect of outlasting them in siege warfare must have seemed
tenuous at best. But to have a chance in decisive field battles against
the Spanish steamroller—a pike and firepower formation know as the
tercio—meant building not just an improved Dutch army, but a radically improved one.

Firepower was the key, and there were two potential approaches: better
and more. At the end of the sixteenth century an experienced musketeer

could still manage no more than one shot every two minutes. In the case of the Dutch this meant the tercio would be upon them after only a few rounds fired at effective ranges. One prospect was to make the shots more accurate by equipping troops with military versions of available hunting rifles. But the tight seal between bullet and bore made rifles even slower to load—literally a matter of hammering the projectile down the barrel. The alternative was to keep the smoothbore with its larger tolerances, and figure out a way for musketeers to fire faster.

On December 8, 1594, William Louis of Nassau wrote his cousin and military colleague Maurice to announce a really good idea. He had just read Aelian's description of Roman drill and believed that rotating ranks of musketeers could approximate the continuous hail of pilum once launched by legionaries. It was a countermarch en masse, though with a much thinner, more stretched out formation. This meant that the lines of troops had to not only fire volleys simultaneously, but load with a much greater precision.

The solution was practice—lots of it. The ratio of officers to troops was doubled for instructional purposes, and during the 1590s Dutch soldiers were almost constantly at their "exercises"—working on the countermarch, forming ranks, and parading—according to one of Maurice's staff members, Anthonis Duyck. By 1599 Maurice had convinced the States-General to

One of 32 pictograms used to train Dutch musketeers in loading and firing.

appropriate the funds to equip the entire Dutch field army with guns of the same size and caliber, further regularizing drill. Meanwhile, another cousin, John, was busy working on an illustrated drill manual, a key innovation. Cousin John analyzed the movements necessary to load and fire a smooth-bore—twenty-five for the arquebus and thirty-two for muskets—and captured them in a series of precise pictograms, which could be practiced over and over until they became automatic. The result was a 100 percent increase in firepower.

Qualitatively, the change was even more fundamental. An experienced soldier himself, Maurice understood that at the core of his reforms was the necessity for absolute discipline and control, and this meant turning troops into battlefield automata. To do this the Dutch came to rely almost completely on mercenary troops, paid with absolute regularity to eliminate mutinies. In return, Maurice's troops were required to dig the field fortifications, work that mercenaries had previously shirked. But most of all they had to drill.

Relentless drill, historian William McNeill theorizes, had an effect that Maurice and his colleagues probably barely understood. When groups of humans moved their large muscles in unison for prolonged periods, a primitive but significant social bond emerged—a bond strong enough to weather issues of life and death. Thus, if the small-group loyalties formed by our past as hunters can be analogized to the force that holds protons and neutrons together, this drill-induced affinity can be viewed as the social equivalent of the force that binds molecules—the stuff of companies and battalions. An army that drilled together stuck together.

Yet in another sense there was nothing primitive or visceral about the effects of Maurice's drills. They were a mélange of highly abstracted gestures, largely alien to the heroic standards of the previous era. Instead of intimate physical contact and the direct infliction of injury, everything turned on the precise and automatic repetition of a complicated routine of second-order movements on a killing instrument optimized for long range. To a degree this had been true of the bow, but the associated motions were simple enough to constitute a difference in kind. The abstract moves needed to load a musket anticipated the esoteric military rituals of the technological future, and would dominate the next 250 years of military history.

The reforms of Maurice and his family provided a palliative for a great deal that had gone wrong in this mangled stretch of military history. The very close control of troops, the regularization of their existence, and the abstraction of aggression all pointed the way to a more benign brand of war. If, in the long run, it was not foolproof, then at least it provided a metastable means of using firearms without totally brutalizing warfare. Yet Maurice and his well-trained army never really won a decisive victory in the field; instead they simply learned to hold their own against the tercio. So the reforms spread slowly, and Europe would not be spared another horrific spate of bloodletting.

There remained a real sense that the Western way of doing things was brought to the brink during the Thirty Years War, that the anarchy and suffering inflicted on Germany and Bohemia were products of forces intrinsic to the politico-military system itself. In the end the war provided a stark example of what happened when these forces were not closely controlled. It was a terrible price, especially considering that the lesson was eventually forgotten.

The slow progress of the *trace italienne*'s bastion defenses in poorer, more rural Central Europe insured there were many more pitched battles in the Thirty Years War; but they were generally indecisive. Armies were simply rebuilt, their size creeping up to a theoretical preindustrial limit of 100,000. Although individual combat persisted, this was mass warfare with the fighting weighted heavily in the direction of predatory aggression. Pistol-wielding heavy cavalry assumed an enhanced importance largely because of their ability to run down and dispatch defeated infantry. As elsewhere, refusal of quarter was increasingly the norm, with mass executions becoming regular features of the aftermath of battle. Although sieges were plainly less dominant than in the Netherlands, they nonetheless reached new levels of barbarity, escalating the same pattern of looting, mass rape, and indiscriminate killing into a frenzy of violence that was the grotesque fate of Magdeburg. History seldom witnessed a scene so perverted as the victorious Te Deum sung in the city's redeemed cathedral amidst the 20,000 shallow graves and blackened ruins of the formerly Lutheran city.

War's passage from Italy to France and the Low Countries and then to Germany had been a steady decline into an abyss. In all of this the gun played a role. Whether serving as ready-made thumbscrews for peasants bent on concealing their food or women, or impelling soldiers to melt church steeples for bullets, firearms technology contributed directly and indirectly to the war's burdens. As all other forms of supply collapsed, guns and powder remained plentiful, holding out the promise of turning peasants into soldiers within a few weeks or months. So the fighting staggered on, choreographed by the staccato crackle of gunfire. Yet the dance

The Miseries of War

Jacques Callot was a great artist driven to despair by what he had seen. Born in the city of Nancy in 1592, where his father was master of ceremonies to the Court of Lorraine, Callot began drawing at an early age, and at sixteen was permitted to go to Rome, where he studied engraving. A prodigious worker, he quickly abandoned this medium for the more rapid technique of etching. Soon he was in the service of Cosimo II de Medici of Florence, during which time he developed the "hard grounds" technique, which allowed him to turn his small etched compositions into richly detailed microcosms. His virtuosity combined with patronage from the Court of Lorraine earned him wide renown and numerous commissions from the powerful, hungry for images of their military triumphs, such as the sieges of Rochelle and Breda. Callot obediently produced them, often from Olympian heights that captured the entire tableau, but also reduced the participants to insectlike proportions and sanitized the violence. Yet Callot had plainly witnessed these events at close hand, and what he saw deeply disturbed him.

Haunted by the images of modern war and afflicted with incurable stomach ulcers, Callot retired to Nancy to await death. There he managed his best and most lasting work—an unflinching vision, executed in eighteen hair-raising scenes, of the depths to which war had sunk: *Les Misères et les Malheurs de la Guerre*. There was nothing sterilized about the butchery here. Not only was the viewer treated to the carnage of battle, but special emphasis was placed on gratuitous violence. Troops were shown alternately running wild,

Enrollment of troops from Jacques Callot's Les Misères et les Malheurs de la Guerre.

then suffering for their crimes. In two particularly horrifying scenes soldiers are portrayed ripping apart the great room of a rich farm—raping, looting, torturing, stabbing, and even burning the inhabitants—and then facing the consequences, twenty of them hanged from a single tree. Meanwhile, guns were everywhere—arquebuses handed out to new recruits, vast clouds of gunsmoke hanging over a battlefield, pistols alternately slaughtering cavalry and peaceful travelers, muskets performing the grisly work of firing squads, or small arms simply gracing the shoulders of troops in formation. Their effects were similarly evident in a crowd of penitent amputees gathered in front of a hospital. *The Miseries of War* ended with two particularly trenchant plates—one depicting a mass of peasants exacting their revenge on a group of soldiers, some shot with their own guns, and the other "Distribution des récompenses," depicting the prince and his courtiers sharing the spoils in a setting far removed from anything vaguely suggesting the suffering they have inflicted.

Some have suggested that Callot's series was prompted by Louis XIII's invasion of Lorraine, or at least events confined to France. This was contradicted not just by the absence of identifiable flags or insignia, but by the universality of the theme. Rampaging troops, killing of the most callous sort, the greed of rulers and their backers, the suffering of civilians, and the lethality of firearms—Callot had seen this going on all over Europe, and it is probable that he meant to depict it as such. This is why his dissection of conflict during the adolescence of firepower has endured and taken a place among antiwar art second only to Goya's *Desastres de la guerra.* Seeing was believing. The gun was behind everything. It had to be tamed.

was increasingly a lumbering exhausted shuffle, whose most apparent purpose was to grind the lives of innocents underfoot, and then bury them.

Living in the midst of the nightmare, thoughtful individuals witnessed what was happening with growing disbelief and horror. "I saw prevailing throughout Europe a license in making of war of which even barbarous nations would have been ashamed," wrote Hugo Grotius, the Dutch legal philosopher.[7]

Chapter 8

GUNS AWAY

T he gun was not a Western invention. Nor was it true, after the Chinese lost interest, that the West held a monopoly on firepower. Several societies, most notably the Ottoman Turks and the Japanese, not only adopted guns quickly, but learned to use them to great military advantage. Unlike the peoples of the Americas, Siberia, and elsewhere who were overwhelmed quickly, they were able to hold the West at bay until the late nineteenth century. Yet even one of the most outspoken of the military revisionists, Jeremy Black, concedes that it was the West that expanded eastward, not vice versa: "the Portuguese were on Deshima, the Japanese were not on the Isle of Wight."[1]

Central here was the development of the big-gunned, full-rigged, oceanic warship. As an implement of power projection no other society had anything close. It could go practically anywhere with a seacoast, mounting the armament of a good-sized fortress. It could carry goods to trade, along with sufficient foot soldiers to establish a base. If things went wrong, it provided a ready avenue of escape, or, alternately, a means of resupply and communications. Stretching terminology only a bit, the shipwrights of Europe had come up with what amounted to an Elizabethan space cruiser.

The sailing gunship evolved in a complex process of north–south cross fertilization, with ideas flowing back and forth from the Mediterranean and the states facing the North Atlantic during fourteenth, fifteenth, and sixteenth centuries. Although the final product visually most resembled the high-sided merchant cogs first brought into the Mediterranean by Basques from northern Spain, it had in fact come from both nautical worlds.

As in the Hellenistic era, a key institutional change was the centralization of arms development, in this case the establishment of naval arsenals consciously dedicated to improving ship types. While France, England, Portugal, and Spain would set up government-operated wharves, the model for all was the Arsenal of Venice. Its first major success with warship design took place around 1300, when it managed radical improvements on the Byzantine dromon, the Mediterranean's standard battle galley. By placing three rowers on the same level seated in echelon, all using sweeps of different lengths, the Venetians produced the trireme *alla zenzile* (literally "in the simple manner"), which quickly came to dominate galley warfare. It was slower than the Greek trireme (seven knots versus ten), but by keeping all the oarsmen on a single deck it lowered the center of gravity, making for a more stable platform, an important consideration once guns were added.

The arsenal was managed to encourage innovation. Engineering talent was attracted and then allowed to work with a minimum of technical interference. A cadre of master shipwrights was developed. Advances were carefully recorded and incorporated, leading to the first treatises on ship-building, the stockpiling of interchangeable parts, and the beginnings of standardized naval design. As this transpired, the arsenal staff learned more and more about hull construction, and were therefore an important factor in a new and radically improved means of putting a ship together.

The ancient scheme of shell construction, joining planks edge to edge with mortise and tenon joints and then adding a light internal frame, worked well enough; but it consumed large amounts of carpentry skill, time, and money. An alternative was to construct a strong internal frame made up of numerous vertical ribs, and then attaching outer planking to them. The transition to this skeletal construction took place gradually over several hundred years driven by a shortage of skilled carpenters after the collapse of Rome. The first all-frame ships were probably built early in the eleventh century.

Shaping and positioning the frames demanded highly trained craftsmen; but fastening the planking to the skeleton could be managed by almost anyone capable of handling a hammer. This fit perfectly with Venetian naval construction strategy. Galleys were short-lived, fragile craft. Rather than keeping a large fleet in being, stockpiling complex parts at the arsenal allowed the state to maintain a core of skilled shipwrights, fully able to direct a dramatically enlarged workforce to rapidly build a brand-new fleet in an emergency. But as good as was the technique (known subsequently as

Dead-End Galleys

The *alla zenzile* arrangement was an advance, but still retained the fundamental short-comings of the galley as an instrument of war. Its hull was 136 feet long and around eighteen feet wide, with twenty-four clusters of oars lining the side. This configuration was not just manpower- and skill-consumptive—a single rower at each sweep implied considerable individual expertise—but left the crews exposed, since there was no deck above them.

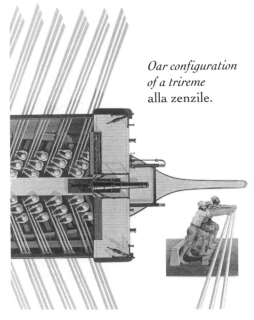

Oar configuration of a trireme alla zenzile.

Because triremes *alla zenzile* were slower and less handy than their Classical counterparts, the tactical tendency was to avoid individual maneuver, advancing instead side by side, or line abreast, hoping to create disorder and vulnerability in the opposing line. The addition of guns reinforced this approach. Because galleys remained flimsy and vulnerable from the sides, the obvious mounting point was the bow, where they could be firmly secured and essentially aimed by the helmsman. By the end of the fifteenth century, very heavy main centerline guns firing balls of fifty pounds or an equivalent amount of scattershot were common in the Western Mediterranean. These were frequently flanked by two smaller cannon and a variety of swivel pieces, all basically forward firing. There was but one way to attack: ahead.

Since a massive round of inflation during the sixteenth century put the wages of citizens well beyond what states could pay them to man oars, slaves, war captives, and criminals were the only alternative. But there was only so much that could be done with desperadoes at sea. The skills needed to properly handle an individual sweep floated beyond the reach of the Mediterranean's naval powers, leaving a stately advance about the only tactical choice. The *alla zenzile* configuration was replaced by the *scaloccio* galleys, in which five or more drudges pulled a single large oar—an archetype of job dissatisfaction.

To prevent mutiny or simply suicide, oarsmen were chained to their benches, forced to relieve themselves where they sat, and placed in the most extreme form of peril should the galley begin to founder. Guns made this a certainty. For what better way to disable a galley than to rake its immobilized crew? Guns even helped to insure a multiplicity of targets, since as their number and size increased, still more wretches found themselves permanently attached to oars.

This had strategic implications. Captive oarsmen still required a lot of food and water to keep going. Provisions took up space, exactly what the slender, shallow-draft, tightly packed vessels lacked. So as galleys grew, their radius of operations shrank. At the very time Europe was looking outward, these floating torture chambers were being pulled inward. It was a death sentence.

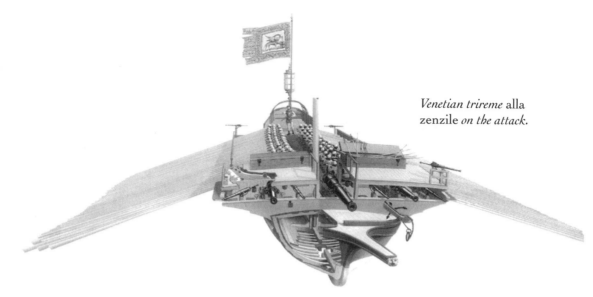

Venetian trireme alla zenzile *on the attack.*

carvel or carvel-built), the product to which it was devoted was headed for an impasse.

An alternative was taking shape far north. Shipbuilders here were faced with an array of challenges not present in the Mediterranean—inexorable tides, more fickle prevailing winds, and a greater probability of killer storms. Just to survive, ships had to be built for heavy seas.

Between 1000 and 1250 the humble cog, heretofore a flat-bottomed, though high-sided, coaster, began to evolve in ways that would enable it to take on the fury of the North Sea and North Atlantic. The essential change was the addition of a keel, which greatly stiffened the hull form, and anchored the mast firmly in place. Cogs were clinker-built, made up of overlapping planking nailed together at the point of double thickness, in effect resembling clapboard siding on a house. Their primary strength stemmed from this outer shell, although an internal frame was added for further rigidity—a combination resulting in one of the cog's key advantages . . . it could be built bigger. As trade in bulk goods like grain grew, so did cogs, until by the late thirteenth century the largest measured ninety feet long, with a draft of up to sixteen feet. They also grew upward, with the already high freeboard increasing along with the cogs' other dimensions.

The net effect was a much more defensible ship. On it superstructures were built fore and aft, in time coming to resemble the battlements of a castle. From here archers and other soldiers had a critical advantage over attackers, such as low-slung Viking longboats.

Hansa cog.

Height mattered in other ways. On such a tall ship the twin steering oars traditionally mounted on either side of the hull's rear had to be made excessively long just to reach the water, especially when the ship was heeled over. They were replaced by a single sternpost rudder, which considerably improved maneuverability. Height also precluded rowing. Cogs were strictly wind-driven, propelled by a single large square sail, its mast stepped slightly forward for more efficient running before the breeze. They were hardly speed merchants, but sail power meant smaller crews than galleys, and in times of danger more hands free to defend the ship.

Guns were slow to make their mark in northern waters. After 1360 there is evidence of firearms at sea here, but they were limited to small man killers. Larger guns were available, but they were impossibly heavy iron monsters used for siege work on land. Then there was the more general question of where to position the firepower aboard a sailing ship. The obvious answer, at least for light pieces, was in the castles. Around 1480, however, they began to be supplemented by naval versions of the new iron-ball-firing bronze cannon that were revolutionizing warfare on land. These were compact enough for sea duty, and could shoot through an enemy ship's timbers. Soon they could be seen protruding from the weather deck or waist (the area between the castles), clustered in lines of three or so on either side.

These were still hefty weapons, and they remained high above the waterline—a top-heavy recipe for disaster, which allowed even light weather conditions to capsize a number of prominent vessels. The alternative, mounting the cannon lower and shooting them through ports in the hull, was confounded by the fact that northern ships were clinker-built, and breaching their outer shell of planks meant dangerously weakening the entire structure.

The solution came with the arrival of the cog in Mediterranean waters around the turn of the fourteenth century. The typical two-masted lateen-rigged (based on a triangular sail) merchantman steered by side oars was gradually abandoned in favor of a larger version of the cog with a single square sail and stern rudder. Clinker construction was dropped in favor of the carvel-built system with a full internal skeleton, and about 1370 a second mast with a so-called lateen mizzen was added in the rear, considerably improving maneuverability.

The resulting *cocha* soon sailed north, where they were labeled carracks and incrementally copied. Carracks had a number of important features as merchantmen, but their singular advantage from a military perspective had to do with structure. The outer hulls were not load-bearing, and could therefore be pierced with ports for heavy guns, mounted low enough to add rather than subtract from stability. This logical step—traditionally attributed to a French shipwright named, appropriately enough, Descharges, in 1501—was not simply a matter of sawing holes in the side of a ship. Equally important was the addition of hinged waterproof lids, which could be sealed tight when not in use . . . a key operational precaution. (In 1545 some English tar aboard the *Mary Rose,* pride of Henry VIII's fleet, forgot to shut the gun ports as it sailed down the Solent to thwart a French invasion. When the ship heeled over, it took on water so fast it sank like a stone.)

The introduction of the gun port made possible rows of heavy cannon on both sides of the militarized sailing ship, capable of firing the broadsides that determined the nature of naval warfare all the way up to the mid-nineteenth century. Historians have noted that non-Western powers also added firepower to their ships. In 1637 traveler Peter Mundy sketched a Chinese junk in the Pearl River, near Canton, with gun ports. But it is important to realize that such ships lacked the stout internal construction capable of withstanding the recoil generated by a truly heavy cannon, much less the simultaneous discharge of a row of them. Europeans alone were on the way to creating true bruisers of the high seas. But putting the gun in the ship was one thing, getting it where you wanted it was another.

Nowhere was a clearer understanding of shipbuilding being gained more methodically than at the Arsenal of Venice, where hulls were beginning to be built according to mathematically and geometrically derived principles known eventually as whole molding. Although the system was later shown to be faulted hydrodynamically, it was based on a notion of fixed ratios between major parts such as the keel, stem, sternpost, and especially the ribs as they narrowed toward the ends of the hull. There was not yet a means of generating precise drawings of a ship prior to construction; yet these geometric techniques could be applied at the dockyard, marking on the timbers exactly the shape into which they were to be sawed. This meant dimensions could be kept and successful designs replicated.

The arsenal's major military preoccupation was the development of galleasses—hybrid vessels combining the galley's slim lines and oar power with the size and broadside gun placement of a sailing vessel. Galleasses were a great success against the fragile Turkish galleys at Lepanto in 1571, the last major battle between oared warships; but against armed sailing vessels they were clumsy, short-ranged due to their huge crews, and outgunned. By the end of the sixteenth century they had disappeared.

Other designs were continually being tested. In 1527 a new type of pure sailing vessel was laid down in Venice called the galleon. Based on the ideas of arsenal supervisor Matteo Bresan, it was full-rigged, and built for running

Rigging the Fight

During the fifteenth and early sixteenth centuries, European sailing vessels underwent remarkable development, transforming their mode of propulsion from one not much advanced from those of the ancient Mycenaean Greeks and Phoenicians to a sophisticated array of sails capable of taking maximum advantage of virtually any weather condition likely to be encountered around the globe. It was this process that made such ships into interoceanic vessels for all seasons.

To the naked eye, the most obvious thing about the process was the proliferation of sails. Common sense argued that more sails meant more speed; but it really meant more control. A single sail could be reefed, or partially taken down, but it was better to have smaller more specialized versions. The sequence began around 1465 with addition of masts from which to hang extra canvas—first, the lateen-rigged mizzen, then a square-sailed foremast, a bowsprit protruding at a diagonal from the front, and finally a second lateen-rigged bonaventure at the stern, destined to be dropped by 1640. Around 1500, in order to capitalize upon elevated wind currents, masts were built upward. First came secondary extensions—square-sailed main and fore topmasts and a lateen equivalent for the mizzen and bonaventure. Finally, tertiary appendages were added to the fore, main, and mizzen masts known as topgallants.

By 1514, one basic sail had grown to around twelve along with all the necessary appurtenances. While the English were leaders, the process was going on all over coastal Europe. Shipwrights had traditionally been illiterate and secretive, but new sails were hard to hide and the dramatic increase in their products' general sophistication pointed to an equivalent growth in their overall knowledge. European shipbuilders were gaining a clearer understanding of why vessels sailed well, what made them strong, and how they could be turned into even more useful mechanisms

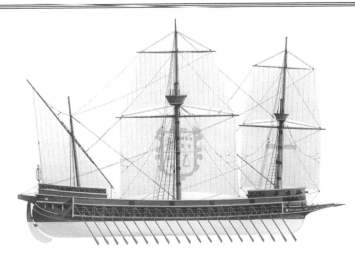

Galleass.

down pirates. The hull still had high castles, but it had a long straight keel, its stem was shortened, and it was significantly narrower than previous models. It had a lengthy and successful career, and before it was broken up in 1547 the Senate of Venice decreed that its measurements be carefully recorded so they could be used for future construction. The shipwrights overstepped the mark. The next galleon was too narrow to safely carry its heavy guns, which were still positioned fairly high off the water. It capsized in 1558. Nonetheless, the concept was sound, and it would soon find a home in northern waters where it would make history of the most fundamental sort.

The naval future would be contested by the states facing the Atlantic— Catholic Spain on one side and the Protestant Dutch and English on the other. Spain, however, took to the water with the motives and inclinations of an imperial land power. As with Rome, the sea represented a means of coming to grips with her enemies and a conduit for colonial tribute. Although commerce was involved, Spanish naval power was primarily an instrument of imperial integration, specifically a means of delivering gold and silver from her Latin American possessions to finance Philip II's efforts at suppressing the revolt in the Netherlands. Conversely as early as 1569 William the Silent was commissioning privateers to directly attack Spanish convoys for profitable ends. Three years later the great Lisbon fleet was intercepted off Walcheren, yielding a haul of half a million gold crowns, enough to keep the Dutch war effort going for two years. Windfalls like these did not go unnoticed by their even more aggressive fellow Protestants and naval collaborators, the English.

The roguish career of Sir Francis Drake epitomized the shortcomings of the Spanish naval approach. A driven man, dedicated practically from adolescence to wreaking personal havoc on Philip's empire, "El Draque" became virtually the devil incarnate to Spaniards who plied the seas, intercepting plate ship after plate ship and driving the frustrated Habsburg even further into debt. By 1586 not a single bar of Peruvian or Mexican silver safely crossed the Atlantic. Further humiliation followed the next year when the audacious Drake directly attacked the Spaniards at their naval base in the bay of Cadiz, in the process demonstrating the complete superiority of gun-based sailing warships over galleys even in sheltered waters.

This was not a random attack. The English were well aware that Philip was planning his revenge and that his implement of retribution was a huge fleet, the Armada, taking shape in Lisbon. It included more ships and munitions than had ever been collected in a single European port. The vessels involved were not necessarily outmoded, but neither were they of the latest design. Many were converted merchantmen, large carracks with massive fore and aft castles. The Armada also included a number of true warships, galleons, but they too were built high along Venetian lines. Still more suggestive was their armament. The standard guns for both carracks and galleons were elongated muzzle-loaded cannon mounted on two-wheeled carriages with long trails—designs indistinguishable from land artillery. They went into battle fully charged. But because of their length, the only way to reload such weapons was to lower a crew member over the side, leaving him both exposed to enemy fire and in an exceedingly difficult position to clean the barrel, ram in the powder, and hoist the heavy ball—virtually an impossibility in any but the calmest seas. The very configuration of the Armada's vessels made their tactics obvious: sail in close, let loose one devastating fusillade, and then board using the height advantage of their lofty superstructures.

The English had no intention of being boarded, and the superior design of their warships and armaments would insure it did not happen. Around 1570 British shipwrights came up with a sleek new galleon design, appropriately termed "race-built." Fore and aft castles were radically cut down, not simply for stability in heavy winds, but because they could mount only light, secondary batteries. Instead, heavy guns were concentrated on an extended gun deck, barely six feet off the waterline. The resulting race-built galleons were not only better sailors—faster and more maneuverable

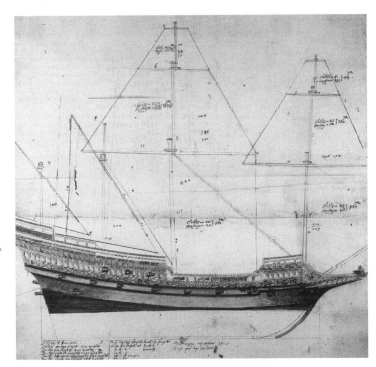

English race-built galleon.

than their Spanish counterparts—but had a considerable advantage in fire-power. The first of the race-builts, launched in 1573 and propitiously dubbed *Dreadnought,*[2] carried nearly 5 percent of its 700-ton displacement in ordnance, an unprecedented figure. Nor was the program just a matter of new ships. The mastermind of the England's anti-Armada, Drake's privateer colleague Sir John Hawkins, began in 1577 to "reform" older vessels, cutting them down to race-built standards, and packing them with heavy guns.

Errant Armada

Nothing went as planned. When word of the Armada's sailing in May 1588 reached England, the Lord High Admiral Howard and Drake hatched a plan to intercept it well before it got near English waters. But ill-winds pushed the Spaniards onward and drove the anti-Armada back to Plymouth. Medina Sidonia sailed into the Channel near the end of July, his intent being to link up at Calais with the Duke of Parma, who had assembled an invasion force of 27,000 men, and escort them to the shores of Kent. Legend has it that news of the Armada's presence reached Drake and Howard on July 29 in the midst of a game of bowls, causing the former to comment that they had time to finish the game and beat the Spanish too. If true, it was bravado. "Fleets like these were a new thing in the world," explains Garrett Mattingly, the great historian of the Armada. "Nobody knew what the new weapons would do, or what tactics would make them most effective. This was the beginning of a new era of naval warfare, of the long day in which the ship-of-the-line, wooden walled and sail driven, and armed with smooth bore cannon, was to be queen of battles, a day for which the armor-plated steam-powered battleship with rifled cannon merely marked the evening."[3]

Fighting erupted on July 31, and the race-builts' combination of speed and gunpower did succeed in keeping the Spanish off of them, but were not able to do much structural damage to their adversaries at long ranges. (The same combination did not work well three and a quarter centuries later for Sir John Fisher, Hawkins's naval reincarnation and inventor of the modern dreadnought.) Two substantial Spanish vessels, *San Salvador* and *Rosario,* were captured, but more by good fortune than anything else. "We pluck their feathers little by little," wrote Lord Howard.

The sputtering series of engagements did little to stop Medina Sidonia's leisurely procession toward Parma. He managed to shepherd nearly 130 ships to Calais, anchoring them there for a day and a half. Parma succeeded in loading his invasion force into barges, also in a day and a half. But each task occupied a different thirty-six

The ordnance buried deep along the sides of the race-builts was as superior to the Spaniard's weapons as the ship's design. Although firing a heavier charge, their barrels were reduced from thirteen feet to about nine. Moreover, they rode on four-wheeled truck carriages that effectively halved the length of the whole package. Such pieces were compact enough to be fired, unhitched, and hauled inboard for reloading (rigs controlling the guns' recoil to bring them in automatically were not perfected until the seventeenth century). They could then be pushed

hours. Medina Sidonia reached Calais on August 6, but Parma, just twenty-five miles away, did not know until the next day and it took him until the 8th to get his troops embarked. By that time the English had succeeded in rooting out the Armada, but not with the heavy guns of their race-builts.

Instead they scared them away with nothing more than eight merchantmen, hastily purchased for around 5,000 pounds—"the cheapest national investment that this country has ever made," wrote one English historian. They were burning at the time. The British filled the fire ships with flammables, and double-loaded their guns to discharge spontaneously when the flames reached them. The attack began around midnight. Initially the Spanish kept their composure, even intercepting two of them. But when the guns of the remaining six began discharging randomly, most captains gave in to panic. Rather than hauling in their anchors, they simply cut the lines and fled into the strong tidal rip of the Channel. Without anchors there would be no going back, or safely approaching any coast for that matter.

The invasion threat was over, and as dawn broke the English galleons moved in, this time discarding any pretense of long-range gunnery. Instead, they subjected the Spanish to the withering point-blank firepower that would characterize Royal Navy tactics for the next two centuries. Medina Sidonia's flagship took over a hundred hits, while his second-in-command, Recalde, believed something like a thousand rounds had been fired against him.

Nonetheless, the British effectively ran out of ammunition before they managed to break up the Armada. More than gunfire, it was the privation they encountered in the stormy North Atlantic during their hegira around Scotland and Ireland that humbled the Spanish fleet. Conversely, the English were energized. In a very real sense the Armada campaign brought the Royal Navy into being, from this point until nearly the middle of the twentieth century, Britannia ruled the waves, and she did so largely with battleships.

back and resecured—a time- and labor-consuming process, but one that gave the English ships the key capacity for multiple broadsides.

The galleon had room for growth in terms of firepower. While even the strongest wooden ships could not be lengthened beyond a certain point due to hogging, they could be built higher—an ironic fate for the immediate successors to the race-builts.

This was not a matter of reintroducing lofty armed castles, but of adding a second row of heavy ordnance along the hull. As far back as 1577 there is a depiction of a galleon with two continuous battery decks, but this was a rarity for the time. By the turn of the seventeenth century, however, such ships were being built in England with some regularity—though over the objections of those who favored nimbleness and speed.

The controversy heated up in 1609 when master shipwright Phineas Pett laid down *Prince,* first of the triple-deckers. William Monson, a champion of the classic galleon, was beside himself, writing, "a ship with three decks . . . is very inconvenient, dangerous, and unserviceable; the number and weight of the ordnance wrings her sides and weakens her."[4] Pett had been experimenting with new proportions to better bear the load, but his critics, not knowing, petitioned King James I, claiming shoddy construction. A monarchic technical inspection was ordered. As a rule of thumb, sovereigns could be depended on to like big ships, and after a flurry of tests, the king had the charges against Pett dropped and *Prince* completed.

Actually there were problems with load bearing due largely to the limitations of whole molding as a construction guide, but Pett continued to work on them, and in 1637 his work bore fruit with the launching of the 1,500-ton, 104-gun *Sovereign of the Seas*—a design similar in all but detail to every English capital ship built until 1860. The technical limits of sailing warships firing solid shot had been reached, and henceforth military naval architecture would be defined by one-, two-, and three-decked ships. The Dutch, for one, would veer away from the heaviest units; but in the end they conformed to the standards set by the British shipwrights.

As fighting platforms, these sailing battleships had very definite implications for the future of warfare. Like Maurice's automata infantry, tactically they were most naturally used in a formalized and contrived manner. Although tactical formations gravitated toward multiple-unit lines, combat within the line was individualized ship versus ship. Moreover, the physical characteristics of wood, wind, and water tended to enforce symmetrical engagement by class. In effect, big ships (those mounting more gun decks) could always defeat smaller ships but, due to their bulk, could not catch them. So like fought like.

Arriving on Santo Domingo in 1492, Christopher Columbus wasted little time before ordering two gunpowder weapons, a cannon and a forerunner to the arquebus, the spingard, to be fired. "When the king saw the effect of their force and what they penetrated he was astonished. And when his people heard the shots they all fell to the ground."[5] The reaction was simply a prelude to a thunderous injection of military power without precedent in human history.

The Americas constituted another world for the taking, and Europeans brought war-making techniques and weapons that allowed them to capture it with astonishing ease. Archetypical was the fate of the Aztecs and Tawantinsuyu, the empire of those we call the Incas. The statistics speak for themselves. Together these two imperial domains had ruled around a fifth of the world's population; but in less than fifty years after Columbus's arrival in the New World they were gone, brought down by improbably small Spanish forces. In the case of the Aztecs, Hernán Cortés managed with a band of around 500; Francisco Pizarro's force that conquered the Incas was a third that size, a total of 168.

The nature of Aztec society did little to insure its survival. The Incas were more viable. The traditional view of the "conquering Incas"—a sort of high-altitude, native American version of Rome—has largely been dispelled. To a considerable degree this was a matter of the lofty environment. Survival here—where only one harvest in three was bountiful, where niches literally from top to bottom had to be exploited, where massive food storage against bad times was a must—demanded cooperation and accommodation. There may have been rivalries at the top, but the social structure was sound and had the basic loyalty of the people. In terms of territory this was probably the largest empire on the face of the earth at the time of the Spanish conquest. It was a realm put together very rapidly through force by some of the best armies the region had ever seen, and held together by superior logistics and a web of carefully engineered highways capable of bringing reinforcements quickly to any trouble spot. But the Inca troops sent against the Spanish proved utterly unable to cope.

The Incas, like the Aztecs, were without walls around their major population centers, making them readily accessible. And although their war making was more purposeful than that of the Aztecs, they were no more prepared for the ruthlessness of the Spaniards. Not only did the invaders represent a martial tradition founded on the harsh and uniquely Old World premises of the original agro-pastoral split; but they employed force with the state-of-the-art pragmatism embodied in the thinking of Niccolò Machiavelli.

Yet what the Spaniards fought with was of equal or greater consequence than how they fought. Nothing was more symbolic of the invaders' advantage in weaponry than the strange fire tubes they used to wreak long-range havoc. The physical destruction wrought by firearms in the two campaigns was considerable, but the psychological effect appears to have been equally devastating. There are numerous indications that the gun, at critical junctures, utterly demoralized Indian warriors and robbed them of their power to resist. Just as Bernardino de Sahagún's compilation of contemporary sources repeatedly cites examples of Aztec warriors fainting away at the reports of discharging cannon, there are similarly well-documented examples of gunfire-induced passivity on the part of the Incas. Unquestionably, the great majority of Indian casualties were not inflicted by guns. Yet firearms seemed to contain the thunder from the very skies, imparting to it a magical quality that made resistance so much more manifestly hopeless. Overawed, the Indians had no recourse to the traditional response to the underarmed, asymmetric counters. There were none. Instead, they had only to submit.

Non-Western societies of the Old World proved far less passive in the face of European guns. None managed to assemble quite the combination of enabling factors that rendered Western firepower so devastating, nor were they able to sustain their initial momentum. Nonetheless, several came very close, generating highly effective gun-based units reached through unique institutional accommodations, and demonstrating long before the twentieth century that there was nothing exclusively Western about advanced weapons technology.

Consider the Turk. The Ottoman dynasty was a unique combination of institutions, a number of them Islamic, but exaggerated to the point that the whole took on the cast of an insect society. Royal reproduction consisted of a single sultan, surrounded by an endless supply of slave girls, recruited not from Turks but from subject Greeks, Balkan dwellers, and Circassians, watched over by a troop of neutered bodyguards. Critical to enforcing the sultan's will was his corps of drones. Also recruited from Christian children in the same areas through a human tax known as the *devsirme,* these Janissaries (in Turkish *yeni ceri,* or "new troops") were a slave army of soldiers, bureaucrats, and technicians who, though not eunuchs, were forbidden to marry and were generally sequestered from women.

So long as the *devsirme* was in force (until 1648), the Janissaries constituted an exquisitely disciplined force. Upon arrival they were instantly sequestered, converted to Islam, and then subjected to years of training under a code of conduct so strict that they emerged as virtual automatons—"obeying their commanders without question, and seeming to care nothing at all for their lives in battle," according to contemporary observer Paolo Giovio.[6] They were troops virtually without social ties except to the Sultan and to themselves as a body. "In them rests the force and firmness of the army of the Turk," observed sixteenth-century Italian traveler Benedetto Ramberti. "Because they all exercise and live together, they become as it were a single body, and in truth they are terrible." They were also malleable, particularly when it came to weapons.

Though Sultan Orphan armed the Janissaries with traditional close-in weapons when he founded them in 1320, the scimitar and the battle-ax had only a limited time to penetrate to the heart of the corps before the possibility of firearms became apparent to Murad II around 1420. He and his two immediate successors, Mehmed II and Bayezid II, rapidly and efficiently grafted both cannon and small arms, manned by their personal slave

Ottoman Janissaries on parade.

troopers, onto the traditional Turkish military forms. What emerged was probably the deadliest fighting mechanism yet created, a swirling mass of horse archers, with a lethal core of firepower. The key was the superb discipline of the relatively small Janissary corps, which, serving as a proxy for Maurice's loading and drill manuals, allowed them to operate both classes of fire weapons with an effect out of all proportion to their numbers.

Although the Ottomans also took guns to sea and eventually challenged Christendom for the entire Mediterranean basin, it was the specter of Turkey as a land power that seemed to be tilting history in the direction of the Sublime Porte. In the twelve years following 1514, Ottoman armies dealt Safavid Persia a crippling blow, conquered the Mamluks of Egypt, and then invaded Hungary, annihilating the defending army at Mohacs and taking over the country. In all cases superior Turkish firepower provided the impetus for victory. A thrill of dread spread across Christendom as the Habsburgs under Charles V and Suleyman I, the greatest of Ottoman sultans, engaged in a mortal struggle that would find Vienna itself twice under siege. Christian armies would gradually come to prevail, but this was hardly apparent at the time. Nor were the Turks alone among Muslims in their adoption of firepower.

Although the Turks and the Moroccans proved they could use guns effectively, the Islamic adoption of firepower proved unequal to the challenge. The Turkish experience highlights some of the reasons why. Unlike the purely European powers, they chose size over mobility and numbers in their artillery. Doubtless big guns appealed to archetypical human notions about weaponry. Geoffrey Parker also points out they were easier to produce for an industrial base incapable of anything approaching mass production. But their sheer bulk had a negative impact on several critical battles, when Ottoman troops proved unable to move them quickly enough to fill tactical voids. There was also an apparent military-industrial weakness in metallurgy, so much so that contemporary sources universally claimed that weapons captured from Ottoman forces were useless. Nor was there anything resembling sustained military innovation. Once Turkish craftsmen managed to copy Western small arms the designs became frozen, causing them to miss even so modest a change as the bayonet. Operationally and tactically they were similarly glacial, never learning to deploy in the thin lines most effective for firefighting in the field, and never mastering the complexity of the offensive siege and, defensively, the *trace italienne*. Cumulatively, it spelled the end of military progress for the Turks— who were observed to have fought in the late nineteenth century just "as in the days of Sulyman I."

Moroccan Amnesia

Portugal long had its eye on Morocco as a client state. As its influence and colonies expanded here, the Watasids of Fez were effectively replaced by the more aggressive Sa'dis, who were determined to stand up to the Christians. Between 1536 and 1549 they took the steps necessary—broadening European sources of munitions, integrating small arms and field artillery into their forces, and developing combined-arms tactics featuring mounted arquebusiers. During these years the Moroccans drove the Portuguese from several strongholds, with the capture of Santa Cruz constituting their first artillery victory and a good sign that the technology gap had all but closed. More success came in the 1550s when the Moroccans thwarted the advancing Ottomans, switching successfully from siege tactics to rapid maneuver.

The Portuguese remained oblivious to the implications, and in 1578 their young king Sebastian rashly invaded, compounding his error by leading his poorly trained army of around 20,000 into the Moroccan interior, largely without benefit of cavalry. Sebastian sought battle in the heat of August 4 at Alcazarquivir, deploying his infantry in the standard Iberian tercio, with artillery in front, and what few cavalry he had on the flanks. The Moroccans picked them apart, one piece at a time, largely with their horse arquebusiers. First they overran the unsupported artillery, next they swept away the cavalry on both wings. Finally, when Sebastian's infantry moved forward, the Moroccan horse arquebusiers threw themselves into a gap that formed, and dismantled the Portuguese rear right flank. The rest was inevitable, and the day ended with Sebastian dead and his entire army either killed or captured. It was, in historian Jeremy Black's words, "the single most important defeat for European expansion during the sixteenth century,"7 and saved Morocco from invasion until 1844.

During the first half of the twentieth century the Japanese were similarly labeled copiers, not creators, only to prove the truth was a great deal more complex. The naysayers might have been more circumspect had they considered seriously Japan's appropriation of the gun four centuries earlier. Of all the East Asian powers in the sixteenth century, the Japanese made the most of firearms. First introduced by Portuguese castaways in 1543, they were rapidly duplicated by local metalworkers and taken up by warlords locked in a seemingly perpetual battle for supremacy. There is some debate over their early effectiveness; but on May 21, 1575, guns exploded onto the stage at the battle of Nagashino, where 3,000 peasants turned musketeers under Oda Nobunaga annihilated the opposing Takeda cavalry.

These devastating fire effects were a long time in the making. From the beginning, the Japanese placed heavy emphasis on accuracy, as opposed to the Europeans, who focused on getting off more shots. After 1550, Japanese military treatises concentrated on instructing how best to aim, the results being notably well-directed fire considering the nature of smoothbores. Rather than rationalize the loading process, which was the central focus of European military manuals, the Japanese concentrated on improved troop deployment. As early as 1560 Nobunaga began experimenting with arrangements conducive to salvo fire, drawing his musketeers into separate lines and teaching each to fire in volleys, while the others were loading. By the time of Nagashino he was able to effectively deploy his troops in just three lines and still maintain regular fire—a remarkable achievement not matched in Europe until the eighteenth century. Indeed, Nobunaga's final solution proved itself on the battlefield nearly two decades before William Louis even thought of volley fire and wrote his brother Maurice.

The Japanese approach to fortification was similarly original. Because Nobunaga and his contemporaries were quick to grasp the potential of artillery as well as small arms, they realized it would render every castle in Japan indefensible. In response they scrapped curtain walls, and began building huge fortifications designed to fit around hills and stone outcroppings like caps on teeth. The result consisted of cyclopean stone-faced foundations backed by soil and living rock, topped off by massive pagoda-shaped superstructures of up to seven stories—quite possibly the most cannon-proof structures existing anywhere in the early modern world. These defensive complexes were truly huge, with the cores of two (one in Osaka and the other in Kumamato) exceeding seven miles in circumference, with the circle of safety extended still further by a ring of supporting fortifications—like the Europeans, the Japanese understood the importance of outliers.

The Japanese responded brilliantly to the arrival of the gun, with solutions that were both thorough and original. Ironically, almost immediately they fell prey to a major technological surprise.

It is always a surprise when a long-standing balance-of-power scheme is short-circuited and replaced by a hegemon, but that appears to be one of history's basic political progressions. This is what happened in Japan with the brilliant one-two combination of Nobunaga and Hideyoshi, followed by Tokugawa Ieyasu, who consolidated the settlement, which was destined to persist until 1867.

Iron Turtle

In 1592, two years after he succeeded in uniting Japan, Toyotomi Hideyoshi, Nobunaga's successor, sought to extend his power still further by invading Korea. After trying unsuccessfully to hire two Portuguese galleons, he increased the size of his invading fleet to 700 ships, assuming the Koreans would resort to close-in boarding tactics and be overwhelmed. Instead, the invading force was met by no one, landing unopposed at Pusan and striking out immediately for Seoul, which it reached in twenty days on May 2.

All seemed to be well—but it wasn't. While these troops were preoccupied, the Korean fleet under Yi Sun-sin went into action, inflicting several minor defeats on the waiting Japanese ships, and later, on July 8, crushing their main body in Hansan Bay. The Japanese crews were not the best, but of equal importance were the Korean ships facing them.

As their adversaries rowed out to battle, the Japanese might well have thought they were being confronted by a herd of giant turtles. About 100 feet long, the sides of the Korean ships were pierced by twelve gun ports and twenty-two loopholes for small arms. The remainder was encased in a massive carapace of hexagonal metal plates—history's first ironclad, more than two and a half centuries before the *Monitor* and CSS *Virginia* sized each other up in Chesapeake Bay. Impossible to board or hole, the avenging terrapins moved in close to fire broadsides with armament outweighing their opponents by forty to one, and then rammed, leaving a trail of mortally wounded Japanese warships to be finished off by more conventional Korean cohorts.

Hideyoshi reacted quickly and symmetrically: his remaining ships took on heavier guns and hid clustered below the harbor's fortifications. Meanwhile, he ordered his nobles to send him metal plates, doubtless intending his own ironclads.

Instead, there was a break in the war until 1597, when the Japanese invaded again. After suffering an initial loss in July, Yi Sun-sin managed to lure them into a tidal race, where the oar-powered turtles once again had their way with their opponents. The Korean admiral was killed, but the Japanese naval arm was broken and the invasion had to be called off in 1598, the same year Hideyoshi died. The turtles would never see action again, and this short but brilliant episode of naval innovation stopped dead in the water, then went into reverse.

Such stability among the fractious Japanese required extraordinary measures. In a process begun by Hideyoshi and ratified by Tokugawa, Japan's new rulers disarmed the country, focusing on the gun. The peasant-musketeers, so studiously trained for firefighting, were disbanded almost immediately. All guns were confiscated and removed to government arsenals, and for the remainder of the seventeenth century the fabrication of firearms was strictly controlled and steadily decreased. There was more. Successive central administrations even conducted a string of "sword hunts" with the intent of seizing every weapon from peasants, the townspeople, and the temples, until only hereditary warriors, or samurai, were left with their blades . . . but not much else.

Even as it was inventing the new cannon-proof fortifications, the future regime was tearing down the great forts of the nobles they had thus far defeated. Ultimately, in 1615 Tokugawa Ieyasu decreed that each of the great lords could keep only one castle; the rest had to go. An analogous process took place at sea. Not only was the Japanese fleet allowed to wither, but guns were removed from all merchant ships—replaced until 1635 (when overseas trade was forbidden) by the shogun's Red-Seal, a license of free passage enforced by nothing more than bribery and commercial pressure. Ultimately, the disarmament process reached the point of attacking the very memory of the gun. Beginning in 1671 and continuing for several decades, all foreign books concerned with military issues were banned from import; while in 1737 the *Honcho Gunki-ko* (Investigation of the Military Weapons of Japan) limited itself to a single chapter on firearms—a short one.

The gun's demise in Japan only underscores its fundamental importance. While it was not allowed to ignite the kind of firestorm that burned over Europe for a century and a half, its impact here was sufficiently powerful that the governing classes concluded that stability required its removal. Their solution worked. The gun was not much missed, and war making was allowed to reclaim its heroic past and implements. Firepower was dangerous, not just to this society but to all societies. Europe would not give up the gun; but it would also find ways to circumscribe it . . . at least for a while.

Chapter 9

GUN CONTROL

Firearms revolutionized the conduct of warfare in Europe. Although politics, economics, and religion certainly exacerbated the resulting conflicts, the role of guns as facilitators for violence was profound. Unlike Japan, the possibility of elimination was never seriously entertained—the European power structure remained too fragmented and the Machiavellian code of conduct too opportunistic.

But by 1648 it was clear to many observers that something had to be done. John Locke, writing shortly after his own country's Civil War, was well aware that the problem transcended national boundaries—"All those flames that have made such havoc and desolation in Europe . . . have not been quenched but with the blood of so many millions."[1] The only alternative was to control the gun.

To do it, the politico-military system rigidly enforced limitation of ends until 1789 and again for some time after 1815. Excepting the Napoleonic interlude, the behavior of Europe's emergent nation-states came to resemble the conduct of players in the city-state-dominated balance-of-power schemes of Sumer and Hellenic Greece. They too had to grapple with a particularly deadly instrument of war, the phalanx. Their solution was based on the ritualization of combat. Control of the gun demanded an equivalent resort to rules. The outlines of such a system had already been drawn by Maurice of Orange and the architects of the sailing battleship; it only remained for it to be applied across Europe. Much of this had to do with the imposition of limited warfare through training and operational ritual. But implicit in everything was the removal of weapons improvement as a fundamental dynamic of victory or defeat. Arms technology did not disappear; it was relegated to perfecting what already existed . . . essentially a carefully channeled intensification of the status quo. As a result, for nearly two centuries after 1648 European civilization managed to conduct its martial affairs largely without recourse to more deadly weaponry. That this stasis was supported by an increasingly anachronistic vision of social and political reality, and persisted in the face of what became a deluge of innovation in the civilian sphere, only serves to underline its vigor.

The core of land warfare during this period was infantry. Naturally, it was also the focus of much of the effort that went into controlling the gun and regularizing its use. Deploying foot soldiers in elongated formations to maximize their firepower left them terribly vulnerable. Once a line like this was breached or flanked so it could be rolled up, the whole was likely to crumble. Even orderly retreat under fire was difficult and dangerous. All too often precipitous flight was the infantryman's only viable means of getting off the battlefield.

Alternately he could stand and fight. It was, in essence, a psychological proposition, a thin line of men staking everything on their ability to perform a prescribed ritual while subjected to the most terrifying possibilities—opposing infantry blasting away, cannonballs whizzing by, and the ever present danger of cavalry, their thundering horses and razor-sharp sabers—all designed to push the foot soldier over the edge. Against this concentrated assault on the senses, infantrymen had only their training.

It was no accident that the name of the French itinerant inspector of training, Lieutenant Colonel Martinet, passed into our language as synonymous with the authoritarian insistence on obedience down to the last detail. His intent and those of his colleagues—the steadily growing cadre of officer-aristocrats and senior enlisted grades responsible for the system's imposition—was the total submission of recruits to an utterly contrived and counterintuitive behavioral sequence. The process required at least a year (five to yield a full-fledged veteran), and the means applied, as befit the agricultural origins of those involved, were based on the time-honored rituals of animal training—breaking the spirit through a healthy measure of negative reinforcement and, above all, repetition.

The pervasiveness of the mental state induced through long periods of performing precisely the same activities is not to be underestimated. This was especially true of marching, one of the two basic aspects of infantry training. A substantial measure of drilling's effectiveness can be attributed to the primeval human penchant for dance, with shared patterns of movement helping to weld a military force into an amalgamated entity. Years of marching along at exactly seventy-five paces a minute, each step precisely seventy-four centimeters in length, practicing the same evolutions designed solely to fold and unfold a body of men on the battlefield, had the effect of blotting or at least dimming the individual consciousness sufficiently to enable soldiers to perform their mechanical ballet while being blown apart.

This oblivious demeanor was further reinforced by the other focus of training ritual, weapons drill. The graphically represented procedures necessary to load and fire a musket in the first Dutch training manuals became institutionalized in Europe's armies—a routine of fine-motor functions roughly equivalent to the impact of marching on the large muscle groups. Through endless reiteration these basic motions formed a pattern of behavior so strong that the foot soldier became an extension of his gun, an automaton of firepower. Troops trained in this manner consistently exhibited great discipline and fortitude under the most appalling circumstances, but seldom showed initiative or enthusiasm.

In the infantry firepower equation, men were assigned the essential variables, and their weapons acted as constants. Troops might be taught to fire fast (though aiming was not encouraged), but their guns were not seen as a fundamental source of improvement. This does not mean that there were no technical developments. There were several innovations of lasting import, but their net effect was to reinforce the system, not change it.

The muzzle-loading musket was finalized from a design perspective around the turn of the eighteenth century with several key features, the bayonet and flintlock firing mechanism being the most important.

The need to protect musketeers while loading had led to the perpetuation of the pike. Yet as the sheer volume of fire grew more critical in the eyes of commanders, every hand on a pike was seen as a hand not loading or shooting. The elongated shape of small arms suggested an alternative. If a blade could simply be added to the end of the barrel, then the musket would become, in effect, both pike and firearm. But before an infantryman could have his pike and shoot it too, a practical means had to be found to attach a blade and keep it there. After early attempts at tying knife to barrel came to nothing, someone thought of simply ramming the blade, or at least its tapered wooden handle, down the muzzle, thereby creating the

Flintlock.

"plug" bayonet. This terminology revealed its chief weakness; as long as the barrel was plugged, the musket could be neither loaded nor fired.

It wasn't until 1687 that a true solution appeared in the form of a "socket" bayonet, based on a sleeve that could be slid over the muzzle and secured with a lug and slot. The source of this concept was the great fortress architect Marshal Sébastien Le Prestre de Vauban, one of the age's few great military-technical innovators. Like so much of his work, however, the socket bayonet was applied to conservative ends. Contrary to military myth, it was employed primarily as a defensive weapon, intended more for deterrence than actual fighting. The essential idea was to keep attackers away, as infantry formations loaded and fired at each other. There were frequent bayonet charges, but they seldom resulted in bloody melees. Instead, they were ordered when an opposing line was thought to be sufficiently degraded by gunfire to be near breaking. When this judgment proved correct—as it usually did—the object of the charge almost always surrendered or ran away. Otherwise, the approaching line was shot to pieces.

The bayonet contributed to the functioning of the system in another important way; it homogenized infantry. Eliminating the pike insured symmetry among foot soldiers, generating standardized fighting modules capable of being plugged into any number of military combinations, a feature of no small import to a political system whose primary balance wheel was coalition.

The replacement of the cranky matchlock with a triggering device based on striker-generated flint sparks was also a marked advance, but a conservative one. Lowering misfires from one in two to one in three was both advantageous and acceptable in a regime more eager to increase the rate of fire than effectiveness. The flintlock's simplified loading procedures were also welcome in that they allowed more men to be trained in a shorter time to fire faster. The effect of the iron ramrod was similar.

Introduced in Prussia around 1720 and quickly copied elsewhere, it offered nothing in the way of ballistic performance, only a faster, more reliable way of loading. This was true of the entire sequence of improvement. Tactically it further stretched what were already elongated formations by permitting a reduction of the ranks from five to three; but a paucity of gains in hitting power and accuracy insured that the distances and conditions of engagement would remain stable. At this juncture, development of smoothbore small arms basically ceased.

Brown Bess and Her Sisters

What had emerged was a sort of universal musket. The primary weapon of Europe's various foot soldiers differed little more than the spears of Hellenic Greece's phalangites. This symmetry of arms would persist for nearly a century and a half. Archetypical was the long life of the English sidelock .78 caliber piece known as Brown Bess. This Methuselah among firearms first saw action around 1703, and continued to blast away at her equally conservative Continental sisters in the midst of the Industrial Revolution. During this period Brown Bess reproduced an estimated 7.8 million times, and was copied around the world. Thus, Committee of Safety muskets fabricated by the colonies during the American Revolution were basically replicas of Brown Bess. In fact, every foot soldier in Europe could have taken Brown Bess into his arms without a significant change in his shooting.

The key to Brown Bess's tactical capabilities was found in the .73 caliber lead ball it fired. While the .05-inch space, or "windage," between ball and barrel was crucial to fast loading (bullets could simply be dropped down the muzzle), it severely limited the gun's accuracy. When Bess was fired, the ball rattled up the barrel, its final direction determined by the last side it hit before emerging. At very short ranges this deflection had only a minor effect. It was possible to hit a foot-square target at forty yards almost every time. Beyond this range results deteriorated quickly. Colonel George Hanger, a British officer, wrote in 1814 that "as for firing at a man at 200 yards with a common musket, you might just as well fire at the moon." "In very poor visibility," he might have added, since a force of several thousand men shooting volley after volley was sure to generate a smoke cloud sufficient to obscure most every target. It followed that Brown Bess lacked even the most rudimentary sighting devices, and the rate of hits was accordingly low.

Brown Bess.

Why was this tolerated? Didn't commanders realize how inaccurate their men's guns really were? The evidence indicates that they didn't spend much time worrying about it. Although most shots were bound to miss, the infernal atmospherics, concussive sound, and the grisly effects on those unlucky enough to be hit were sufficient. In their own way, military muskets like Brown Bess were formidable weapons, and the style of fighting they epitomized was sufficiently deadly to be decisive. Beyond that, their fundamental weakness—inaccuracy—was something only a few skeptics like Colonel Hanger cared to question. Even given available metallurgy and fabricating techniques, there was a good deal of room for improvement in firearms. Yet fundamental development in weaponry was precisely what the system could not stand. Really accurate small arms used by troops out in the open, lined up like so many clay pigeons, would have meant almost immediate slaughter. It was critical to avoid the question of how it was possible to march two lines of foot soldiers to within a hundred yards' range and then, disdaining all cover, have them fire over and over at each other. Had this question been pursued systematically, the military musket would have been found at the center of the riddle: Brown Bess and her sisters, having cohabited for so long with infantrymen, refused to kill them promiscuously. If this was a marriage of convenience, then so be it.

The dynamics of warfare were much more complex than two lines of infantry blazing away at point-blank range. From a command perspective, it was an activity that demanded a rare ability to juggle both variables and unknowns, and to do it on the fly. Consequently, the eighteenth century was a time for generals. The degree of control that had to be exercised over this brand of firepower demanded almost total centralization of authority and extraordinary responsibility placed ultimately on the shoulders of a single individual. It has been estimated that without electronic communications, 80,000 to 100,000 was the maximum number of troops that could be directed by a lone commander. As armies steadily grew, the art of generalship came to exist at and beyond this ragged barrier.

Strategically, this meant delivering a very large body of men to the battlefield in a timely fashion and in a position that put the adversary under maximum disadvantage. A daunting task under any circumstances, it was compounded by another fundamental characteristic of the military system—huge support structures. Given the demands of training and the generally miserable conditions of service, all major European armies suffered continuous attrition through desertion. This was especially true while on campaign. Obviously, such troops could not be allowed to forage on their own. Instead, they had to be regularly fed as they moved forward.

Into the Breech . . . Prematurely

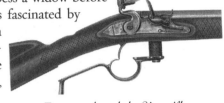

Ferguson breech-loading rifle.

One persistent individual nearly made Bess a widow before her time. Major Patrick Ferguson was fascinated by firearms and an inveterate tinkerer—a combination that led him to produce a breech-loading rifle so apparently lethal that even the hierarchy of the British army could not ignore it, at least initially.

Capitalizing on an earlier French concept, Ferguson drilled a hole down through the breech of a standard musket, removing the trigger guard. He then replaced it with a threaded plug attached at the bottom to a new trigger guard. Loading was simply a matter of unscrewing the stopper and exposing the inner portion of the barrel, thereby allowing a lead ball and charge to be inserted and then sealed. Because the ball was loaded at the rear, he could add rifling and, suggestively, an adjustable rear sight.

The gun was ready in June 1776, and tests were conducted before Lord Townsend, the master general of ordnance, and Lord Amherst, the army's commander-in-chief. The results were astonishing. Ferguson kept up a steady rate of four shots per minute for over five minutes, missing a target 200 yards away only three times—all of this without misfires despite a steady rain. Characteristically, British authorities were attracted to the gun's rapidity of fire, not its accuracy. The crisis in North America was reaching the boiling point, and the army required every bit of firepower it could muster. The master of ordnance ordered 100 of the breech-loaders to be manufactured, while Ferguson was ordered to form and train a special corps of riflemen.

On September 11, 1777, Ferguson's rifles had their first and only combat test at Brandywine Creek, where the British caught George Washington trying to cross at Chadd's Ford. Ferguson and his tiny rifle corps repeatedly flanked larger bodies of Americans, subjecting them to lethally accurate fire. In return, they took few casualties, since they were able to fire on the move and from a prone position, neither of which was possible with muzzle-loaders. Infantry tactics over a century in the making were wrenched in new directions in the space of minutes by Ferguson's gun.

At one point an opposing officer rode within Ferguson's range, thinking himself immune from normal musket fire. "I could have lodged a half dozen balls in or about him," wrote Ferguson.[2] It was George Washington, and a less chivalrous rifleman might have crippled the American Revolution with a pull of the trigger.

Ferguson himself was not so lucky. In the process of extricating his men from an untenable position, the major had his right elbow shattered by an American ball. He would eventually recover, but the shot proved fatal to his gun. During his convalescence, British commanding general Howe took the opportunity to disband and disarm his rifle corps. The guns were packed up and placed in storage for good.

Armies were tied to huge, slow-moving supply trains running into thousands of vehicles. Thus fighting forces were removed from the circumstances that had led to the worst depredations of the Thirty Years War. Regular nutrition became a factor in the system of control to which the soldier and his weapon were subjected. Yet armies, further laden with artillery, ammunition, and masses of animals and camp followers, were left as shackled entities, capable of only the most ponderous strategic maneuver. Campaigns not only unfolded at an elephantine pace, but mistakes once made could not quickly be rectified. So great were the difficulties involved in maneuvering armies effectively, along with the rewards to be gained from doing so, that the Comte de Saxe, a leading military theorist of the day, suggested that a commander of sufficient skill might make war successfully all his life without ever being forced into a pitched battle. This accidental paraphrase of the ancient Chinese military sage Sun Tzu also implied that not fighting was in some sense preferable to fighting—an outlook rare in European military history, but characteristic of the eighteenth century.

Set-piece engagements occurred, but with less frequency. And when they did, tactics were equivalently complicated. Battle became an exercise in opportunism—seeking a vulnerability and exploiting it, while taking care to avoid a similar fate. Amidst the smoke and confusion, keeping track of what was going on, and reacting quickly and appropriately, were so difficult that it came to be described in near mystical terms. *Coup d'oeil* was the phrase used to connote this rare gift of insight, which enabled a commander to take in at a glance all tactical probability and know exactly the right moment to apply overwhelming pressure. The general had become battle manager, essentially a cerebral figure reflecting the new ethic of courage previewed at the Duke of Parma's lunch table. Rather than leading a charge of flailing broadswords, the classic military pose now revealed the commander scrutinizing enemy dispositions from afar, perhaps snapping a spyglass shut with a decisive flick of the wrist before issuing his final orders.

Aristocratic dominance of the officer corps and the extreme centralization of such armies insured that very few ever got the opportunity for command. Even fewer had much talent for it. But those who did had a major advantage. With the homogenization of arms and forces, only numbers and leadership really mattered. Faced with a truly clever commander, on the order of a Turenne or Marlborough, even numbers could be neutralized. So generals could and did win wars.

They had, besides front-line infantry, three implements to dismantle the enemy—extra infantry, cavalry, and artillery. How and when a commander committed them was critical. The injection of the fresh troops was frequently enough to collapse an opponent's wavering line. The longer he waited, the greater the disparity in fire generation between his fresh and the enemy's used-up troops. Should he wait too long, however, his own line might disintegrate or fall to opposing reserves. Thus the decisive use of reserves was a key element separating the merely mediocre from true captains.

Cavalry served a similar function, but less effectively. Mounted troops continued to be held in high esteem by commanders, but there were already signs that they were galloping toward obsolescence. At the heart of the problem was a lingering inability to come to grips with firepower. While cavalry retained its importance strategically for reconnaissance and in pursuit (actually, until Napoleon it was seldom used ruthlessly during this period to chase down defeated armies) on the battlefield, its role became largely psychological. Bodies of horse were increasingly employed to flank extended infantry lines with the aim of rolling them up, or to catch foot soldiers when disorganized or isolated. But in response ground troops were taught to form hollow squares, so cavalry units could not ride around and attack from the rear. Cavalry relied on an ability to induce panic; otherwise, direct confrontation was highly problematic. As always, formations of horse could not overwhelm massed and disciplined infantry projecting a row of points—now muskets with bayonets. Lines were certainly thinner and more brittle, but firepower made up for it—one reason musket slugs were so big was to enable foot solders to bring down horses at close range, where the animals presented hard-to-miss targets. Cavalry also could prey on artillery, especially if they were on the move or otherwise unprepared. But the penalties for failure were even greater, especially if a charge was bungled and a line of horse was met by a hail of subprojectiles at short range.

There were other signs of weakness. Military historian David Chandler points to a dramatic drop in the proportional representation of cavalry in force structures over the two centuries separating the Thirty Years War from the end of the Napoleonic conflicts. These figures varied from army to army, and can be explained in part by the high cost of horse. The fact remained, however, that as armies grew, cavalry shrank. The rise of the dragoons also pointed to the compromised state of cavalry. These were essentially foot soldiers put on the back of a horse for the sake of mobility and trained to fight dismounted—handymen of

Horse Artillery

The possibility of being caught in the open by a sustained artillery barrage posed one of the greatest dangers to a body of horsemen. A single ball could easily pulverize multiple members of a closely packed cavalry or dragoon formation. The only safety was to be found in the shelter of equivalent cannon power, but given the ponderous nature of field pieces this came at the cost of mobility, the cavalry's only true advantage on the battlefield.

Falcon: early light artillery piece.

Gradually, however, the horsemen's predicament was relieved by the evolution of much more portable artillery. First and most important was the appearance of lighter artillery tubes. Since the days of the composite leather gun, the Swedes had been experimenting with very light pieces, eventually developing a .17 caliber 4-pounder with a useful range of nearly 600 yards that weighed only 600 pounds. This type of gun was either ignored or adopted slowly; nonetheless, it held the potential of being drawn at a gallop by a team of as few as four horses. This, in turn, was made possible by the steady improvement of gun carriages, and the eventual redistribution of the gross weight of such light pieces from two to four wheels, allowing them to be pulled much more efficiently at high speeds even over rough ground. Closely related was the introduction of a wheeled caisson, or chest, to hold ammunition and spare parts, which could be hitched behind the cannon to generate a truly self-contained fire-support package.

With cannoneers and loaders individually mounted, the resultant horse artillery, or "galloper guns," were capable of keeping up with cavalry and dragoons, or moving independently to provide fire support on demand—roles that accurately foresaw the function of self-propelled artillery on the modern battlefield. Nevertheless, the entire process required in excess of fifty years, none of the constituent improvements was the least bit revolutionary, and each step met with considerable opposition . . . the epitome of the Enlightenment's quiet but apparent disinclination to foster new arms.

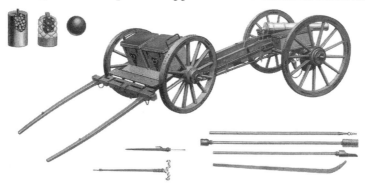

Galloper gun and accessories.

battle who could be used quickly to plug a hole or exploit a weakness with gunfire.

This was decidedly not the case with pure cavalry. In armament the view from their saddles was backward. With some exceptions most armies followed up on Gustav Adolph's deemphasis of pistols during the Thirty Years War, and the return of cold steel to the hands of cavalrymen. This may have been psychologically satisfying, but at base it could be traced to a continuing inability to reload while on horseback. Cavalry charges right up through the Napoleonic era came to be dominated by flashing sabers and lances at the ready, reducing the horseman's range of killing power to approximately three feet. While multishot pistols and carbines would eventually restore some ability to strike from a distance, through it all horsemen remained large and vulnerable targets committed to a confrontational style of warfare in contravention to the logic of fire-power. There was, however, one successful blending of horseflesh and gunpowder, which pointed directly toward a future of truly mobile fire-power, horse artillery.

Much the same thing could be said about the evolution of artillery as a whole in Europe during and well past the eighteenth century. Keeping cannon on a tight leash presented special problems, however. Unlike the skills associated with other combat arms, gunnery was increasingly understood to be a scientific and mathematical pursuit not ordinarily the province of the officer-aristocrats who ran armies. Instead, the profession tended to draw technically oriented and potentially inventive types. While the militarizing of artillery units begun by Gustav Adolph gained momentum steadily, the persistence of massive ordnance establishments left bastions of independence and potential innovation beyond the reach of strictly uniformed authority.

They were not beyond the control of princes, however. Ordnance centers, whose purview extended from the manufacture, testing, and servicing of the cannon themselves to the education and training of their crews, were dependent upon the direct patronage of monarchs. When Louis XIV had his cannon inscribed with the motto *Ultima Ratio Regis* (the last argument of the King) he meant it literally. Cannon were the most deadly concentration of power; monarchs hoarded them like gold. Nevertheless, a political system based on the limitation of objectives and war by the rules also demanded that the development of weapons with the lethal potential of artillery be carefully circumscribed.

The trajectory of artillery, at least until 1776, was largely retrograde. Numerically, this was represented by a drop in the ratio of cannon to infantry from a high of four per thousand in the later armies of Gustav Adolph and Wallenstein to an average of 1.5 per thousand in twenty-one selected engagements between 1690 and 1745. With the exception of the galloper guns, artillery units were conceived of and used as defensive assets, and mobility was of relatively little consequence. Up to the middle of the eighteenth century most field pieces weighed in at a hefty three tons, if their carriages and trails were included. This despite the discovery by one Antonio Gonzales, around 1680, that the substitution of a spherical powder chamber for the traditional cylindrical version allowed a great deal smaller charge to generate the same range and penetration, opening the way for dramatically lighter pieces.

The discovery went largely unexploited. When the energetic French lieutenant general of artillery, Marquis de la Frezeliere, promoted chamber designs based on the new concept, he not only ran into bitter opposition, but upon his death all of his improved pieces were recast in the old manner. Half a century later, when Joseph-Florent de Valliere reduced the number of French pieces to five standardized calibers, he too based them on the older, heavier designs. Royal approbation can be assumed, since Valliere's king, Louis XV, was reputed to have rejected the use of a newly perfected gunpowder on the grounds that it was "too destructive to human life."

Nor were French authorities out of step in their conservatism. Field pieces employed by all the major powers up to the late 1760s differed only in detail; the archetype consisted of the traditional bronze tube cast in a cored mole, a process that insured both excessive weight and considerable imprecision in dimensions and, therefore, accuracy. Even after 1740 when Swiss engineer Jean Maritz and his son of the same name perfected a machine capable of boring a barrel out of a solid billet, thereby producing a much truer product, Europeans were slow to capitalize on the possibilities.

The key technical factor in stabilizing the conditions of warfare, at sea as well as on the battlefield, was the disinclination to use anything but solid shot in strictly military engagements. Hollow, explosive-filled shells were certainly available; mortars already used them. Howitzers too fired a bursting round. Yet direct-fire pieces, plainly the dominant artillery weapon, were fed no explosive projectiles. Even after 1740, when it was known that the initial blast of a cannon could be relied upon to light the fuse, there was no rush to employ ammunition that promised to blow apart exposed infantry formations, or the sides of oaken warships. Grape and case shot, employing a variety of subprojectiles ranging from nails to musket balls, were used to rake infantry and cavalry at close ranges but without secondary charges. Until 1850 the use of explosives as a destructive agent rather than as a propellant remained highly circumscribed. It was a deliberate choice, or, more accurately, a shared disinclination to pursue certain military-technical possibilities.

Reform at Revolution's Doorstep

One man who did capitalize on the possibilities of bored barrels was Jean Baptiste Vacquette de Gribeauval. An artillerist with long experience in both theory and practice, between 1763 and 1767 he presided over the systematic redesign of all calibers of field artillery by a team of French ordnance experts. Employing advanced chambers, machine-bored tubes, and careful testing, Gribeauval's team was able to substantially reduce both charges and the amount of metal used in fabrication, producing a much more exact, significantly lighter family of guns that were also reduced in length. Efficiency was further enhanced by truly spherical balls of the correct diameter, decreasing windage and improving the potential for accuracy, while preserving range and hitting power. Lighter, more compact, tubes allowed Gribeauval's team to reengineer virtually all the associated elements needed for field artillery, in the process injecting an element of tactical mobility. With characteristic attention to detail, elevation wedges were replaced with a much more precise screw-based mechanism, sights with adjustable hairlines were added, and shot was combined with uniform charges in cartridges—all of which increased the rate of accurate fire significantly. Finally, everything—tubes, carriages, and accessories—while not truly interchangeable, was basically standardized, greatly facilitating repair and maintenance in devices particularly subject to the rigors of warfare.

Despite their obvious superiority, Gribeauval's first prototypes stirred immediate opposition from conservatives, including Valliere. So intense was their campaign that final approval of the new designs did not come until 1776, upward of a decade after they had been vetted initially. Even then, continued opposition from within the army along with production problems slowed deployment to the point that Gribeauval's new artillery pieces were not available until the very eve of the Revolution.

Under the new regime, the cannon made a key contribution at the pivotal battle of Valmay in 1792, and were embraced by the Jacobins in the person of Lazare Carnot, who oversaw their production and proliferation throughout the Revolutionary Army. Bonaparte, trained as an artillerist, subsequently inherited Gribeauval's guns and they remained standard issue throughout his wars, lingering in French service until 1829, when they were finally phased out. At this point the designs were nearly sixty-five years old.

The major exception was siege warfare, where not only shell-firing mortars but large explosive mines were used routinely. Here technology was far from stagnant. Rather it was applied with an increasingly systematic orientation to a variety of military problems facing both besieger and besieged. Yet this was a process dominated by a single towering figure, and his ultimate commitment was to moderation, not predation.

Had Horatio Alger been interested in military history he might have written about Sébastien Le Prestre de Vauban. "Fortune caused me to be born to one of the poorest gentlemen in France,"[3] Vauban once lamented. Compounding this handicap, he joined the army as a military engineer, a corps where no officer had ever risen above captain. Yet in 1703 his patron, Louis XIV, the Sun King, elevated him to the supreme rank, Marshal of France. In a world of ascribed status and aristocracy, pure talent and, as Alger would have added, befriending the boss, won out. Yet in another sense Vauban was very much a man of his times both militarily and in his relationship with his prince.

Early in his reign, Louis, if not exactly on the rampage, was nonetheless in an expansive phase. Vauban devoted himself to perfecting means to attack the many fortifications that littered the Sun King's way into the Low Countries. To more effectively drive defenders from the outer ramparts of a trace, Vauban was the first to exploit the possibilities of the ricochet. Because the flat trajectory of direct-fire cannon propelled balls harmlessly over troops huddled behind low-lying parapets, Vauban took to using reduced charges so that balls bounced short and then arced down on the defenders. He employed the technique with considerable success, but characteristically objected to the phrase *tir a ricochet,* which he thought implied trickery.

To approach fortifications with an attacking party Vauban devised a system of parallel trenches minutely calculated to provide the best firing angles and maximum shelter for his troops. Vauban's trench works required enormous effort, but parallels were considerably safer than the earlier system of zigzags, providing shelter all the way to the point of attack.

But Vauban's true calling became apparent when his master turned from aggrandizement to what historian John Lynn characterized as an obsessive pursuit of security and defense of his acquisitions. Vauban responded with a plan to ring France's border to the north and east with high-tech fortifications—an enemy-tight seal that in its conception and price tag was the Enlightenment equivalent of Ronald Reagan's Star Wars. The ensuing building program established Vauban as among the greatest military architects of all time, and set France on a course destined to culminate in the ill-fated Maginot Line.

Much of the stone and earth has crumbled, but we know exactly what each gem in Vauban's strategic necklace looked like, since detailed scale-models were lovingly constructed for His Majesty to display in the Hall of Mirrors at Versailles and are now preserved in the Musée de l'Armée in Paris. From these it can be discerned that Vauban's designs were based on the practical application of plane geometry to insure that any attacker, no matter what his angle of approach, would be subject to a withering cross-fire. Not a square inch of cover or dead ground was tolerated within the effective range of defending guns. Due to variations in topography, regular polygons—the classic star-shaped trace, for example—were either impossible to construct or insufficient in their coverage. Hence, separate bastions, themselves supported by outworks, were added to close off extraneous angles. In addressing this and other military problems, Vauban's creations went through several separate phases, each one more elaborate, until they became virtual military mazes extending as much as 300 yards outward. The costs were equivalently enormous, rising from an average of 350,000 French livres a year at the outset of the program to a high of nearly 12 million twenty-eight years later. But the strength of these fortifications, along with their positioning on what Vauban dubbed the *pre carre*, or dueling field, caused them to act as strategic magnets, drawing in, then pinning down huge attacking armies. Until Napoleon finally reversed the trend, the ratio of sieges to field engagements grew even greater, further thwarting the offense and employing the fortified gun to further domesticate the gun on the loose.

Paradoxically, we rarely hear of an unsuccessful siege. The continued usefulness of this type of fortification, employed in some cases as late as World War I, argued that this high success rate was not due to design shortcomings. Rather, it was attributable to the general inclination to surrender—itself a manifestation of the mind-set under which the defenses were undertaken in the first place. Sieges, traditionally the most brutal and predatory form of warfare, logically came under particularly close control. The precise conditions of submission were formalized and, in large measure, adhered to. Most influential was a letter dated April 6, 1705, in which Louis XIV ruled that henceforth a commander might surrender honorably after suffering one small breach and repulsing a single assault. This departed significantly from the previous criteria of one large breach along with multiple assaults, and remained the standard until French Revolutionaries guillotined a commander and his wife for what they considered pro forma resistance. "What was taught in the military schools was no longer the art of defending strong places," huffed Lazare Carnot, "but that of surrendering them honorably after certain conventional formalities."[4]

He was right but missed the point. The only alternative to such contrived circumstances was the wretched fate of cities like Magdeburg. So even in the face of fortifications improvements that technically facilitated die-hard resistance, the inclination toward moderation was fortified. There were instances when sieges were carried to their wanton conclusion; but these were usually provoked by atrocity or the pure stubbornness of the defenders. By the same token, even though explosives were used more promiscuously in sieges, solid shot fired out of standard bronze muzzle-loaders remained easily the dominant projectile-gun combination employed against fortified places. The instinct here as elsewhere was to preserve, not destroy.

The Enlightenment prism through which weapons innovation was filtered was reflected by events at sea. Navies were the best balanced and most enduring of the age's accommodations with weaponry. The physics of wood, wind, and water conspired in moderation's behalf, so bolstering the emergent naval order that key elements persisted in the minds of naval officers even a century after sailing warships had disappeared. Yet so long as they existed, the great stabilizer was the persistence of the solid-shot-firing muzzle-loader as the standard engine of naval destruction.

So armed, all men-of-war differed mainly in degree—the more guns, the more fighting power. Length being limited, this meant piling gun decks one upon the other to a maximum of three. Because guns were mounted at regular intervals along each side, only a finite number could be carried per deck. Offensive potency, therefore, turned on whether a warship had one, two, or three such levels. In determining speed, the situation was reversed. Since all rates, or classes, of ships were essentially full-rigged and shared relatively similar amounts of canvas, vessels with more layers of guns were by necessity slower than their less-well-endowed kin. The result, noted naval historian Julian Corbett, was a very clear segregation of function in naval warfare. Smaller ships, inevitably outgunned, zealously avoided combat with bigger ones, while the very plenitude of armament that made larger vessels more powerful prevented them from catching smaller-fry. Sea battles, therefore, unfolded with Homeric decorum—equals seeking out equals to engage symmetrically by class. Generally, the swifter, lighter rates, used primarily for reconnaissance, made contact first, with the action progressing upward until the slow-moving heavyweights finally engaged.

As with musket formations, such actions generally took place at very short range. "How much nearer, so much the better," observed Richard Hawkins in 1622, a statement that characterized the ever aggressive Royal Navy up to and beyond Trafalgar, and, it followed, the behavior of her enemies when forced to fight. It only made sense. The extreme inaccuracies imparted by the windage necessary for efficient loading, along with the natural rocking of the hull, insured that effective ranges did not exceed 300 yards, a mere tenth of the potential range. The use of solid shot limited the structural damage inflicted even at close range, but the impact of cannonballs on wood hulls generated a barrage of splinters capable of inflicting horrific casualties. Close ranges also created openings for effective musket fire, while resistance to structural damage frequently demanded bloody boarding melees to take men-of-war out of action. Easy on ships but hard on their crews, war at sea remained a gory rite of manhood, with combatants slugging it out "yardarm to yardarm."

As with land engagements during this period, pitched naval battles, much less decisive ones, remained relatively infrequent. Also, despite the fact that these ships were built for roaming the high seas, almost all fights took place near the coast, frequently within sight of land. In part this was due to the difficulties of locating a fleet in mid-ocean, but it was also deeply influenced by the preferred approach of the dominant Royal Navy, blockade. As wielded by the English, blockade centered on the placement of the heaviest fleet units athwart an adversary's lines of trade and communication. While the maintenance of a blockade, and conversely blockade running, were left to smaller, swifter classes, ultimately such a siege at sea could only be broken by a physical confrontation between battleships.

These engagements were determined largely by the nature of the weaponry and placed heavy emphasis on control. Since almost all firepower was located along the sides of ships, it was important that a sailing battleship confronted its adversary lengthwise whenever possible. Only one formation allowed multiple ships to satisfy this requirement, and that was line-ahead. This had been known for a long time. Back in 1502, the

Portuguese admiral Vasco da Gama scored a victory over a large Islamic fleet off the coast of Mulberry by deploying his ships "one astern of the other in a line." Yet line-ahead was employed only sporadically (ships in the Armada fight operated in bunches) until 1653 when "Instructions for the Better Ordering of the Fleet in Fighting" were promulgated by the Royal Navy. "Each squadron shall take the best advantage they can to engage with the enemy next unto them; and in order thereunto the ships of every squadron shall endeavor to keep in line with the chief."5 After a series of pitched battles between the Royal Navy and the Dutch over the next thirteen years hammered this home, line-ahead became the standard for battleship engagements for nearly three centuries.

Given this, the ultimate goal of an advancing file of sailing warships became crossing an enemy's line of bearing. Actually, this was extremely difficult and seldom accomplished, but its theoretical advantages captured the naval imagination. As ships proceeded past the opposing file, they could concentrate their entire broadsides on targets that might only reply with a few bow guns. Also, firing down a line maximized the chances of hitting something. Not coincidentally, this tactic ("crossing the T") demanded all units follow the lead ship closely and exactly. Overall the centralization of command and the degree of control necessary to keep ships in formation set the tone of subsequent naval engagements, or lack of them. In particular it encouraged a follow-the-leader mentality, which stifled initiative even among the audacious British.

Nevertheless, the structure of command and control was highly satisfactory in other ways, particularly in the way it was paralleled, and therefore reinforced, by the nature of the weaponry—a hierarchy of ships dictated by the provisos of wood, wind, and solid shot. In a scheme where an admiral commanded a line of battleships ruled by captains, surrounded by a screen of progressively smaller craft controlled by a descending order of officers, the idea of a lesser warship defeating a larger one was not just implausible, it was insubordinate!

Lethality with decorum, power precisely equated to size, the system was a nearly ideal embodiment of combat predicated by intraspecific aggression, and its crown jewel the sailing battleship was the essence of all it implied—large (made to look larger by its sails), stately to the point of arrogance, and capable of generating unprecedented sound and fury with guns that in full roar could be heard fifty miles away.

So satisfactory was the state of naval architecture that it ground to a halt. As on land, improvements were made to hasten firing—pulleys to control recoil and flintlocks rather than fuses on cannon. Rigging was further

The Paragon

HMS Victory.

Horatio Nelson finally had the skittish Villeneuve where he wanted him; after nearly seven months of cat and mouse, the French admiral was leading his Combined Fleet of French and Spanish battleships out for a showdown. What followed became a monument to the system and a vision of decisiveness that inspired naval officers and naval architecture until December 7, 1941.

Actually Nelson was an iconoclast. His battle plan involved letting his own T be crossed by attacking the enemy perpendicular to their line. Also, he deliberately encouraged initiative in his officers . . . the right kind of initiative. "In case signals cannot be seen, or clearly understood, no captain can do very wrong if he places his ship alongside that of an enemy."[6]

Preparations aboard each warship left little to the imagination. Below, benches and tables were stowed or tossed. Hammocks were brought up on deck to provide cover from small arms. Provisions were thrown overboard, including live cattle. Magazines were readied, guns precharged, and watered-down canvas was hung as fire curtains. Far above, sails were reefed and netting hung to thwart boarding; rigging was reinforced with chains to help keep shot-up yards from crashing down on those below. Decks were soaked to inhibit fire and then coated with sand for traction and to absorb blood—some may have even been painted red. The crews were not fooled. As they ate a last cold meal—all stoves had been extinguished—they knew how bloody it would be.

As Nelson's forty-year-old flagship *Victory* approached, the French line erupted, shooting away the main topmast, shredding rigging, and shattering her wheel before the English vessel fired its first broadside. When its port guns were finally in position to rake *Bucentaure,* the results were devastating, taking out twenty guns and inflicting 400 casualties. The French, like the naval pretenders aboard the Armada and those of Rome, emphasized what amounted to land tactics, making heavy use of small arms fire, grenades, and even trying to board *Victory* at a critical moment. They succeeded in immortalizing Nelson— shot fatally through the spine by a musketeer in the mizzen top. But as they did so, *Victory*'s main guns and those of the other closely engaged British battleships literally shot the French and Spanish ships out from under their crews. Trafalgar and the gale that followed left the Combined Fleet shattered, twenty-two of its thirty-three capital ships captured or destroyed. Never again would France challenge for the crown of seapower.

But even more important for the naval future was the pure decisiveness of the clash. It was Armageddon. As such it became the guiding light of naval commanders, and the ultimate prop for the capital ship. However becalmed, they could look back on Trafalgar and dream of its reenactment.

improved and substantial canvas area was added, but not in a way that changed the relationship between classes. As a concept and institution the battleship was perfect. Nothing so stirred the naval imagination. But dreams ultimately must be anchored in reality, and this explained the importance of the clash of behemoths that took place off Cape Trafalgar on October 21, 1805.

Power was a matter of counting. Like homogenized infantry and artillery on land, ships-of-the-line were the universal tokens of naval might. At the close of the Napoleonic Wars in 1814, England reached an all-time high with 100 sailing battleships actually in commission, and a force of sixty fully seaworthy craft held in reserve. Never had the crown of seapower been planted more firmly.

The most extraordinary part of the fading era's accommodation with weapons development was its longevity. In the face of the transformative American and French revolutions, the outbreak of general war, and Napoleon, virtually every aspect of the system was swept away or profoundly compromised: monarchy would never be the same; professional armies of trained automatons were overtaken by citizen legions numbering in the millions; and limited ends were blotted out by visions of total overthrow. Yet weapons changed only marginally, a fact made even more remarkable by the onrush of the Industrial Revolution.

The case of France was instructive. To shield the Revolution, Carnot and the Committee of Public Safety staged a massive program of defense-industrial mobilization—labor and resources were drawn together on a national scale, scientific talent was recruited, and great public weapons factories were set up in the parks and gardens of Paris. By 1794 the city was the largest small arms producer on earth, turning out up to 750 muskets a day, when all of Europe had never before exceeded 1,000. Artillery took longer, but the standardized calibers and parts of Gribeauval's guns eventually allowed production to ramp up dramatically. Yet the factories and foundries were still producing the same basic designs, just a great deal more of them.

Nor did Bonaparte have much impact. Unquestionably one of the greatest soldiers in history, a good measure of his success was due nonetheless to his ruthless opportunism and refusal to obey the rules. Following Revolutionary precedent he capitalized on his troops' loyalty and enthusiasm by cutting their logistical shackles and allowing them to forage, thereby introducing an element of strategic mobility that was all too frequently a dagger in the heart of his lumbering adversaries. He did not so much fight his enemies as he hunted them. Victory on the battlefield was merely an intermediate step. He pursued the survivors relentlessly, using Murat's cavalry to track them and crush their will to resist.

When it came to weapons, though, Napoleon barely deviated from the norm. Trained as an artillery officer, he was the first commander to fully comprehend the latent killing power of cannon, employing them to inflict over 50 percent of the battle casualties suffered by his opponents. He made good use of horse artillery, Gribeauval's system, even howitzers firing explosive shells; but the triumvirate of solid-shot, smoothbores, and muzzle loaders still held sway over Bonaparte's battlefields.

This outlook was even more pronounced in the manner he armed French infantry and cavalry. Despite his heavy reliance on skirmishers and sharpshooters, Napoleon retained the musket as the primary weapon of light infantry as well as troops-of-the-line. Instead it was the British, drawing on their experience in the American Revolution, who began furnishing skirmishers with rifles, using subcaliber balls when a high rate of fire was needed. Napoleon's favorite infantry weapon was the bayonet, the use of which, though selective, very effectively demonstrated the moti-

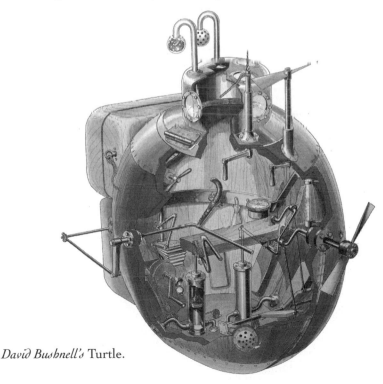

David Bushnell's Turtle.

vation gap between French troops and those they fought. This attraction to
the archaic was also seen in his resurrection of very heavy cavalry units, the
cuirassiers, equipped with breastplates, steel helmets, and mounted on huge
steeds. But Bonaparte's penchant for cold steel was secondary and rooted in
the psychology of fear. Like his contemporaries, Napoleon understood that
"it was by fire and not shock that battles are decided today."7 Not, however,
by new kinds of fire. Weapons solutions that went beyond the conventional
did not interest him. Not only was the infant balloon corps, founded by
Revolutionaries in 1793, disbanded; but the Emperor of the French
displayed a similar lack of imagination when it came to naval armaments.

Bonaparte was tremendously frustrated by British seapower, but not
enough to seriously assault the hierarchy of ships. Back in 1773 Yale under-
graduate David Bushnell had designed and built the first successful submarine, the *Turtle,* a diving-bell-like craft, which he used during the American
Revolution to attack several British warships at anchor. Bushnell failed, but
around twenty-five years later a fellow American, the famed inventor and
engineer Robert Fulton, took up the cause. In 1797 he journeyed to Paris
where he proposed the use of a submersible, the *Nautilus,* against Britain,
only to have the short-lived Directorate reject the idea as an atrocious and
dishonorable way to fight. Undaunted, Fulton constructed the *Nautilus* at his
own expense. In 1801 he approached Napoleon, who had by that time
replaced the Directorate, interesting him only briefly before he turned to
other things.

Fulton subsequently went to England, where William Pitt encouraged
the inventor to test his device against the French squadron at Boulogne. He
did so on the night of October 2, 1805—less than three weeks before
Nelson's fateful meeting with Villeneuve—managing to sink a pinnace, first
blood for what would become the deadliest naval weapon in history. But
further experimentation was quickly dropped. The Earl of St. Vincent, First
Lord of the Admiralty, left no doubt why. "Pitt was the greatest fool that
ever existed to encourage a mode of war which they who commanded the
seas did not want, and which, if successful, would deprive them of it."8

There were other such starts and stops. In 1784 General Henry
Shrapnel developed the first explosive shell containing subprojectiles
designed to be scattered to cut down infantry and cavalry. Foreshadow-
ing the introduction of later highly lethal arms, the first use of Shrapnel's shells was postponed for two decades and employed in Suriname,
not Europe, where they made only a cameo appearance in Spain during
the Napoleonic Wars.

In 1799, Sir William Congreve became interested in the primitive rockets he saw in India, and produced an improved model with which he managed to set the entire city of Boulogne afire in 1806. Soon after he pronounced the rocket "an arm by which the whole system of military tactics is destined to change."[9] The British ordnance establishment even responded by developing an iron-headed version with a contact-fused explosive charge and a two-mile range. But it proved a short-lived revolution. Although Francis Scott Key immortalized Congreve's "rockets' red glare," they failed in their primary purpose ("our flag was still there"), and faded out of existence for the rest of the nineteenth century.

The disinclination to improve weaponry beyond a certain point was a subtle phenomenon, which, as times changed, grew even harder to explain. It would be more reassuring if the era of gun control could be clarified by a prescriptive body of thought or a scrupulously executed plan; but all that remained was a sort of collective somnolence.

The aristocratic guardians of the system, the officers, were men of action, not introspection—soldiers, not technologists. Weapons were an

Philosopher's Lapse

Carl von Clausewitz ranks as one of the greatest thinkers to consider the issues of organized violence. His unfinished masterpiece, *On War,* was so opaquely phrased yet packed with insights that generations of military scholars continued to perseverate on its ponderous pages, like monks poring over a sacred text. Virtually every war and politico-military maneuver ever undertaken has been explained by someone in Clausewitzian terms. Like Darwinian thinking, it seems almost infinitely applicable. So rich were the veins of thought, that sometimes it seemed as though his commentators read different books. In part this is because Clausewitz had something to say on almost every topic even vaguely related to war . . . with one exception.

When it came to weapons, the philosopher was silent. The tools of the trade were barely touched. Clearly, this was not a matter of ignorance, since in asides Clausewitz revealed himself to be thoroughly conversant with the capabilities of contemporary weaponry. Rather, it seemed as if weapons were taken as a given, a sort of base condition, which, like human nature, was so fundamental as to not require explanation. This was hardly the case and it seemed that a man of Clausewitz's intellect, writing as late as the 1830s, should have recognized the possibility that arms might be improved decisively. Yet the thought apparently never fought its way to the surface of his consciousness. In this he was a man of a certain age; but that age had passed.

everyday part of their lives, but they were not by nature inclined to tinker or even think speculatively about them. Psychologically, tactics were designed to accommodate humans to a specific level of violence; the possibility of increasing that level could only painfully disrupt the equilibrium, so it was ignored. There was certainly always the lure of victory; but never out of the context of survival. Thus the anonymous late-eighteenth-century author of *The Reign of George VI: 1900–1925,* the first fantasy of future war—a genre destined to predict arms innovations with uncanny accuracy—foresaw no weapons developments at all. John Ruskin could still proclaim in 1830: "Taken as a whole the ship-of-the-line is the most honorable thing that man, as a gregarious animal, has produced." Yet no one more tellingly betrayed this outlook than the era's most profound student of war, Carl von Clausewitz.

Chapter 10

DEATH MACHINES

At last the Promethean force of arms slipped its chains, free again to run rampant across the affairs of humanity. This process, like all good jail breaks, came from an unexpected quarter. The record indicates that the initial avenue of escape was located in the newly emergent United States and driven essentially by a fascination with the processes of production. Certainly a desire to improve the efficiency of weaponry was involved, but it appears that the earliest and most dynamic participants in changing the course of arms were industrialists first and arms innovators secondarily. In Europe the process took place slightly later and was driven more frankly by international competition and device technology aimed at increased lethality; yet the same entrepreneurial spirit was apparent here also.

As with the origin of so many things American, Thomas Jefferson's fingerprints can be found on the roots of the gun's liberation in the New World. As U.S. minister to France, Jefferson became acquainted with Honoré Blanc, a protege of Gribeauval's, who was seeking to extend the artillerist's concept of parts standardization to small arms. Jefferson was sufficiently taken with the idea that in 1785 he wrote John Jay, describing "making every part of them so exactly alike, that what belongs to any one may be used for every other musket in the magazine."[1]

This was plainly a useful idea. Few things were harder on mechanical devices than military service; and, since small arms were handcrafted, every piece was sufficiently different from its equivalents that breakage meant the custom fabrication of a replacement. Standardization, on the other hand, implied universally applicable spare parts, a particularly attractive concept in early republican America, where inadequate storage and maintenance of muskets by local militias led to vast wastage.

Among those who picked up on the call for interchangeable parts in small arms was the archetypical Connecticut Yankee, Eli Whitney. He had to do something. Already under a small mountain of debt accumulated producing his famous cotton gins, in 1798 Whitney had been awarded a contract by the War Department to produce 10,000 muskets. Ten months later, after receiving a relatively huge advance of $5,000, he had made next to no progress. Then the secretary of the treasury sent him a foreign pamphlet on arms manufacture—probably Blanc's report to the National Assembly—and Whitney saw a way out. With the date of contract fulfill-ment looming in the middle of 1800 and not having delivered so much as a single musket, he pleaded for more time, arguing that he was not just making muskets, he was manufacturing them in a unique and highly advanced manner, with uniform parts made largely by machinery.

To drive the point home, in January 1801, Whitney staged a demonstra-tion for John Adams and President-elect Jefferson in which he fitted ten different gun locks to the same musket using only a screwdriver. Actually, each had been carefully hand-filed to make sure it fit. Jefferson, however, saw what he wanted to see, writing James Monroe that Whitney had "invented molds and machines for making all the pieces of his locks so exactly equal, that take 100 locks to pieces and mingle their parts and the hundred locks may be put together as well by taking the first pieces which come to hand."[2]

Whitney's show-and-tell brought him further concessions in time and money, a pattern that stretched all the way through late 1809, when the last of the muskets were finally delivered at a total cost of $134,000—all but $2,400 in the form of advances. The equipment list for Whitney's Mill Rock armory, which showed little movement toward mechanization, indicated that the entire lot was basically handmade. It followed that the resulting muskets were far from interchangeable; in fact, their quality was miserable, leading to further disputes between Whitney and the government. Still, it would be a mistake to label him a total failure. His advanced cost accounting procedures insured a healthy profit, and his twin conceptions of mechaniza-tion and interchangeability were reflective of something very important taking place among American arms makers—just not in his factory.

Guns by their nature demanded both integrity in the way they fit together and large-scale production—heretofore an impossible combination. Because gunpowder generated thousands of pounds of pressure per square inch in the firing chamber, simple designs, especially at the breech, seemed mandatory. Also, hand-fabricating the numbers necessary to arm forces that had grown dramatically during the Napoleonic Wars argued that qualitative change, if it came at all, had to come gradually. Yet mass production offered an alternative. Precision parts promised the very small tolerances necessary for design flexibility in such a demanding regime, while machine-based production not only implied extreme uniformity in parts, but lots of them. Together, they prophesied a clear path to rapid change—both qualitative and quantitative.

Why the engineers and mechanics of New England became so obsessed with mechanization remains elusive. Nevertheless, perceptions (both real and imagined) of chronic labor shortages, a tradition of literacy and mechanical ingenuity, and a culture of entrepreneurship all brought to fruition between 1820 and 1850 a revolutionary way of making things. Although the womb of the resultant military-industrial complex was located in the highly progressive U.S. arsenal at Springfield, Massachusetts, and among the private small arms manufacturers that sprang up along the Connecticut River valley, the initial birthing actually took place far to the south, in the other national armory, at Harpers Ferry, Virginia—though unquestionably in the hands of a very capable New Englander.

Only recently rediscovered by historians, John H. Hall was a central figure in the evolution of mass production. His story began during the second decade of the nineteenth century when this cabinetmaker from Portland, Maine, with no prior experience in gunsmithing, developed a viable breech-loading rifle. It was based on a simple hinged block, which pivoted above the firing chamber when released by a spring-operated catch, allowing it to be conveniently loaded and then snapped back into place for firing. Despite patent problems, the gun itself worked well from the beginning, easily outperforming breech- and muzzle-loading rivals in a series of tests. Unless very carefully made, however, the rifles were subject to gas leakage at the breech. Hall's preferred solution was precise manufacture, "to make every similar part of every gun so much alike that" they would become interchangeable. These were magic words at the U.S. Ordnance Department, and Hall soon found himself in the arsenal at Harpers Ferry charged with developing a special factory to produce his rifles in limited lots that averaged between 1,000 and 2,000 a year through 1844, funded by a series of rolling government contracts.

The gun itself was soon overshadowed by the production technology. Over the space of years Hall devised and installed a series of water-powered cutting and milling machines along with large and small drop forges, which steadily removed the handwork. Meanwhile, he instituted a system of careful jigging and relentless measurement with precision gauges that resulted in true interchangeability. He employed at least as many men building tools and machinery as he did in actually making the firearms, or rather attending the machines that made the firearms . . . a concept Henry Ford would have grasped immediately. In 1827 a congressionally mandated technical committee reported that "Hall has formed & adopted a system, in the manufacture of small arms, entirely *novel* & which no doubt, may be attended with the most beneficial results to the Country, especially if carried into effect on a large scale."[3] John Hall was continually frustrated by the relatively small lots stipulated in his contracts, since he understood his system was intrinsically capable of a breakout to true mass production. He didn't have to worry. In word and deed his achievements rebounded back on the Connecticut River valley at just the right moment.

The manufacturing nexus there was a model for the defense industry as it would emerge in the twentieth century—a combination of government sponsorship and entrepreneurial capitalism playing off each other. The focal point was the Springfield arsenal. Its longtime superintendent, Roswell Lee, never forgot the War of 1812, when vast numbers of muskets had been damaged beyond repair for want of interchangeable parts. Lee worked zealously at Springfield to introduce advanced practices such as precision gauging and jigging, along with machines that helped take the human error out of the gun-making process. Because John Hall worked under government contract, his production technology was in the public domain; hence Lee felt free to introduce it into his own establishment, literally "lock, stock, and barrel," until by the late 1830s he had made Springfield armory the largest and most mechanized manufacturing facility in the United States.

Among Lee's most fruitful management practices was the use of "inside contractors." An inventor with a viable and applicable production process was provided with shop space, free use of tools and waterpower, and raw materials to produce gun parts at Springfield for a certain set fee. In this way Thomas Blanchard generated a series of fourteen lathes and other woodworking machines that virtually eliminated hand labor in gun stock manufacture. There were other innovations of comparable magnitude, but of equal importance was the aura of originality fostered here. In a climate akin to those of the Hellenistic armories or the arsenal at Venice, Springfield became a center for synergy and best practices, which then spun off cutting-edge technologists to further spread the seeds of progressive industry and support a new class of gun makers, the patent arms manufacturers.

Innovation at Springfield was focused on production technology. The guns themselves were held to a much more conservative path. This created an opening for others. The system fostered by Hall and Lee was not just capable of generating interchangeable parts and vast production runs, it

Revolver Man

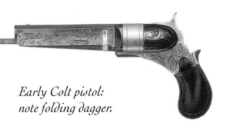

Early Colt pistol:
note folding dagger.

Samuel Colt epitomized the spirit of a whole class of military industrialists. Born in Hartford, son of a well-connected but notably unsuccessful businessman, Colt shipped out to sea young and came home with a taste for travel and a whittled wooden model of a rotating multichambered breech. It proved the template for his life's work.

Colt went through several failed attempts to produce his five- and six-shot revolvers. But ever enterprising and self-promoting, Colt capitalized on the Mexican War and moved back to Connecticut in 1846 with a government contract to produce revolvers, which he successfully subcontracted out to the son of Eli Whitney.

During the next year, he set up shop in Hartford, determined to successfully produce the guns himself. With the help of famed Texas Ranger Samuel Walker, who had been saved from Comanches by the revolver's sheer volume of fire, Colt made several important product modifications. But looking back on his earlier failures, Colt realized his problems centered on production. "With hand labour it was not possible to obtain that amount of uniformity, or accuracy in several parts which is so desirable."4 Colt resolved the issue by hiring Elisha Root, "probably the best mechanic in New England," as his production chief. Root generated an integrated package of special-purpose machinery, including patented advances in drop forging, rifling, cylinder boring, and slide lathes. Before long Colt's shops were crammed with machinery, and production shot up to 40,000 revolvers in 1851. Parts still required some filing, since Colt's gauging system lacked the rigor of Springfield's. But he steadily improved the finish and quality of materials, switching from cast iron to steel and introducing bluing (a chemical rust inhibitor that dramatically reduced the need for constant gun care), while cost penalties were reversed by further mechanization. By 1855 a reporter counted nearly 400 machines at work.

Colt was a relentless salesman, personally spreading word of his revolvers around the globe. In a pioneering gesture of military-industrialism, Colt helped arm the participants of the Crimean War—first selling the Russians 5,000 revolvers by telling them the Turks were buying his pistols, and then journeying to Constantinople, where he used the same ploy in reverse to unload another 5,000. Yet Colt's international reputation was really made at London's 1851 Great Exhibition, where he mounted such a dazzling display of revolvers that he was invited to give a lecture before the prestigious Institution of Civil Engineers. Although he dwelled on the virtues of his firearms, the British were more interested in his production technology, which they labeled "the American system." It was quite a compliment from the inventors of the Industrial Revolution.

could also sustain dramatic changes in device technology. In this case it brought forth a legion of brilliantly designed breech-loading small arms, each one backed by high-tech capitalists who turned the towns lining the shores of the Connecticut River into Gunmetal Valley.

In truth, the guns themselves left considerable room for improvement. The revolver's obvious advantage, the ability to deliver repeating fire, was somewhat compromised by its rotating multichambered breech. Even with close machining, it was prone to occasional multiple ignition. With a pistol, where the hand was behind the action, this was only an annoyance. But Colt's revolver rifles were considered dangerous, since the hand extended to steady the barrel might be shot off by just such an occurrence. Yet among self-inflicted wounds this was minor compared to the one Colt suffered by ignoring a development that took place virtually under his nose.

The Connecticut valley was primed for weapons innovation. Of those involved, Francis Pratt and Amos Whitney (machine tool wizards and founders of the firm that would one day dominate the military jet engine market), George A. Fairfield, Christopher Spencer, and Charles Billings had worked as inside contractors at Colt Firearms, while Benjamin Tyler Henry and a host of lesser lights had been trained at Springfield. All had been exposed to and were duly committed to the latest production technology, but the firms they founded or worked for needed unique and advanced

Better Bullets

Heretofore, each chamber of a Colt revolver had to be loaded individually with powder and ball, and then primed with the recently invented percussion cap. This took time and resulted in power-robbing gas leakage and even multiple ignitions. Colt employee Rolin White had a better idea. The idea of integrating ball and charge into a cartridge was not new, but White took the concept to its logical conclusion, combining the newly developed conoidal bullet with a copper powder casing

All-metal cartridge — an ideal little death-dealing module.

employing a percussion cap embedded in its base. The result was a complete little death-dealing module, handy to load and remarkably efficient thermodynamically, since the copper casing expanded tightly against the breech, completely sealing it. It was the perfect solution to Colt's revolver problems, but he couldn't see it and allowed White to shop his bullets around. Hartford was a town full of potential gun manufacturers, and very soon Samuel Colt was surrounded by competitors.

products to survive. So their energies turned to the guns, and break-throughs in lethality followed at a breathtaking pace.

Self-contained bullets proved the catalyst. In 1852 Hartford gunsmiths Horace Smith and Daniel Wesson formed the Volcanic Repeating Arms Company to produce a multishot breech-loader featuring a spring-fed tubular magazine and a trigger guard lever that operated the loading mechanism. Three years later Smith and Wesson sold their interest in the firm to concentrate on adapting the revolver to metallic bullets, which they began manufacturing when Colt's patent ran out. Meanwhile, Volcanic Arms fell into the hands of a syndicate that included a New Haven shirtmaker, Oliver Fisher Winchester. He knew little of firearms, but Winchester had the sense to leave the factory and product development in the hands of the gifted mechanic Benjamin Tyler Henry, who concentrated on cartridge development. The result was an innovative metal-cased .44 rim fire with a light bullet and charge, but a muzzle velocity of fully 1,200 feet per second. Best of all, sixteen of the compact slugs fit into the rifle's tubular magazine, making Henry's gun truly volcanic. "Beyond all competitors by the rapidity of its execution," enthused the *Scientific American* in 1858. "Thirty shots can be fired in less than one minute."[5]

Actually, an even better rifle was in the works. Whereas the Henry was fed by a magazine in an exposed position below the barrel, Colt apostate Christopher Spencer thought to drill a hole lengthwise into the stock of his repeater from the butt end to the breech, and load via a seven-cartridge metal tube inserted in the cavity. The lever-action Spencer, simpler and more rugged, could discharge all seven shots in twelve seconds, and then be reloaded with a spare tube just as quickly. Patented seven months earlier than the Henry, both guns were ready for the Civil War. The firepower they represented was unprecedented, and they could have been manufactured in huge numbers. The fact that they were not remains symptomatic of the spirit in which this Americanized version of the weapons revolution took place.

The men involved had, for the most part, no significant military experi-

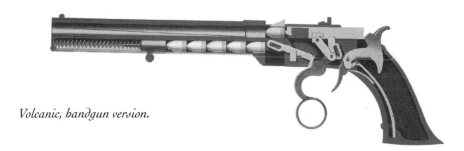

Volcanic, handgun version.

ence. Occasionally they took on titles such as "Captain" or "Colonel," but these were essentially affectations stemming from experience with the decrepit militia. Instead, they were engineers first and businessmen second; weapons were simply their chosen product. When the lethality of their wares was considered at all, it was rationalized through expedient arguments. For the individual consumer, improved firearms meant a greater measure of "self-defense." At the military level it was argued that enhanced killing power, by making wars more terrible, would diminish their frequency, destructiveness, and length. Hence the term "peacemaker" was applied to a number of weapons, most famously by Samuel Colt. The sole context in which more deadly guns were extolled was on the frontier against Indians.

At base, the patent arms manufacturers' pride in development was derived from their own ingenuity and the larger phenomena of mechanization and industrialization. Weapons as implements symbolizing power probably played some role in their outlook. More profound, however, was the sheer intoxication of techniques suddenly capable of casting, beating, and bending metal precisely into unheard of shapes and in quantities previously unimagined.

For all their revolutionary capabilities, repeating rifles did not come to immediately dominate the military market, even during the Civil War. Of the four million small arms purchased for Northern soldiers, just 1,730 Henrys and 75,000 Spencers were included—most of those during the latter stages of the conflict. To get the Ordnance Department to start buying repeating rifles took the direct intervention of President Lincoln, who had personally tested both the Spencer and the Henry.

The Civil War was already proving deadly enough, deadly beyond all expectations. Beneath all the bloodshed was a new gun. Not nearly so advanced as the repeaters, it was not even fed from the breech. Still, its impact was sufficient to change war forever.

As a military weapon the traditional muzzle-loading small arm had two major problem areas: loading and firing. While better than matchlocks, flintlocks were still severely handicapped by inclement weather. Yet around the turn of the nineteenth century, fulminate of mercury, a practical explosive capable of detonation upon impact, was discovered. Within a few years the Reverend Alexander Forsyth, an avid wildfowler, successfully applied the principle to a hammer-fired hunting rifle. This led directly to the development of a copper percussion cap that could be fitted to any flintlock and became available for military use around 1820. During the following two decades, the flintlock mechanism was replaced by a simple hammer, which struck a hollow receptacle or nipple upon which the percussion cap was mounted. The flame was vented down to the main charge, where ignition was completed. Although seating a cap for each shot could be difficult, especially in cold weather, the system was far more dependable than that of its predecessors. For the first time pulling the trigger almost always meant a shot would be fired.

Meanwhile, other significant changes were taking place within the barrel of the muzzle-loader. Traditionally, the necessity of maintaining a tight bore seal made rifled small arms extremely slow to load. This began to change around 1832, when a Captain Norton of the British 34th Regiment became interested in the blowgun darts he saw during his tour in southern India. Crafted with a base of lotus pith that expanded when blown to seal the inner surface of the tube, these darts planted the seed of an idea in Norton's mind. The principle might be applied to a rifle bullet. He designed one with a hollow base shaped to deform on firing. It created a tight seal with the gun barrel, while still allowing it to be dropped easily through the bore.

Norton's concept set off parallel chains of development on both sides of the Atlantic. Best known was the work of French army officer Claude Etienne Minié, who gave his name to the perfected expanding-base cylindroconoidal bullet that appeared in 1849. Of equal importance was the bullet of James Henry Burton, a master mechanic at the Harpers Ferry armory. Burton's bullet was cheaper and had an ogival point rather than the straight-sided cone of the French design, making it somewhat more stable in flight. Both were immediately recognizable advances.

England's Enfield Arsenal reacted quickly, hiring Burton and putting their newly imported American machinery to work manufacturing the Pattern 1853 .577 caliber rifled musket, featuring three-grooved "progressive" rifling that increased in depth from muzzle to breech. Only a few saw

Enfield 1853 rifle.

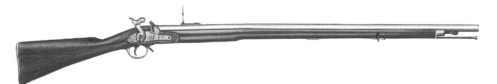

action in the later stages of the Crimean War. But by the late 1850s they were being churned out by the thousands, while the Royal Arsenal was producing over 250,000 Minié bullets per day to feed them—all in time for the American Civil War.

Back in Springfield, a new rifled musket was in the works. With Burton's superior bullet in hand, the U.S. Ordnance Department set about to design a gun to shoot it. What emerged, the U.S. Model 1855, was exceedingly similar to the Enfield, firing a .58 caliber expanding-base slug down a barrel that also boasted three-groove progressive rifling. The key difference in the American gun was an ingenious tape primer, which operated like a children's cap gun to automatically feed a dab of explosive over the nipple when the hammer was cocked. Reliability proved dismal, however, and the primer was replaced by percussion caps in subsequent models. The decision to adopt the Model 1855 (pushed ironically by then Secretary of War Jefferson Davis) proved a momentous one. The rifled musket was not only vastly more accurate, but the increased power generated from the tight bore seal combined with the streamlined cylindroconoidal bullet allowed it to reach out much further—roughly tripling the engagement range of military small arms, and maintaining lethality to nearly 1,000 yards. And the hitting power of the .58 caliber slug was massive—tearing huge holes, pulverizing bone, and destroying tissue and organs. A battlefield filled with such weapons would be a deadly place indeed.

By the end of the Civil War Springfield had produced 802,000 rifled muskets, generating nearly a quarter of a million in 1864 alone. Meanwhile, Enfield, using American technology, added another 436,000 to the Union ranks, and many to the Confederacy as well. It took some time for rifled small arms to filter down to the troops. In 1861 each side relied heavily on smoothbores. But during 1862 most Union regiments got new Springfields and Enfields, and by 1863 nearly all infantrymen on both sides carried rifles. The Killer Angels were armed; the stage was set for slaughter.

The American Civil War was the first modern conflict, and like wars of this nature, its course was heavily influenced by weapons technology. The fighting produced extraordinarily high casualties. Out of a total population North and South of around 31 million, more than 620,000 soldiers lost their lives during the four years of battle—360,000 Unionists and 260,000 Confederates, a figure as great as the cost incurred by all the nation's other wars combined.

Historians Grady McWhiney and Perry Jamieson argue convincingly that rifled small arms were at the root of the carnage—their inherent lethality certainly, but, in particular, commanders' inability to adjust tactically to the new weapons. The careers of numerous Northern and Southern military leaders, including Grant, Lee, McClellan, Beauregard, Hancock, Longstreet, Hooker, A. S. Johnston, Thomas, Jefferson Davis, Meade, and Bragg were shaped by the Mexican War, the last conflict in which smoothbore muskets vastly outnumbered rifles in all decisive engagements. Here, close-ranked linear formations and frequent bayonet charges scattered the weakly motivated Mexicans. It was an experience the participants long remembered. There were some accommodations to the rifle in William Hardee's update of Winfield Scott's basic tactical manual, but most commanders, particularly the Southerners, began the Civil War assuming that what had worked against Mexicans would work against Americans.[6]

It didn't. As rifled muskets filtered into the ranks, the opposing forces found themselves locked in a massive bloodletting. The aggressive Confederates, who were slower to receive such arms, suffered grievously. During eight of the first twelve big battles, the Southerners took the tactical offensive, and lost 97,000 men doing so. "It was not war—it was murder," concluded General Daniel H. Hill after his division lost 2,000 of its 6,500 men attacking Union positions at Malvern Hill.[7] Actually, it was team suicide. In the process of winning a series of Pyrrhic victories, the South bled itself white during the first three years of warfare.

Few commanders had much idea what had happened. When forced to fight under cover in trenches at Petersburg, Robert E. Lee bemoaned losing the tactical initiative, while Jefferson Davis continued to sack generals like Joseph Johnston, who saw virtue in defense and calculated retreat. Northerners were similarly suspicious of anything but head-to-head encounters. Cautious generals scrupulous of their men's lives like George McClellan frequently lost their jobs, while William Tecumseh Sherman, who learned to avoid battle and wage war on Southern property, was often viewed as either half-mad or brutish. Instead, Ulysses S. Grant became the exemplar among Union commanders. "He fights," Abraham Lincoln described him admiringly.

Fight he did—accumulating 65,000 battle casualties in the sledgehammer Wilderness campaign and taking as an important sign of victory that a battle with Lee's troops "outside of their entrenchments cannot be had." Here the Southerners remained, unbeaten until 1865. Meanwhile, a better indication of the true tactical situation was provided Grant by his own troops, many of whom pinned their names on their uniforms before the disastrous June 3 assault on fixed rebel positions at Cold Harbor, so their bodies might be identified later.

Unlike their officers, foot soldiers took an altogether more practical stance, and sought cover whenever possible. Archetypical was Sherman's Army of Tennessee, which historian Bruce Catton described as not having drilled for two years, but refusing to let wagons carry their spades so they might be handy for emergencies. Colonel Theodore Lyman made much the same point about Confederates: "It is a rule that when the Rebels halt, the first day gives them a good rifle pit; the second a regular infantry parapet. . . . You would be amazed how this country is intersected with field works."[8]

Within limits, field fortifications were the correct response to the new rifles. Yet trench warfare was inherently inconclusive, as the siege of Richmond-Petersburg indicated. To win, troops still had to overcome opposing field fortifications, which meant ultimately a massed assault made all the

Sherman's End Run

In the fall of 1864, William Tecumseh Sherman and his army sat in Atlanta, gradually becoming less a sword and more a target. The conventional approach, advocated by Grant and the military establishment in Washington, argued for staying there to secure the hard-won territory of northern Georgia. Meanwhile, as an adjunct he could continue to chase the elusive John Bell Hood and his army of 40,000. Yet he had already tried bird-dogging Hood, only to see him rip up his rail lines and slip away.

He therefore burned Atlanta, ignored Hood, and took off with his long-limbed army of 64,000 to wage war against the economic base of the Confederacy—in other words, its citizens. "I propose to demonstrate the vulnerability of the South," Sherman announced, "and make its inhabitants feel that war and individual ruin are synonymous terms."[9] Famous for pronouncing war "hell," Sherman purposefully shifted its venue from the battlefield to society at large.

Soldiers had wreaked havoc on civilian populations from the beginning of war among humans; but Sherman's reasoning was different and entirely more ominous, a grim strain of logic that eight decades later fueled the firestorms over Hamburg, Dresden, and Tokyo, and eventually begat Mutually Assured Destruction. Battlefields were fed by societies as much as by soldiers, so it made sense to go to the source, to target the means of production and the people who ran them. In actual fact Sherman's march was far less cruel than the depredations of the Thirty Years War or even Napoleon; but its implications would lead to unrestrained war and civilization to the brink of destruction.

more costly by the prepared nature of the objective. Very few fixed fortifications were taken by frontal attack during the Civil War. So, long-range small arms left two basic alternatives to the tradition of close-order fire-fighting: dig in or charge. The first remained the proverbial choice of the foot soldier, the second, the preference of his officers. Neither worked.

Meanwhile, the sheer dominance of long-range small arms was startling enough. Of a representative sample of 144,000 Civil War casualties, 108,000 were caused by cylindroconoidal bullets and only around 13,000 by cannon-fired ball and shell—a major shift, when it is considered that artillery had traditionally inflicted around one half of combat injuries. Such figures indicated that in the space of a very few years the long-standing and balanced relationship between the major battlefield components, infantry, artillery, and cavalry had undergone a major realignment because of the introduction of a single weapon. And this was only a prelude.

Yet the Civil War could have been far more deadly. The technology was plainly available to have proliferated a far more lethal generation of small arms. The seven-shot Spencer was ready for production in 1862, and could have been massively deployed in time for most of the major battles. Only the intercession of Chief of Ordnance James Wolfe Ripley, obsessed by excessive ammunition consumption, postponed the repeater's acquisition until late in the war, when it was used with great effect. The single-shot breech-loading Sharps and the Henry repeater were similarly thwarted. Of even more significance, in 1862 J. D. Mill and then Richard Jordan Gatling came forward with workable prototypes of machine guns, only to meet the opposition of the indefatigable Ripley.

The latter weapon, in particular, held great promise. Billed by its inventor as "a labor-saving device for warfare," Gatling's gun fired a metal-cased rim-fire cartridge at the rate of 200 shots a minute, and had the penetrating power of a Springfield musket. Extensively deployed both in the United States and abroad during the 1870s, Gatling met nothing but frustration in having it adopted while the Civil War still raged. At one point he was even jailed as a Southern sympathizer. The arguments against such advanced death machinery appeared rational—cost, logistical problems, production challenges—but most were more on the order of excuses than true roadblocks. Given the obvious pressures to win the war, it is difficult to document the deeper motivation of the dissenters—almost all career military men—but there is a clear undercurrent of disapproval, a sentiment for leaving well enough . . . or, given the horrific bloodshed, bad-enough alone.

On the other hand, the chorus of those in favor of advanced weaponry, led by Abraham Lincoln, viewed them as panaceas, means by which the war might be shortened dramatically through superiority of arms. And considering the tiny industrial base of the Confederacy, they were right.

But the opponents were right too: war was bad enough. For the first time since Napoleon, it had been waged by mass armies composed of troops with a real stake in the fighting. In this case the forces also confronted each other with arms radically more deadly. Among Greek city-states and in eighteenth-century Europe, excessive lethality had been dealt with through ritualization and by freezing weapons development. The changes of the mid-nineteenth century came upon the soldiers so rapidly that they had no choice but to bear up in the face of the slaughter. In the near term this meant compulsively digging trenches to hide in. In the longer run there were signs of deep resentment. These were free men who had mostly volunteered to fight. When conscription was begun in the North, serious rioting occurred. And once the war was over, Americans collectively turned their backs on things military. Free men had not necessarily liked being cannon, or more properly, rifle fodder. Nonetheless, the very qualities of freedom, initiative, and enterprise that propelled Americans through their daily lives liberated death-dealing.

The arms breakout in Europe started about a decade later than in America, was directed more consistently by governments as a matter of high policy, and was aimed more frankly at lethality than production.

In Europe competition among states rather than private companies was responsible for the pace of arms improvements, propelled more by arms racing than arms marketing. Nonetheless, as time went by, the need to manufacture weapons on a massive scale did foster the growth of large and important private arms enterprises like Krupp, Armstrong, and Schneider-Creusot, who, driven by the need to keep their capital-intensive production bases steadily employed, came to sell their products to anyone who would buy them . . . though these customers habitually wore uniforms. What emerged on the continent during the 1870s and 1880s was the essence of modern military-industrialism—state enterprises sponsoring and cajoling technical advances from corporate arms manufacturers, who, recognizing their dependency, at once worked with their primary clients and hedged their bets through other customers. When America emerged from its state of military quiescence during the last decade of the nineteenth century, it was this model that was adopted.

There was one aspect of the process that varied dramatically. That was the impression it left on the respective participants. The American Civil War provided a starkly accurate picture of its lethality. Europe, on the other hand, drew its lessons from a series of short and decisive wars that made it seem that better weapons were essentially a benign force on the battlefield. It was an illusion of the most dangerous sort.

At the center of Europe's story was Prussia, that proverbial army masquerading as a state. Smashed by Napoleon at Jena-Auerstedt, Prussia emerged out of its own ashes—not chastened by war but committed to the concept of a nation-at-arms and a productive mechanism that beat swords into better swords. Geography was central. Prussia remained a patchwork of disparate holdings stretching from the Rhine to the Vistula, still subject to attack by one or more of the three great Continental powers: France, Russia, and Austria. As in the eighteenth century, the army was the sole institution capable of protecting the state and welding it into something approaching a whole. So the military planning and control mechanism, the *Grosser Generalstab* (Great General Staff) evolved into a self-conscious elite, which progressively seized the helm of state.

Success was predicated on technology. War plans, in the Prussian idiom, were preoccupied with communications and troop movement—only through superior planning could the army concentrate quickly enough to maintain the initiative over the state's many potential adversaries. By the 1860s two new inventions, the telegraph and the railroad, were beginning to work miracles in just these areas.

Telegraphy revolutionized command and control. For the first time an advancing army might maintain instantaneous communications with a central headquarters merely by spinning out copper wire in its wake. In practice there were inevitably breakdowns and delays; but, to a previously undreamed of degree, a command element might monitor and modify the deployment of its forces on a real-time basis.

Meanwhile, the velocity and magnitude of these deployments were being dramatically amplified by the railroad. With the introduction of rail transport, a single train could carry the equivalent of 1,000 horse-drawn carts at ten times the speed. Suddenly supplies in vast quantities could be drawn from hundreds of miles away, while huge numbers of men might be transported similar distances virtually overnight. Less apparent but of equal significance, such troops were to be spared the hardening process of the march, allowing what amounted to civilians taking their place along the front. Force structures might now metastasize into the multiple millions, fed by short-term conscripts cascading into massive pools of reservists. In recruiting, Prussia also led the way, and when it began relentlessly enlarging available manpower pools through conscription, the rest of the Continental powers had to follow suit. But quantity, as the truism goes, has a quality all its own, and opportunities afforded by vast manpower also raised fundamental and unavoidable problems.

How could short-term conscripts and reservists of these million-man armies be expected to fight with lethal efficiency? The obvious answer was with better weapons, weapons that were easier to use and faster firing. Again Prussia turned to technology.

A rare liberal exception to his state's conservative leadership, Frederick William IV was a true Prussian when it came to matters of military preparedness. No sooner had he assumed the throne in 1840 than he moved to equip his soldiers with a radical new firearm. The technology was immature, the production base minuscule, but the tactical possibilities overwhelmed caution.

Nicholas von Dreyse had been working on better guns for over two decades, and had been subsidized by the Prussian government since 1833. His focus was on a self-contained cartridge. Yet Dreyse's solution was nowhere near as practical as those in America. In fact, it defied logic. Rather than at the base of the cartridge, he placed the primer ahead of the propellant and just in back of the bullet. This necessitated a triggering mechanism in the form of a long spring-loaded firing pin, which had to be driven through the paper-coated charge to reach the primer. Similarly novel was his mechanism for loading a cartridge at the breech, which looked and worked like the turn-bolt of a door, an ancient device that had never before been applied to gun design, but provided the starting point for a long line of bolt-action guns.

In Dreyse's version it leaked so much hot gas that it robbed the gun of power and left Prussian soldiers inclined to fire from the hip. The gun's elongated firing pin also had an embarrassing tendency to break or bend, so it missed the primer. Yet the appropriately named "needle" gun's nine-round-per-minute rate of fire and rifled accuracy overwhelmed all opposition. Frederick William ordered 60,000, the first installment of his plan to completely reequip the army. Upon its official acceptance, a royal pronouncement called the Dreyse gun "a special dispensation of Providence for the strengthening of our national resources," and put forward the hope that "the system may be kept secret until the great part it is destined to play in history may couple it with the glory of Prussian arms and extension of empire."[10]

That would take time. Dreyse's workshops never managed to turn out more than 10,000 guns a year. Since the Prussian army and its reserves numbered around 320,000, conversion promised to stretch over three decades. Three state arsenals were converted to produce needle guns, and production inched up to a maximum 22,000 per annum. In the end it took until 1866 to fully equip every unit in the Prussian army, ready at last for a full-scale test in battle.

After winning a ten-week conflict with the Danes in 1864, Prussia picked a fight over the spoils with Austria, which led to war two years later. For the first time the *Grosser Generalstab* received authority to issue direct orders in the field and largely directed the fighting. Its logistical tables misfired badly during the seven-week campaign, and the rapid capitulation of the Austrians rested as much on the Habsburg tradition of seeking peace after one or two lost battles as it did on Prussian force of arms.

Nonetheless, the needle gun impressed the other European powers. One great advantage of breech-action weapons like the needle gun was that they could be loaded and fired crouching or lying down under cover, which is exactly what Prussian troops did to great advantage. The Dreyse gun still leaked at the breech, broke its firing pins, and had to be loaded one shot at a time, but it took the competitive European military environment by storm, touching off a two-decades-long race to build better small arms. Although private firms and American technology clearly played a role, it remained primarily an interstate competition.

The British reacted rapidly and very practically. Taking advantage of its American-derived production technology and a clever side-hinged breech block submitted by U.S. citizen Jacob Snider, they began rapidly converting their large stock of .577 Enfield rifled muskets to breech-loaders at a cost of about one pound apiece. To this they added an all-metal cartridge designed by the superintendent of the Royal Laboratories, Colonel Boxer. It was crude, but it provided a fully effective breach seal and it was cheap. Having plugged the initial gap, the British then moved to a series of more advanced designs.

The French, sharing a border with Prussia's Rhineland, acted more precipitously, introducing their own bolt-action derivative of the needle gun in 1866, the Chassepot. It too employed a paper cartridge and leaked gas, but it had a much shorter, more robust firing pin detonating a primer set in the rear. Yet the charge it ignited and the bullet constituted the real differences—an oversized slug of 386 grams propelled by a hefty eighty-five grains of powder—enough powder to give it a 20 percent flatter trajectory than the already powerful Snider, and completely out-ranging and out-

Mitrailleuse Unmasked

Mitrailleuse machine gun.

I n the spring of 1868 Richard Jordan Gatling was a worried man. "It is whispered around," he wrote an associate, "that it is the intention of the French Gov't to go to making my guns from the samples they have without leave or license from me."[11] The French had no such intention. The two Gatling guns they had purchased were simply for testing against their most secret of weapons, the Mitrailleuse machine gun.

The gun was actually a slightly modified version of an older Belgian design produced in small numbers by ordnance engineers Joseph Montigny and Louis Christophe to arm fortifications. The French government had snapped up the entire program, cloaked it in a veil of the utmost secrecy, and by 1867 the Mitrailleuse had been surreptitiously adopted, its manufacture begun by secret imperial order.

No one not directly concerned with its production was even allowed to see the finished Mitrailleuse. They might have been disappointed. A cluster of twenty-five stationary barrels, it had to be loaded manually at the breech with a plate drilled to accommodate twenty-five similarly placed cartridges, as opposed to the continuous feed of the Gatling, whose barrels rotated. If slow firing—around 120 shots a minute—the Mitrailleuse design was at least blessed with simplicity.

The combination of secrecy and constant government intervention limited the number of guns on hand at the outbreak of war with Prussia in 1870 to less than 200. Further, the blackness of the program made adequate training and discussion as to how the gun might be best employed nearly impossible. In one division before the battle of Froeschwiller, only a single noncommissioned officer could be found knowledgeable enough to work the Mitrailleuse. Even worse was the tactical doctrine. Since the guns were fairly large and resembled cannon, most commanders decided to deploy them in batteries out in the open and shot at maximum range as artillery. But the Mitrailleuse was easily out-ranged by real artillery—the Prussians' new Krupp steel cannon—and blown apart.

While all of this said a good deal about secrecy, it did not say much about the possibilities of machine guns. In the wake of the Franco-Prussian War many military authorities used the record of the Mitrailleuse to conclude that such rapid-fire weapons had no future on the battlefield. It was wishful thinking.

hitting the Dreyse. Also, France managed a complete conversion to the Chassepot in four short years, just in time for the Franco-Prussian War of 1870. The Emperor and his generals entered the fray with a great deal of confidence in their weapons, and it was not simply a matter of their new and very powerful rifle.

Meanwhile, the race for better rifles continued apace. In 1869 the British adopted the very fast-firing hinged-lock Martini-Henry. It still used the Boxer cartridge, now better made from a single piece of brass. The French reacted quickly after the war, converting the Chassepot into the more efficient metallic cartridge Gras. The Prussians, realizing the Dreyse was obsolete, switched in 1871 to a bolt-action designed by Peter Paul Mauser, the first in a long line. Mauser's rifle was an instant success, incorporating a "cock-on-opening" feature that withdrew the firing pin as the bolt handle was lifted, finally removing the danger of accidental discharge upon closing. In short order, the other European powers followed suit with one or another English, American, German, or local design. All remained single-shot weapons, however.

The Prussians were first out of the block with a repeater. They simply took their excellent Mauser Model 71 and added a tubular magazine below the barrel, on the order of the Henry-Winchester, to create the Model 71/84. But storing rounds nose to tail in a military small arm was not optimal, since there was a potential for the primer cap at the base of the cartridge to be detonated by the bullet behind with a hard jolt. Then in England James Paris Lee created a stir with a bolt-action fitted with a box magazine below the receiver in which could be placed five cartridges, spring-loaded one on top of the other. This system became the heart of the excellent Lee-Metford, Britain's first general-issue bolt-action repeater. In 1885, however, Ferdinand Ritter von Mannlicher added another wrinkle—a separate metal clip holding five cartridges, which could be loaded at once into the magazine and then dropped out the bottom after the last bullet was fired, making way for a new clip of five—truly a fumble-free means of loading and the basic ammunition feeder for subsequent military rifles.

The very next year the French army fired a shot that resounded across Europe, with the introduction of the Lebel. The rifle itself was nothing remarkable, simply a strengthened Gras with a tubular magazine. But as with the Chassepot, it was the cartridge that mattered. This time the French reversed the equation, using a very small 8mm 216-grain steel-jacketed bullet, barely half the weight of the Chassepot's slug, but shot at the extraordinary velocity of just over 2,000 feet per second. The secret was a revolutionary propellant, Poudre B, one of the first successful applications of nitrocellulose, or guncotton, a compound known since 1845, but heretofore too unstable to use in firearms. That ancient staple of mayhem, black powder, was finished; within twenty years it would disappear from Europe's battlefields, and with it the pall of smoke that had enveloped armies since the introduction of the gun. Of more importance to the French was the resulting flat trajectory that so greatly expanded the Lebel cartridge's zone of lethality, and the extreme velocity, which allowed the hard-jacketed bullet to punch through brick walls and stout trees where once there had been safe cover.

At long ranges especially, the battlefield became a far more perilous place, and not just for France's enemies. Almost immediately Germany set her chemical industry to copying Poudre B. The rest of Europe followed suit, and by 1892 all of the major powers were arming or armed with small-bore high-velocity bolt-action repeaters. Barely realizing it, their beneficiaries and users, the conscripts of Europe's mass armies, were also becoming targets and victims. And it was not just a matter of small arms.

Artillery's takeoff reached back to the late 1850s and did include Americans such as Parrott and Rodman, who had experimented with reinforcement techniques that led to stronger gun barrels; but their impact on the Civil War was not significant. During the postbellum decline in U.S. military procurement, the revolution in artillery was spearheaded by Europeans.

Artillery pieces, being subject to enormous physical stresses, had much to gain from developments in metallurgy and chemistry. Because these advances were initiated by civilian "smokestack" industries, the revolution in artillery development would be touched off and primarily conducted by a class of engineer-capitalists, the equivalents of Hall and Colt. What differed was the more overtly militaristic and nationalistic environment in which they worked. Krupp and his colleagues would sell to virtually anyone, but they also remained essentially in league with their own militaries, pursuing development with the aim of giving them a war-winning edge. Hence the process had the characteristics of modern arms racing—secrecy, espionage, technical leapfrogging, zero-sum mentality, and sleight-of-hand financing.

One day in 1854, William Armstrong, a Northumbrian inventor and successful manufacturer of hydraulic equipment, was sitting in his club reading a newspaper account of the battle of Inkerman, won when British troops managed to drag two heavy 18-pound smoothbore pieces into firing position. Realizing the need for a light, accurate, long-range field piece, Armstrong was heard to say that it was "time military engineering was brought up to the level of current engineering practice."[12] These were not idle musings. Within two years he brought forth a revolutionary gun—a 12-pounder with polygroove rifling, and a rather crude breech mechanism consisting of a threaded plug. Yet the gun's true strength was in its barrel. Rather than being cast of homogeneous metal, it was built up from a series of metal strips (later wire) wound around a central core and sheathed with an outer hoop heated to fit around it and then allowed to cool so as to exert tremendous pressure inward. Not only did Armstrong's gun pass its acceptance tests with flying colors, but its strength-to-weight ratio was so superior that it pointed the way to guns vastly more powerful.

The use of rifling was also of major importance, more so even than in small arms. Not only was accuracy increased, but the spin imparted to cylindroconoidal projectiles insured they would hit point first, making it possible to employ fuses actuated on contact. The detonation of a shell was no longer an event controlled by the random burn of a time fuse; instead, explosive-filled projectiles burst when they hit something.

The only significant problem with Armstrong's design was the breech, which was weak and slow to operate. By 1865 his guns were exceeded in performance by a new class of rifled muzzle-loaders, pioneered by the French. Simple, efficient, and inherently cheap, these new guns—when combined with the strength of built-up barrel technology—dominated until approximately 1880, when the demand for increased projectile velocities caused barrels to grow so long they became extremely difficult to load at the muzzle. This dead end would force the Anglo-French mainstream back to the breech. Meanwhile, there was a gentleman in the Rhineland with a different way of doing things, pursued with such enthusiasm and determination that he came to personify the military-industrialist out of control—the proverbial merchant of death.

If there was better living through chemistry, so was there better killing. Although the first of the new propellants (nitrocellulose) and explosives (nitroglycerin) were synthesized before 1850, both remained too unstable to employ in weapons until the late 1880s. Each had tremendous possibilities.

As ballistics and thermodynamics became better understood, it was increasingly apparent that a longer propellant burn would accelerate a

Cannon King

Casting steel was the family business, and casting steel cannon became the dream of Alfred Krupp. Built-up barrels were good but hard to make; casting was simple and straightforward. Steel was immensely strong; yet pouring and cooling it so it was malleable and free of imperfections, in billets large enough for artillery tubes, was no easy matter. But Krupp had a will approximating that of his favorite medium.

He was fourteen when his father, Friedrich, died in despair, leaving young Alfred and his mother, Therese, with seven sullen workers, the crumbling remnants of a cast-steel factory, and a water-powered hammer positioned on a stream that was usually dry. Out of this meager legacy Alfred created the biggest industrial empire in Europe, based on the manufacture of his magnificent instruments of death. Tall and whippet-thin all his life, he was a mass of energy and neuroses—at once an insomniac, a hypochondriac, and a workaholic dynamo of business schemes. Although he did not discover weapons until almost twenty years of his frantic career had passed, he exhibited from the beginning the central contradictions of the industry. Although Krupp was always patriotically insistent that the Prussian and later the German government nurture his infant Krupp *Gusstahlfabrik,* when given the opportunity he would sell to anyone with money.

In the early years this was a matter of necessity. Prussian bureaucrats and army officers remained notably resistant to Krupp's wares. After three years of expensive and frustrating experimentation, Alfred finally managed to deliver a 6.5-centimeter crucible steel cannon to the Spandau arsenal for evaluation. The Artillery Test Commission left the gun out in the weather to rust for nearly two years without even firing it. When Krupp brought pressure on them to do so, they blandly informed him that only bronze pieces were adequate for purposes of field artillery.

Consequently, when Alfred the chauvinist failed, Krupp the huckster took charge—showing off his cannon, polished to a dazzling sheen, at a string of international exhibitions. Nor did his pace slacken when the accession of his patrons, militaristic Wilhelm I and his Iron Chancellor, Otto von Bismarck, assured Krupp a steady domestic market by purchasing 300 of his cast-steel breech-loaders, the guns that blew the French apart in the war of 1870.

Krupp still faced resistance from the snobbish Prussian officer corps. Once

Krupp steel breech-loading field artillery.

again, he turned to demonstrations, setting up a special firing range at Meppen and conducting a series of tests in 1878–79 for both foreign and German observers, which fully demonstrated what giant strides his guns had made, especially the big ones. International orders streamed in, and as late as 1891 fully 87 percent of Krupp's guns were sold abroad. But hereafter Krupp's relationship with the military officialdom of the newly united Germany tightened steadily. Cannon making had become a large-scale industry, full of sunk costs and huge investments in research and development that might or might not pay off. Governments simply could not expect ordnance makers to take risks of this magnitude without substantial fiscal patronage. With Krupp, the German government backed a series of guns designed to take full advantage of the new high explosives. In particular, the Kruppwerk focused on large siege howitzers. These high-angle/steep-descent guns were designed to demolish massive concrete fortifications; but when the time came they would prove ideal for pulverizing trenches and those in them.

projectile much more efficiently, while also lowering the pressures within a cannon. Further, with the invention of the crusher gauge, it became possible to accurately measure pressure variance along the tube, and to tailor strength and thickness accordingly. Initially, a slower, but far from optimal, burn was achieved by compressing black powder into pellets. This allowed barrels to grow longer and more slender, while lower pressures opened the way to more efficient breech locks, culminating in the sliding block or wedge (favored by Krupp and other German designers) and the interrupted screw and Nordenfeldt rotating screw concepts (pioneered by the French and British). Nevertheless, even with the enormous strength of built-up and cast-steel-barrel technology, gun performance was limited by the uneven burn of even pelletized black powder's combustion.

By 1890 the search for better energizers had produced spectacular results. In Sweden Alfred Nobel produced ballistite, one of the earliest nitroglycerin-derived propellants; and in England, cordite (a mixture of nitroglycerin, guncotton, and acetone) was so slow-burning that a strand could be ignited and safely held in the hand like a match. Because grain size could be controlled by the chemical process, it was possible even to customize propellants for different gun designs and uses. The gains in energetics were dramatic—significantly heavier projectiles could be shot from the same bore cannon simply by making them longer. At the end of the century muzzle velocities were nearing 3,000 feet per second, with equivalent increases in range.

Because the new propellants were largely smoke-free, it became possible for artillery to be effectively hidden—not a bad idea given the increased reach and hitting power of small arms. But long-range firing from concealed positions ordinarily made direct sighting impossible. Instead, an aiming point was selected, often on the crest of a hill, which was approximately in line with the target. At the crest was a spotter with a direct view of the enemy's position and the landing point of the initial shells. He then signaled back to the artillery position with corrections, which allowed them to zero in on the target. For those on the receiving end, frequently their first warning came with a barrage of explosions erupting with a terrifying suddenness.

By the turn of the twentieth century explosive shells had become a great deal more dangerous; though this, like so much else having to do with the weapons revolution, was not fully recognized. Key was the development of high explosives stable enough to withstand a ride in a cannon shell—nitric acid–based compounds like picric acid, the essence of lyddite, melinite, and the other high-explosive shell fills destined to dominate World War I.

In a conventional sense these substances were extraordinarily resistant to accidental detonation—you could hit them with a hammer, for instance. Unlike gunpowder and other conventional charges that were lit, high explosives were set off by exposure to a detonating shock wave. The resulting blast wave was extraordinarily energetic—TNT's wave, for example, traveled at 15,600 miles per hour—and had a shattering impact on whatever it encountered. In the case of humans, high explosives could kill simply by shock effect—rupturing organs, inducing internal bleeding, and bringing about the eerie phenomenon of downed troops unmarked but no less dead. Yet the implications were not fully understood until World War I was well underway. Previously, it was thought the main danger artillery posed to infantry was through shrapnel, which demanded that victims actually be hit by one of the balls scattered when such a shell exploded above them. High explosives certainly killed through fragmentation of their shell casings, but the blast effect made them doubly dangerous.

Only a single hurdle for artillery remained; but it was a law of nature—Newton's Third . . . every action resulted in an equal and opposite reaction. For artillery this meant recoil, the inevitable tendency to leap backward with every shot. Traditionally, cannon had been hauled back into place by the muscle power of their crews. But as guns grew and rounds per minute increased, this became an impossibility. With characteristic hardheadedness, Alfred Krupp sought to forbid recoil with an 1875 mounting design so massive as to resist all movement. He failed. Instead, more humble souls turned to a variety of mechanical contrivances aimed at mitigating the effects and returning the gun to its original position—inclined planes, rubber pads, springs, and various other shock absorbers. This trial-and-error process did produce results; but longer ranges and the necessity for greater precision in aiming increased the challenge. By the early 1890s recoil mechanisms reliably delivered gun barrels back to their approximate point of origin, but they still had to be zeroed in after every shot. For the moment, truly fast firing and pinpoint accuracy appeared irreconcilable. Not to the French, however, who were at work on another secret project.

The implements that created the Western Front were largely in place by century's end. Since the final defeat of Napoleon, combat in Europe had been limited to three short decisive wars clustered together in a span of only six years. The much more portentous American Civil War was largely dismissed as amateurish. So Europe's military elite marched lockstep toward the future having little idea what awaited them, or that they were carrying the instruments of their own destruction.

Mademoiselle Soixante-quinze

Having dug itself out of the rubble of the Franco-Prussian War, the French army longed for an instrument of revenge—a "dream gun" capable of delivering up to twenty accurate rounds per minute. Self-contained metal-cased ammunition and the new screw breech mechanism were part of the solution, if only a way could be found to handle the recoil so the gun did not have to be constantly retrained.

In 1892, the director of French artillery, acting on purloined information, asked the heads of the country's arsenals if they could design such a gun around a new long-recoil principle—rather than inches, the barrel would be allowed to move several feet in a special cradle. The answer was no—until the head of the Chantillón-Commentry gun foundry agreed to accept the task. Work began in utmost secrecy, funded with a total of $60 million supposedly earmarked for the purchase of property around Paris.

Everything turned on the development of the recoil mechanism, supervised by Captain Sainte-Claire Deville, a gifted and relentless engineer. The principles were borrowed from a Belgian hydropneumatic brake and recuperator. They seemed sound, but so great were the pressures that the first examples, cast in bronze, oozed hydraulic fluid through the very pores of the metal. This was corrected by using machined steel, and gradually, problem by problem, solutions were found until Deville and his team brought the revolutionary 75mm gun to fruition in 1897.

The performance of the recoil system was astonishing. It was subsequently calculated that the energy released to the barrel upon firing was equal to that of an automobile traveling at 100 miles per hour, which was then stopped within a space of three feet. The action reversed fifteen or more times a minute; yet a glass of water placed on the wheel would not spill.

In one minute, a single gun weighing just over 2,500 pounds was capable of spewing fifteen precisely aimed 75mm high-explosive or shrapnel shells up to a distance of five miles, while a standard battery of four might rain 300 such projectiles on a body of troops at a similar distance in just five minutes. As might be expected the French were determined to protect *Mademoiselle Soixante-quinze*'s privacy. A veil of secrecy descended, ensuring that no figures on ranges, rates of fire, weight of shells, or anything else was released, and long prison terms were provided for those who sought out such information. The curtain that surrounded the recoil mechanism was especially thick, with production of its parts deliberately scattered around France. Still, the French managed to produce and deploy well over 4,000 75s during the fifteen years prior to World War I. The Mademoiselle was no Mitrailleuse.[13]

*French model 1897
75mm gun.*

Chapter 11

STEAMING THROUGH TROUBLED WATERS

If the application of industrial technology to weapons had a profound impact on land warfare, it utterly transformed the world of the sailor. During the second half of the nineteenth century the naval horizon morphed from a vista dotted with puffy expanses of white canvas to one smudged with towering smoke columns belched from coal-fired steam engines; from arks of oak to steel leviathans; from weapons of impact to those of explosive force, and, most unexpectedly, from a battlefield measured in two dimensions to one in which the gravest threats came from below. To add further military uncertainty, there was even less war at sea than there had been on land to help chart the course of development. Lacking more than a few channel beacons of recent combat, planners and shipbuilders were thrown back on an incongruous combination of naval tradition and technological possibility to grope their way toward the future.

The era of self-propelled ships began in 1807, when that nautical provocateur Robert Fulton, having switched from submarines, managed to create a commercially viable steamboat. Nearly four decades passed before the world's navies took up steam propulsion. Besides sheer prejudice there were several very practical reasons for their hesitation. Early steam engines were cumbersome, unreliable, and inefficient devices that produced less horsepower per pound than sail, and consumed so much coal that oceanic steamers still required elaborate standby rigging lest they run out or break down. They were also highly vulnerable. The first steamers were driven by massive paddle wheels, whose high axis demanded that the machinery be placed well above the waterline—all of it presenting a large target and leaving little room for armament.

Still, the steam engine's potential for military application was great. Wind is inexhaustible, but navies had always sailed at its sufferance: becalmed at critical junctures or subject to such heavy blows that fighting was impossible. Mechanical propulsion promised a solution, a means of going anywhere at any time. Therefore, steam's progress in the civilian sphere was closely followed.

It was spectacular. Just thirty years after Fulton's little *Clermont* had proved the viability of steam by chugging up the Hudson River, the *Sirius,* powered by an engine rated at 600 horsepower, crossed the Atlantic. Barely two decades after that, steam engines with a combined horsepower of 11,500 drove the 27,000-ton *Great Eastern* at nearly fifteen knots. But these ships were paddle wheelers, so the basic military drawbacks remained.

The breakthrough came in the form of the screw propeller, derived from principles first described by Swiss mathematician Daniel Bernoulli in the eighteenth century. In 1843 the brilliant naval engineer and inventor John Ericsson applied them in the USS *Princeton,* the first propeller-driven warship in the world. The new screw arrangement removed the ship's propulsion to a far less exposed position at the bottom of the hull, and brought with it most of the steam equipment, which could now operate safely below the waterline. Up above, this opened up room for significant armament. Though masts and sheets stayed aboard for now, there was at least a free vista from which to contemplate the further harnessing of steam power to warlike ends.

The bulky machinery associated with reciprocating power still occupied space belowdecks that traditionally had been reserved for guns. Although it was possible to preserve the broadside configuration in early oceanic warships, the logic of technology pointed to fewer guns. This in turn set off a search for compensatory measures. Some of these looked forward; one looked back to the Homeric urge to close.

A much more realistic response was a continued reliance on guns; but guns greatly increased in size and power. Although naval and land ordnance development was technically similar, the divergent nature of combat at sea raised several novel aspects. One of these was the delay in the replacement of solid shot with explosive shells.

Bursting projectiles had long been launched from high-trajectory siege

Building a Head of Steam

The military career of the reciprocating steam engine (one in which the to-and-fro motion of pistons within cylinders was translated into the rotary motion of a crank shaft) stretched over six decades. The cycle began in earnest once the thermodynamics of steam were better understood, and it became apparent that improved performance of a simple single-expansion engine demanded considerably higher operating pressures than the fifteen to twenty-five pounds per square inch typical in the 1850s and 1860s. This led to the introduction of more efficient cylindrical and oval boilers, which enabled pressures to rise to the neighborhood of sixty to seventy pounds—higher than could be used efficiently during a single expansion due to heat loss and pressure fluctuations. A far better alternative was to handle the steam in two stages in separate cylinders. Such dual-action or compound engines not only reduced wear through smoother operations, but cut coal consumption dramatically—so dramatically that by 1870 it became possible to omit sails completely in warships intended for oceanic use. Instead, twin screws were adopted. Dual propellers enabled the steering gear to be placed in a protected position, enhanced maneuverability, and improved efficiency—all important qualities for warships.

Meanwhile, back among the bunkers and stokers, further developments in steam generation—boilers with corrugated steel furnaces (130 pounds), evaporators (155 pounds), and large and small tube condensers (250 pounds)—created another surfeit of pressure most logically accommodated by a switch around 1885 to triple-expansion, the last major development in reciprocating steam engines. The last of these engines before the steam turbine took over, the four-cylinder, twin-shaft installation in the Royal Navy's *Drake* in 1898, developed over 31,000 horsepower. It was enough to drive the 450-foot-long ship, weighing nearly three million pounds, at a rate of over twenty-seven miles an hour.

mortars and later from howitzers on land. But the rocking action of a ship made anything other than direct-fire (flat-shooting) guns nearly impossible to aim without fairly sophisticated fire control. It was not until 1824 when Henri Paixhans demonstrated the feasibility of firing exploding shells hori-

Ram Redux

Among the strangest effects self-propulsion had upon naval traditionalists was the mania to resurrect the ram. The notion that steam-powered vessels might revert to what they had been for 3,000 oar-powered years—floating projectiles—proved extraordinarily attractive.

It was done with scant evidence of tactical success. During the Civil War the ram achieved little despite many attempts, most famously when the CSS *Virginia* tried and failed to skewer the *Monitor*. Only during the 1866 battle of Lissa, when the Austrian ironclad *Ferdinand Max* rammed and sank the *Re d'Italia*, did the ancient beak draw significant enemy blood. And *Re d'Italia* was already lying dead in the water with a disabled rudder. Thus a detailed examination of seventy-four ramming incidents revealed hardly any success unless the victim was stationary, very little damage in general, and a strong likelihood that it would be suffered by the rammer and not the intended victim. It was suggestive that the ram's primary claim to true lethality was a series of fatal collisions (*Iron Duke–Vanguard, Koenig Wilhelm–Grosser Kurfurst, Camperdown–Victoria*) that major fleet units inflicted on each other during maneuvers—sinkings that, in the words of respected naval engineer and historian William Hovgaard, "confirmed naval men in their high opinion of the ram."[1]

Hence, for four decades after Lissa, this fratricidal weapon continued to be fitted to virtually every new capital ship and cruiser. There were even several specialized classes solely devoted to ramming. Only with the appearance of the all-big-gun battleship did these maladroit beaks finally disappear. While the tale of the ram resurrected retains a certain element of black humor, it also reflects just how disorienting the revolution in naval affairs was. For those figuratively shipwrecked by the transition, it was hardly ridiculous to grasp at tradition in hopes that it might prove a life preserver and not simply debris.

zontally that major navies began to seriously consider the possibilities of such missiles. The prospects were not pleasant, since it took only a rudimentary understanding of energetics to realize that oak was no match for the bursting power of even black-powder-filled shells. For almost three decades an absence of major naval combat postponed the day of reckoning; but on November 30, 1853, shell-firing Russians attacked Sinope and blew a large Turkish squadron and 300 years of naval warfare out of the water. It took some time for exploding projectiles to become universal and for purely wooden warships to disappear, but from this moment the revolution in naval armament was on in earnest.

The bursting effect of the shell suggested its own antidote . . . armor. Since shells tended to detonate upon plunging into the sea, only surfaces near or above the waterline needed protection. A high degree of invulnerability seemed attainable within reasonable weight limitations through the addition of armor belts of sufficient impenetrability. As with the ram's reappearance, the attractiveness of this scheme in a time of extreme change was reinforced by tradition—in this case the penchant for armoring key combatants. Hence the simple logic of offense and defense became a crucial factor in charting the course of military naval architecture for virtually the life span of the modern battleship.

An arms race was on. Muzzle velocities climbed from around 1,200 feet per second, to 1,600 feet per second, and finally to about 2,000 feet per second, where they reached a plateau, limited by the combustive properties of gunpowder. Penetration was aided significantly by the introduction of the chilled iron, or Palliser, projectile, and then, during the mid-1880s, by forged steel shells. Yet the major gains in power were achieved simply by making guns bigger. A representative cycle of British naval guns spiraled relentlessly upward, beginning in 1865 with 7-inch 6.5-ton pieces that threw projectiles of around 115 pounds, and reaching a peak with 16.25-inch 110-ton monsters launching 1,800-pound shells in 1882, whereupon the introduction of smokeless propellants prompted a new design phase based on longer, thinner tubes.

From around 1855 to the mid-1870s, protection consisted of countering each rise in gun power with a proportionately thicker slab of relatively soft, but tough and malleable, wrought iron. But unacceptable increases in weight (by this time belts of armor were over a foot across), manufacturing difficulties, and hardened projectiles made further thickening futile. This led to an indecisive competition between all-steel armor and compound plates of steel backed by wrought iron that stretched through the 1880s. All the while both were gaining in thickness and weight, a phenomenon that, along with the growth of guns, was definitive in shaping the replacement for the sailing battleship.

For those immersed in the process the overall thrust was far from clear. In fact, two separate versions of the warship of the future sailed into view. One was the Royal Navy's studious response to the naval revolution, the other was built in haste by a newcomer navy. One survives, docked at the Portsmouth Navy Yard proudly restored and a magnet for tourists; the other lies in 225 feet of water off Cape Hatteras, rusting. A strange end, since the shape of the future is better seen in the rotting hulk.

Warrior

When it joined the fleet in 1861, HMS *Warrior* was viewed as a revolutionary breakthrough in warship design and a decisive response to Gallic naval pretensions. The year before the French navy had launched the 5,600-ton *Gloire,* the first of the armored capital ships—a single-screw, three-masted vessel powerfully armed with thirty 6.4-inch rifled muzzle-loaders. Built of wood but clad with wrought iron plate, the ship was seen as a gauntlet thrown down to the Royal Navy.

French Gloire, *the ship that inspired the* Warrior.

The answer was already on the way. *Warrior* dwarfed *Gloire*—nearly double the tonnage, half again as long, propelled by a steam engine with twice the horsepower, and studded with thirty-eight 8-inch guns, all behind four and a half inches of armor. The bulk of this crushing superiority was made possible by *Warrior*'s most significant innovation: the ship was constructed of iron not wood. Vastly stronger than organic materials, ferrous structures swept away the wooden warship's inherent limitations on length and displacement, and also allowed for internal subdivision to mitigate flooding. Contemporaries marveled. By all accounts *Warrior* not only put the French in their place, but marked a major milestone in the future of military naval architecture.

Today, one glimpse of *Warrior*'s graceful lines raises doubts. It looks just like any other sailing warship, longer and less tubby, but still the same basic form. Three giant masts predominated, guns were placed at regular intervals along the outer perimeter, the high freeboard of the oceanic sailor preserved. There is the sense of pouring new wine in old bottles.

Warrior—
revolutionary structure, conservative format.

Monitor

It had from the beginning the aura of underdog. Shortly after the start of the Civil War, U.S. Secretary of the Navy Gideon Welles learned the Confederates were planning to raise the burned-out hulk of the USS *Merrimac* and convert it into what became the ironclad central battery ship CSS *Virginia,* or the South's ultimate weapon intended to break single-handedly the Union blockade. In self-defense, but pessimistic about the prospects, Welles's advisory board on ironclads called for proposals.

One of the submissions was radical bordering on ridiculous. John Ericsson, the designer of the *Princeton,* finalized plans for a small, low-slung craft whose most visible feature was a large revolving turret that sheltered two seven-ton Dahlgren smoothbore guns, the ship's only armament. Encouraged by the comments of weapons technology enthusiast Abraham Lincoln ("All I have to say is what the girl said when she put her foot into the stocking, 'It strikes me there's something in it.'"), the navy granted a contract for $275,000 in late 1861, with the provision that the vessel be ready for sea in 100 days.

Monitor was completed in the allotted time, but whether it was ready for sea remained an open question. Some even wondered if it would float. The diminutive 1,200-ton vessel consisted of two separate pieces: a shallow hull (122 feet long and thirty-four feet wide) upon which rested an iron "raft" (172 feet by forty-one feet) reinforced to carry the turret and protect the hull from ramming. Within the hull there was an engine room, coal bunkers, ammunition storage, and cramped space for forty-one crew members. Since the vessel's freeboard was best measured in inches not feet and there were no masts, the only components vulnerable to gunfire were a small pilot house and the turret, which was constructed of eight layers of one-inch iron plate.

Monitor lived up to everyone's expectations—both those of her critics and those of her creators. She nearly sank on the way from New York to Chesapeake Bay to meet CSS *Virginia.* But once she joined battle on March 9, 1862, *Monitor* proved hard to hit; Confederate guns were incapable of penetrating her armor, and when *Virginia* managed to ram, the sharp upper edge of the raft easily blunted the blow, doing the attacker more damage than was inflicted. But when *Monitor* next ventured into the Atlantic bound for blockade duty off Wilmington, North Carolina, she promptly foundered in a storm.

Ericsson's revolutionary Monitor.

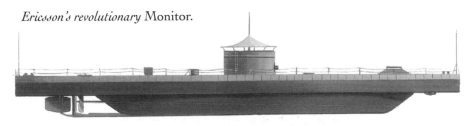

French capital ship Vainquer;
note low freeboard.

Monitor embodied that future—one without sails, a small number of big guns, and only selective protection. Yet *Warrior* too was a harbinger. Its existence was symptomatic of a naval establishment that instinctively sought to domesticate change and place it in a familiar context. If iron sailboats proved problematic, navies would save what they could from their familiar world— the primacy of guns, decorous lines of battle, and the hierarchy of ships.

Americans having turned to other things, the evolution of military naval architecture during the 1870s and 1880s took place almost exclusively in Europe. Yet the ghost of *Monitor* thrived. Capital ships took on a low-slung aspect, maintaining the minimum freeboard it took to make them seaworthy in the rough waters surrounding the Continent. Guns grew truly huge, until main armament consisted of four or even just two of the great naval rifles, mounted on revolving platforms or in heavily protected turrets. There was some secondary armament, but this too was concentrated in armored citadels. The net result was the floating fortress configuration characteristic of the modern battleship. The introduction of compound and triple-expansion engines gave these ships fairly high top speeds of eighteen or nineteen knots in calm waters; but they were far slower in high seas where their low lines proved a distinct handicap. Meanwhile, their relatively small numbers and apparent power encouraged strategists to keep them on a short leash, concentrating them locally to counter rivals.

Yet this was also a time of European commercial and imperial expansion, and steam warships held out a far better means of projecting power, protecting trade, and overawing native regimes. These contradictory ends—concentration and extension—were bridged by another major class of warship, more lightly armed and armored but far better suited to roaming the high seas. Aptly named cruisers, they performed many of the duties of frigates, but were designated by role not configuration since early designs varied widely. With time though, technology and tradition conspired to turn them not simply into useful adjuncts of naval power, but also members of an emergent hierarchy of steamships.

As usual the Royal Navy led the way. The Civil War had demonstrated that even a small number of raiders like the Confederate *Alabama* could inflict heavy damage on merchant marines. Possessing virtually half the world's transport fleet, most of which was unable to exceed nine knots, Britain had every reason to develop warships capable of running down predators. This issue became even more urgent in 1871 when several respected naval authorities testified that the volume of trade and the distances involved had rendered convoys unfeasible—a misguided verdict and one that almost led England to defeat in World War I. A security blanket of cruisers seemed the only alternative.

Endurance, speed, and guns capable of quickly pulverizing an opponent were the dynamic factors that drove cruiser development. Better, more economical steam engines were an obvious boost, but long slender hulls with plenty of freeboard also became characteristic of this class, since they promoted speed and provided room for ample coal supplies. These ships carried large guns, since their capacity to hit at long range was judged a counterbalance to the cruiser's characteristically light protection—generally a combination of shock-absorbing coal bunkers, careful compartmentation, and armored decks curved like the shell of a turtle.

Cruisers demonstrated a propensity toward growth, their hulls stretching to nearly the length of battleships and their displacement growing to

*Cutaway of a
light cruiser.*

around 6,000 tons, around two thirds that of a capital ship. They were also expensive, since speed and long-range-hitting potential required lots of machinery and high-tech guns. Nevertheless, the cost was perceived as money well spent; since these versatile workhorses were sent around the globe to serve a host of functions, not the least of which was "showing the flag"—that combination of diplomacy and intimidation which proved so effective in setting the stage for European aggrandizement. Concomitant to such long-range deployments was the necessity to establish and protect a network of cable and coaling stations, which in itself became a further stimulus to expansion. A worldwide power grab was taking form, and the steam-powered warship was at its core.

Another reality lurked beneath the waves. Traditional naval belief was founded on a profound misconception of water; it was the surface ship's misfortune to float on a paradoxical medium—willing to swallow up leaky vessels, but remarkably unyielding when subjected to sudden stress. This quirk of physics was the fatal flaw in the logical structure that favored surface warships. Due to the immense inertial resistance of water, the entire force of a submerged explosion close to a ship was drawn inward through the hull (which, after all, was filled with very compressible air) along the line of least resistance. A battleship was less the impregnable floating fortress perceived by the naval establishment than a bubble clad in a thin layer of steel. To compound the problem, water was capable of acting as a medium of transport to great depths, allowing surface vessels to be approached unseen from below.

The idea of underwater attack stretched back to the late fifteenth century, but only as a concept. Leonardo proposed a submarine in his notebooks and then decried the possibility in literally the same sentence. The idea that there was something innately ignoble about the concept persisted through the Enlightenment to the dawn of the democratic revolution, when political fervor and a fascination with technology began to overcome the spirit of moderation in the form of David Bushnell's *Turtle* and Fulton's *Nautilus.* By 1843, Samuel Colt, during a brief sabbatical from his revolvers, had designed a mine employing electrical means of discharge, which successfully destroyed a vessel several miles offshore.

It was not until the Civil War that the formidable nature of these weapons was revealed, however. Mines were used, particularly by the blockaded Confederacy, to defend estuaries and harbors, sinking a total of

Kamikaze Fish

The year 1864 marked the conception of what battleship historian Richard Hough called "the most destructive weapon at sea until the arrival of nuclear power."[2] The process began when Giovanni Luppis, a captain in the Austrian navy, designed and built a miniature propeller-driven "boat," powered by clockworks and equipped with a pistol-detonated explosive charge in the nose (as well as the obligatory ram). The nautical projectile's only guidance consisted of lines extending from the launch point to its rudder—a serious hindrance in both range and reliability, and one that caused the Austrian navy to reject it.

Short on solutions, Luppis turned to Robert Whitehead, the manager of a marine engineering firm in Fiume. With the help of his twelve-year-old son, Whitehead constructed the first of his torpedoes in 1866. It proved a rather feeble device. Powered this time by compressed air, it managed a speed of only six knots over a short course, and was highly erratic in its vertical motion—a major shortcoming, since it was likely to dive harmlessly under its target. Two years later Whitehead mastered the problem with a depth regulator based on a hydrostatic valve set to respond to a predetermined water pressure, and known subsequently as "the Secret." That same year Whitehead introduced another key component, the launching tube, also powered by compressed air and featuring self-sealing doors front and rear. By 1870 the first production torpedo was ready—sixteen feet long, capable of only eight knots over a range of 800 yards, but carrying a massive seventy-six-pound charge of guncotton.

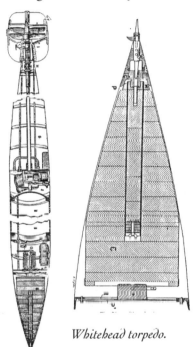

Whitehead torpedo.

Rather than press on alone, the Austrians invited representatives of the major naval powers to Fiume to witness the torpedo in action. As much to keep an eye on Whitehead and give themselves a chance to test defenses, the British subsidized further development with the hefty sum of £15,000. Still, they were in no position to monopolize the concept. Soon virtually all the naval powers had bought into "the Secret." Performance improved steadily, and by 1907 the torpedo's maximum velocity increased to thirty-six knots over a range of 4,000 yards. When it hit, the hole it opened was likely to be the size of a barn door, since its high-explosive warhead not only took advantage of the noncompressibility of water, but accounted for as much as 20 percent of the torpedo's net weight.

twenty-eight vessels. By the beginning of the twentieth century, the quantity and quality of mines had improved to the point of rendering close blockade exceedingly dangerous. Mines were essentially a coastal weapon, effective in constricted areas, but not of much value on the vast expanses of the high seas. They remained passive, needing the unwitting cooperation of surface ships to venture into the danger zone. They were warheads without a delivery system; but they were also just a beginning.

The horrific possibility of the torpedo caused the international establishment to vacillate from near paralysis, to dismissing the threat, to schemes bent on accommodation. Because it was clear from the beginning that torpedoes would always require a platform to transport them to the vicinity of the target, the debate over their future was couched in terms of these means of delivery. There were two candidates.

The first was to mount the mechanical fish on small, fast, and basically dispensable auxiliary surface craft. The introduction of the so-called torpedo boat shortly before 1880 generated such consternation that the navies of Germany, Austria, and Russia temporarily abandoned their battleship programs, while Britain's was dramatically scaled back. The impact was still greater in the French navy, where, after the debacle of Trafalgar, the idea of circumventing the orthodoxy of surface warfare was particularly engaging. Admiral Theophile Aube and his fellow theorists of the Jeune Ecole hailed the torpedo boat as the warship of the future. They argued that England's and, by implication, other European powers' growing dependence on overseas trade had tipped the scales in favor of commerce raiding (*guerre de course*) as the prime means of naval warfare, rendering battle fleets largely superfluous.

Aube's reasoning with respect to commerce raiding was eventually supported by the course of events, but, in the absence of major naval combat during the late nineteenth century, hard to prove. Meanwhile, defenders of naval orthodoxy did what they could to sink the torpedo boat. The development of quick-firing secondary armament, more comprehensive hull subdivision, and bulky torpedo nets gave at least the aura of security for major fleet units. (The nets proved worthless against torpedoes; but their deployment provided a valued substitute for furling and unfurling sails as a means of demonstrating the smartness of crews.)

To deal with the pesky craft more actively, the Royal Navy led the way by introducing boats specifically designed to hunt and kill them. Larger, faster, and better armed with both guns and torpedoes, this useful type came to be known as the destroyer, and was universally adopted as the prime means of screening a battle fleet. The torpedo boat fell quickly into disfavor.

The second bearer of torpedoes was not so easily thwarted. Since the days of Bushnell it had been apparent that a warship could potentially become both invisible and virtually invulnerable if it could dive beneath the surface. A submerged craft was also a particularly good candidate for torpedo armament, since unlike guns they could be discharged underwater.

There had been a number of military experiments with submarines after 1850, most notably the ill-fated Confederate *Hunley*, remembered as the first submersible to sink a man-of-war in combat, and also for repeatedly drowning its own crews. Until the late 1870s, the absence of a reliable power source for underwater operation and the mistaken belief that a submarine must always descend and ascend on an even keel, precluded early models from becoming much more than contraptions and death traps. Subsequently, however, pioneers on both sides of the Atlantic—Lancashire Reverend George W. Garrett; Thorsten Nordenfelt, later famed for his machine guns, a series of Frenchmen inspired by Jules Verne's 20,000 *Leagues Under the Sea;* and American Simon Lake, who built submersibles with wheels to roll along the sea bottom—methodically attacked these obstacles.

Yet the career of John Philip Holland epitomized the combination of ingenuity, tenacity, and angry iconoclasm that enabled these outsiders to bring such a weapon to fruition in the face of nature and those who preferred to fight atop the bounding main. Born in County Clare in 1840, Holland emigrated to New Jersey at age thirty-three and within two years had embarked on his life's work. "The submarine boat is a small ship on the model of the Whitehead torpedo. . . . It is like no other small vessels, compelled to select for its antagonist a vessel of about its own or inferior power; the larger and more powerful its mark, the better its opportunity."[3] At the very outset, he announced his intention of toppling the hierarchy of ships, and his personal bull's-eye never wandered from the hull of a British battleship.

Holland's efforts remained confined to ideas and drawings until the Fenian Brotherhood stepped in with money from its Skirmishing Fund, allowing him to build a one-man pedal-powered experimental craft, followed in 1882 by the much more elaborate *Fenian Ram,* a thirty-foot boat with a primitive torpedo tube and powered by a Brayton petroleum engine. The Fenians scuttled their plan to smuggle a flotilla of Rams to raise havoc in English waters, and instead seized Holland's boat, which they ran aground off New Haven, effectively ending the relationship.

Not exactly the luck of the Irish, but Holland had already discovered one key to success. Although his boats did make effective use of ballast tanks and compressed air to clear them, they did not descend solely by their own weight. Instead, they dove, maneuvered, and rose to the surface by angling large hydroplanes mounted on the outer hull and designed to work in concert with the forward thrust of the propeller and against a small reserve buoyancy, kept as a safety feature in case of power failure. This arrangement helped Holland to gain official support, though his new client proved as headstrong as the Fenians.

There was little enthusiasm for submersible warships in the U.S. Navy. Twice, in 1888 and in 1889, the department opened competition for experimental submarine designs, only to close them down without issuing a contract. Finally, in March 1893, after Congress had passed an appropriation to reopen the contest, the navy called for new designs. Holland's proposal won easily over a field that included both Nordenfelt and Lake, and in early 1895 he was awarded a $150,000 contract to construct a submersible that would be named, inauspiciously, *Plunger.* Holland's plan for the vessel incorporated the revolutionary idea of combining battery-powered electric motors, which were small and consumed no oxygen, with a conventional steam engine for surface propulsion. The cigar-shaped boat was to be much larger than his earlier efforts (eighty-five feet long, nearly twelve feet in diameter, and 168 tons in displacement) and far better armed, with five Whitehead torpedoes launched from twin forward tubes. But the navy insisted on stuffing the hull with two 600-horsepower triple-expansion engines fed by petrol-fired boilers. The dispirited Holland realized that the heat generated by such a setup would render *Plunger* uninhabitable and failure inevitable. Well before construction was completed, Holland had begun planning a new boat.

This time the specifications were purely his own, and the financing came from a private source, which promptly changed its name to Electric Boat Company, to this day the principal supplier of submarines to the U.S. Navy. Holland selected as power source a combination of electric motors and a 160-horsepower version of the new Otto four-cycle gasoline engine, the progenitor of today's internal combustion automobile engine. It was fuel-efficient, compact, and ran cool enough to make serious submarining possible at last. It was also versatile. The engine powered a dynamo that fed the storage batteries, making the eponymous *Holland* the first submarine capable of recharging at sea; it also operated a compressor to refill air tanks with a capacity of forty-two cubic feet (the equivalent of a cubic mile of atmosphere), enough to blow ballast tanks, charge torpedoes and their tubes, and keep the crew breathing.

Holland was a success from the moment it hit the water in 1897. But

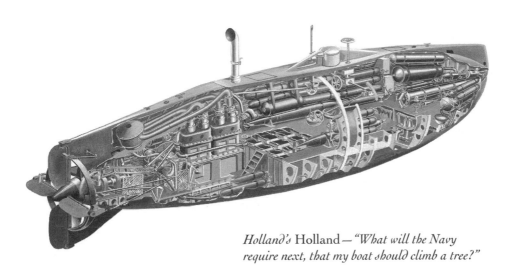

Holland's Holland — *"What will the Navy
require next, that my boat should climb a tree?"*

despite the fact that the new boat had run submerged twice the original
distance stipulated for the *Plunger,* the department hesitated to accept her.
"What will the Navy require next," fumed the exasperated inventor, "that
my boat should climb a tree?"[4] It was not until April 1900, after three
boards of inspection had thoroughly scrutinized the craft, that the department officially accepted it, making the U.S. Navy the second after France to
commission a submarine. Holland's dream was now a practicable, if fragile,
warship, and a number of naval powers lined up to acquire them.

The naval establishment might have been more hesitant had they stopped
to ponder what the submarine implied or its developmental possibilities.
Instead, their unease was tempered by a misplaced confidence that
submarines would never be anything more than coastal defense weapons of
severely limited range. "With two of those in Galveston, all the navies in the
world could not blockade that place," George Dewey, the hero of Manila Bay,
testified before Congress in 1900.[5] Yet the same Dewey failed to consider
what might happen if submarines took to the high seas, and he became
instead a major figure in amassing heavy units for the U.S. surface fleet.

Far from being tied to coastal waters, the submarine had sea-keeping
abilities that amazed even its most ardent supporters. Able to dive below
bad weather, its range and endurance grew at a dramatic pace. So long as
submersibles employed gasoline engines, they remained prone to sudden
explosion and capable of traveling barely a thousand miles before refueling.
But in 1908, as the result of prior experiments, the Royal Navy launched
the D-class submarine, the first equipped with a diesel engine requiring no

ignition apparatus, burning safe oil fuel, and stretching the boat's range to nearly 4,000 miles. Four years later the German navy launched the diesel-powered U-19, with a range of 7,600 miles. The submarine had become interoceanic.

Meanwhile, out on the high seas the torpedo's influence began to complicate plans for surface engagements. Torpedoes worked best when launched at the broadside of a ship, which was particularly ominous for parallel line-ahead tactics. So long as it ran at proper depth, a torpedo would have to be erratic indeed to miss an entire battle line of 400-foot-long ships. The only answer lay in separating the battle fleets beyond the reach of torpedoes. Yet the naval establishment's solution, battleships that engaged at very long ranges, dealt with torpedoes largely without reference to submarines, whose invisibility meant that they might be anywhere between two lines of capital ships.

In actuality, when war came the threat that submarines posed to military targets was secondary to the challenge they presented as commerce raiders. Meanwhile, the naval establishment became fixated on the surface artillery duel, engaging in a competition to build battleships that stands even today as one of history's most monumental arms races . . . all the while dismissing the submarine as it grew ever more lethal beneath their very noses. There were two reasons. The first was simple. The submarine was an unattractive weapon, especially for those steeped in naval tradition.

Early submarines were uncomfortable and hazardous. "To the Navy," wrote one American officer, "the submarine is a 'pig-boat,' and pig boats are pariahs—useful but dangerous craft."[6] Every time a submarine dove underwater there was the possibility that it would not come up again. Even in the best of circumstances the inside of a submarine was a damp, claustrophobic, foul-smelling environment, where the starch of naval custom quickly wilted. John Holland touched a nerve when he complained, "The Navy does not like submarines because there is no deck to strut on."[7]

The surface naval establishment also resented having to serve against submarines. The submarine and its torpedoes flouted the basic values that were associated with proper and traditional conduct of war at sea. Unlike the carefully regulated surface fleet, the submarine traveled alone, free of control by higher authority. It was a nautical terrorist—invisible, capricious, and potentially ubiquitous. Worse yet, it made a mockery of the hierarchy of ships. Torpedo armament afforded the most diminutive of warships the capability to attack and even destroy the largest—a possibility anticipated by Holland with something approaching glee. But to the conventionally oriented officer this must have seemed not only potentially disastrous but tantamount to tactical insubordination.

Even the quiet discharge of the submarine's lethal torpedoes must have seemed incongruous to ears conditioned to the reassuring thunder of great guns. The submarine was simply not a weapon based on confrontation. The whole manner of its operation implied skulking, treachery, and deception—qualities warriors, particularly warriors in the Western tradition, disdained. British Admiral A. K. Wilson spoke for the naval establishment when he described the submarine as "underhanded, unfair, and damned un-English."[8]

The feeling against the submarine was probably greatest at the hub of the naval world, Great Britain. It was not until 1904 that a small development program was begun, ironically by First Sea Lord Sir John Fisher, the iconoclastic chief sponsor of the all-big-gun battleship. In spite of brilliant early work with the introduction of the D-class diesel boats, familiarity in the Royal Navy bred mostly contempt, exemplified by the reaction of a flagship, thrice torpedoed during maneuvers and politely requested by its submarine antagonist to remove itself from the fleet. "You be damned!" was the signal flashed back.

The submarine policies of the other major powers paralleled the Royal Navy's. In the Orient the Imperial Japanese Navy built submersible prototypes, but concentrated on capital ships. American efforts proceeded haphazardly at best. Congress proved supportive, but the navy showed little enthusiasm. After three years of watching submarines dominate combat at sea, America entered World War I without a single submarine fit to fight. Even Germany's emphasis remained squarely on the battleship fleet, which so dominated naval construction that only twenty-one submarines, just nine of them diesels, were ready for service in 1914.

So the submarine languished. Yet it was more than innate dislike that blinded the naval establishment to its possibilities. Submarine advocates were preaching to a congregation already converted by another prophet.

While the naval community led by Great Britain had reacted purposefully to the initial wave of technology that turned their wooden ships to metal, shorn them of their sails, and clad them in armor worthy of Achilles, the makeover had been accomplished ad hoc. The naval establishment was without a guiding light. In part this explains why the appearance of the torpedo and the catcalls from the Jeune Ecole came as such a shock. When danger appeared from an unexpected quarter, there was no explicit body of thought to fall back on. Instead, ships rusted on the building ways while planners dithered and conducted tests.

Things were particularly bad in North America. The U.S. Navy had fallen upon truly hard times. Not only had the citizenry turned its back on things naval after 1865, but its line officers used the service's decline to rid themselves of steam power, newfangled weaponry, and naval engineers. The result was a tenfold contraction in ships and guns in less than a decade, and a deliberate policy of technological retrogression. A combination of poverty and institutional neglect insured that for more than twenty years no seagoing armored vessels were built in the United States.

By the 1880s conditions had deteriorated to the degree that, just when other naval powers were experiencing a crisis of confidence, the U.S. government felt compelled to take the first halting steps toward resurrecting its rotting fleet. A series of inquiries were held, and in 1886 ship construction and its attendant heavy industries received a tremendous boost when Congress stipulated that in the future only domestically manufactured components were to be employed on U.S. warships. The process required the acquisition of some very specialized production technology and substantial entrepreneurial capital investment; but within a decade the American warship industrial complex had the capability to produce naval components equal to any in the world. Yet the key issues of exactly what kinds of ships would be built remained unanswered. The U.S. Navy, like other navies, lacked a central vision.

Naval construction took on renewed energy. Between 1890 and 1905 the capital ships and auxiliaries of all the major powers evolved in parallel directions—a condition that certainly reflected the guiding light of Mahan, but also common technological stimuli. Slow-burning smokeless propellants and high-explosive fills for shells brought a temporary end to the escalation of gun size. As on land, gun barrels became longer and more slender, until the main armament of all the naval powers stabilized right around twelve inches, where it stayed for fifteen years. Testing convinced all participants that the higher velocity and picric-acid-based fills of the shells 12-inchers threw made them so much more destructive that additional size was unnecessary. Meanwhile, projectile development continued until two distinct types had evolved—a thin-walled high-explosive shell that maximized blast effect on unarmored portions of a target, and an armor-piercing round with a thick chrome-steel snout coated with a

Blue Messiah

Alfred Thayer Mahan, a scholarly officer with an urgent desire to "raise the profession in the eyes of its members,"[9] was invited to join the faculty of the newly established Naval War College and prepare a series of lectures. In 1890 they would appear in print as *The Influence of Sea Power upon History: 1660–1783*, one of the most influential American books ever published. Mahan was a historian and he addressed the impact of mechanization with a strategic retreat. "It is doubly necessary to study critically the history of naval warfare in the days of sailing ships," he wrote, "because while these will be found to afford lessons of present application and value, steam navies have as yet made no history."[10]

What Mahan found, or thought he found, was deeply reassuring. The tsunami of technology had left intact the premises of traditional naval strategy. History proved the importance of fleet concentration. Only fights between massed battleships, never raids on commerce, proved decisive. It was unthinkable to Mahan that seapower could ever be anchored to anything but a fleet of numerous and largely homogeneous capital ships—modern equivalents of the great sailing ships-of-the-line.

Unfortunately his analysis turned on historical analogy, not technical reality. He confused the stability of the past with immutability. Much of what was overtaking navies really was unprecedented. But by refusing to accept this, Mahan worked wonders. He argued convincingly that fleets remained the supreme instrument of power and defense. His model was the queen of the naval world and imperial power par excellence, Great Britain. It was her trade, her merchant fleet, her Suez Canal, her net of bases, and above all, her massive fleet that provided the inspiration. But it was Mahan alone who towed the naval establishment and its battleships out of the swamp of technology-induced uncertainty and launched them on the rising tide of imperialism.

For this all the major naval powers (save France) took this unlikely savior into their hearts. In America, not only were several generations of naval thinkers brought up on Mahan, but the sacred text was read and used by Theodore Roosevelt as the central justification for both a battleship-based navy and an expansionist foreign policy, which led directly to the acquisition of the Philippines, Puerto Rico, and the Panama Canal. The effect was similar on Japan and Germany. More of Mahan's books were translated into Japanese than those of any other author, and together they helped point the helm of state on a course set for battleships, aggression, and eventually Pearl Harbor. Kaiser Wilhelm II, for his part, "devoured" *The Influence of Sea Power*, "trying to learn it by heart." He was so impressed that the book became standard equipment on all German ships, serving as both a monument to the past and an inspiration for Germany's ill-fated High Seas Fleet. Meanwhile, Mahan had become, in the words of naval historian Arthur J. Marder, "practically the naval Mohammed of England"—a significant factor in the prophet's acceptability to the worldwide naval community. Dazzled by his vision, the major naval contestants sailed toward the twentieth century building battleships. The Blue Messiah, who never invented a thing nor designed a single ship, exerted more influence on naval acquisition than any man of his age.

wrought iron "cap" and a much smaller charge designed to penetrate to the vitals of a ship before detonating.

The 1890s also saw major improvements in armor. First nickel was alloyed with steel for elasticity and strength, and later American A. H. Harvey and then the Kruppwerke found ways to heat-treat a single plate so it mimicked the qualities of a compound sandwich of wrought iron and steel with none of the tendencies to split. Trials demonstrated that six inches of Krupp "cemented" armor would provide protection considerably better than double that thickness of wrought iron, keeping out armor-piercing (AP) shells of up to six inches at most battle ranges. Such armor could be thinner and therefore lighter, so coverage could be increased and vital areas better protected. The result for battleships was a welcome addition in exposed freeboard and therefore sea keeping, and for cruisers the possibility of more concentrated and effective protection.

By 1905 the capital ships of the world's navies were configured very much alike. England's *Majestics*, France's *Républiques*, America's *Virginias*, and Germany's *Deutschlands* were all vessels of about 15,000 tons displacement, having sixteen to eighteen knots top speed even in high seas. All mounted similar mixed-caliber armament (four 11- or 12-inch main guns in turrets atop barbettes high off the waterline, ten to twelve 6- to 9-inch secondary guns in turrets or casements, and a large number of small quick-firing guns for anti-torpedo-boat work). And all took advantage of advanced armor to improve protection.

Meanwhile, a new class of "armored" cruisers now occupied a space just below the battleship in the increasingly complete pyramid of ships. They mounted main armament of up to nine inches, and were much better protected. While no match for a battleship in a prolonged fight, they relied on speeds of up to twenty-four knots to steer clear of trouble. Such high speeds meant long hulls crammed with boilers and triple-expansion engines, so armored cruisers proved to be large ships ranging between 11,000 and 15,000 tons and equivalently expensive—especially since two per battleship were thought to be needed.

These factors opened the way for the further development of smaller cruisers, ostensibly for trade protection but increasingly enlisted as scouts and support for the battle fleet's destroyer screen. Eventually known as light cruisers, this useful class combined high speeds of up to twenty-nine knots, good endurance, and moderate armament generally based on 6-inch quick-firing guns to perform a number of missions far and wide. Yet like other auxiliaries, they were perceived through the prism of the ships' hierarchy as acolytes of the battleship and servants of Mahan's larger vision.

From this perspective by 1900 the surface naval establishment was in far better shape than just a decade earlier. The pyramid of destroyers, light cruisers, armored cruisers, and battleships was an admirable match for what had come before. Moreover, the entire ziggurat seemed to serve the ends of a modern industrial state looking outward. One major quandary remained: despite the great guns, the crews couldn't shoot straight.

✺✺✺

By this time the surface naval community understood that the battleship had a problem with the torpedo. "If a fleet might otherwise close and reduce the range in an action to increase the effect of gun fire," a panel at the U.S. Naval War College trenchantly observed, "the torpedo would prevent its closing within effective range."[11] As the reach of the lethal fish stretched beyond 5,000 yards, the problem only grew more acute. Mixed-caliber battleships might have had big guns capable of shooting out to the horizon, but the chances of them hitting anything there were not good.

The conclusion was obvious. The best resolution of the problems associated with long-range naval firefighting lay in the construction of a battleship with a single-caliber main battery composed of the largest naval rifles available. With that thought the all-big-gun battleship captured the imagination of the naval world.

By early 1902 Sims's friend Homer G. Poundstone, on leave with severe rheumatoid arthritis, was working on plans for an all-big-gun battleship, the *Possible,* which quickly found its way into the sympathetic hands of Theodore Roosevelt. The next year Vitorio Cuniberti, a distinguished but romantic Italian naval constructor, published a piece in the semiofficial naval annual *Jane's Fighting Ships,* advocating a vessel armed with all 12-inch guns and a "very high speed—superior to that of any existing battleship." Cuniberti reasoned that his hypothetical "Invincible" could not only quickly reduce anything at sea to a smoldering hulk, but had the speed to escape any trap or combination of adversaries.

It didn't take long for this mirage of an ultimate weapon to convince the new First Sea Lord. Admiral John Arbuthnot Fisher was the most determined and charismatic figure in the Royal Navy since Nelson. Robbed by the long peace of immortality in battle, his significance lay in his influence over naval construction. At once a brilliant visionary and

Scientific Gunnery

On November 17, 1901, an obscure commander in the American navy took the chance of his life. He wrote President Theodore Roosevelt informing him of the results of the North Atlantic Squadron's latest target practice. Five battleships had fired at a hulk for five minutes at the relatively short range of around 2,500 yards. They hit the target just twice. But William Sowden Sims knew of a better way, courtesy of an iconoclastic British colleague.

Captain Sir Percy Scott found equally miserable shooting upon arriving at the Royal Navy's Asiatic Squadron in 1898. As commander of HMS *Scylla* and HMS *Terrible* he developed a series of training aids and introduced powerful telescopic sights that, in combination with relentless practice, produced some very respectable results. Scott's methods were contrived—ships fired singly in smooth water at targets unobscured by funnel and other smoke—but they still constituted a major improvement and were accordingly treated as carefully guarded secrets, of which the French and Germans were told nothing.

The American navy was different. The British, faced with an increasingly competitive naval world, had almost unconsciously decided to treat it as an apprentice. So, when Sims visited Scott in Hong Kong during 1901, he was shown everything. No sooner had Sims learned the rudiments of Scott's system than he was proselytizing the U.S. Navy and Theodore Roosevelt, who appointed him inspector of target practice. Working hand in glove, Sims and Scott soon had their respective fleets locked in a friendly competition, at one point conspiring to inflate each other's shooting results to spur on their home authorities.

As the process took hold, a group of ordnance experts formed around the two targeteers. All were line officers and very much a part of the naval mainstream, but they were also innovators. Scientific gunnery exactly fit the bill for what amounted to an Anglo-American naval guild, exchanging visits and comparing notes. But even if they had remained isolated on either side of the Atlantic, ordnance technology would have led them along parallel paths. As naval antagonists drew further and further apart the problem of the gunner became less a matter of aim than of distance.

The basic means of establishing an unknown range was to create an imaginary triangle and find its tangent. But beyond 6,000 yards the angle created became too acute to measure unless one of the legs was lengthened by mounting a spotter high above the ship on a mast, where he might calculate the range. Even with good equipment and powerful optics the angle was still very narrow and difficult to interpret. Shell splashes remained the best means of checking and correcting. Recognizable splash patterns required that rounds be fired in unison, and that they be as identical as possible in charge and mass. It was also observed that the heaviest shells were the steadiest in flight characteristics. Mixed-caliber armament meant confusion and inaccuracy. Scientific gunnery, which had begun as a way of optimizing available weapons, suddenly had major implications for capital ship design.

notorious charmer, Jacky Fisher was also dogmatic, chauvinistic, and utterly pugnacious. "If a man throws a glass of wine in your face, do not throw a glass of wine in his," he once advised. "Throw the decanter stopper!"[12]

Fisher personified English suspicion of the Kaiser's upstart High Seas Fleet. An avid supporter of his friend Scott—"I don't care if he drinks, gambles, and womanizes; he hits targets!"—Fisher was already a convert to scientific gunnery when appointed First Sea Lord in 1904.[13] Soon after, he created a Committee on Design packed with all-big-gun supporters including his biographer Reginald Bacon and John Jellicoe, destined to command the Grand Fleet at Jutland. Their recommendations were a foregone conclusion—a battleship with ten 12-inch guns mounted in five turrets, no secondary armament beyond 12-pounders for antitorpedo work, "adequate armor," and a speed of twenty-one knots made possible by a switch from reciprocating engines to revolutionary Parsons steam turbines. (No ram was included, a first for British steam-powered capital ships, demonstrating the degree to which gunnery had captured the Royal Navy's imagination.)

In early May 1905, the ship, HMS *Dreadnought,* was undertaken in absolute secrecy and rushed to completion in eighteen months. Shortly

HMS Dreadnought.

after work had begun, news reached England of the spectacular Japanese victory in the straits of Tsushima, during which Admiral Togo had managed to cross the T of the Russians and virtually annihilated their fleet with gunfire. This convinced the British and left the rest of the naval world equivalently impressed, but more impressive in hindsight was the fact that fully half of the vessels sunk during the Russo-Japanese War had fallen prey to mines.

Fisher was in no mood for temporizing. He knew that the all-big-gun battleship would render every mixed-caliber battleship in Germany's fleet

The Full Cuniberti

Jacky Fisher already had the Committee on Design working on three new dreadnoughts, *Invincible, Inflexible,* and *Indomitable,* but they were not battleships. They were cruisers, very big cruisers. When some members proposed a uniform armament of 9.2-inch guns, Fisher overruled them. They would have 12-inch guns as befit a capital ship, eight of them in four turrets, but far less armor—six inches in the main belt, the minimum to gain immunity from an ordinary cruiser, the only warship swift enough to keep up. The weight saved in protection was invested in extra turbines and horsepower—45,000 to the *Dreadnought's* 25,000. "It's no use one or two knots superiority of speed—a dirty bottom brings that down!" he wrote a friend. "It's a d——d big six or seven knots surplus that does the trick! THEN you can fight HOW you like, WHEN you like, and WHERE you like!"[14] It amounted to nothing less than the Full Cuniberti, the ultimate weapon at sea. Fisher got so carried away with the concept he wanted to build nothing else, and told the committee as much. It was a good thing he was overruled. These "large armored cruisers"—soon to be dubbed battle cruisers—sacrificed too much protection for sheer speed, and in the heat of battle blew apart. Three centuries earlier Jacky Fisher's spiritual ancestor John Hawkins had been similarly mesmerized by speed and big guns; both were disappointed.

Nevertheless, dreadnought battle cruisers symbolized what weapons and machines became for many in the twentieth century. In this sense they were the equivalents of sports cars for the nautically inclined—very fast and powerful, thunderously loud, and lightly built—and even given names like *Lion* and *Tiger,* if not Cougar and Jaguar. These were weapons as men would have liked them to be, culturally ideal compromises between technology, tradition, and the martial imagination. Unfortunately, the marriage between technology and preferred weapons was an uneasy one, with the wandering utilitarian eye of the former seeking and finding less desirable partners . . . like the lowdown submarine.

hopelessly obsolete. That it did the same to the British fleet was of secondary importance. He was aware the Americans were already working on two all-big-gun battleships, *Michigan* and *South Carolina,* and in December 1906 Sims was brought in civilian clothes to the Portsmouth Navy Yard by Scott and Jellicoe and given a thorough tour of *Dreadnought.* Nobody there was thinking in terms of single units. The ordnance experts and their political supporters on both sides of the Atlantic were already bent on rebuilding their entire battle lines with dreadnoughts. Still, the man who matched salvos of wine with decanter stoppers had bigger plans.

British naval prestige and the logic of scientific gunnery set the world's naval powers building dreadnoughts at an astonishing rate—though largely the heftier battleship versions. Not since the macro-galley-building orgy of the Hellenistic monarchies had history seen a naval arms race of such intensity with such a wholesale commitment to large ships.

Germany's response was instructive. None of the other naval powers were more inconvenienced by HMS *Dreadnought.* Here, it was not simply a matter of the Kaiser's brand-new fleet of mixed-caliber battleships being instantly outdated; the Kiel Canal—a sixty-one-mile shortcut between the Baltic and the North Sea—was rendered strategically obsolete for lack of locks big enough to handle such ships. It was rebuilt—it had to be.

The Kaiser had come to look upon his fleet as a cure for his frustrations and those of his adolescent nation. For the Kaiser his pack of battleships implied not just power, but acceptance. Most of all, wrote one intimate, the "Kaiser wanted a fleet like that of England."[15] German officers were no less wed to the example of the Royal Navy, and the logic they used to justify the birth of the High Seas Fleet demanded that Germany follow Britain's lead in constructing dreadnoughts. "Risk fleet" was the term invented by the program's architect, Admiral Alfred P. von Tirpitz, to describe his oblique attempt to paralyze Britain at sea by building only enough battleships to prevent the Royal Navy from challenging its German rival without fatally weakening itself in the face of the remaining naval powers. As strategy it was profoundly flawed. Not only were the other naval powers—America, Japan, and France—allied or well disposed toward the English, but risk fleet condemned Germany to perpetual second place. Yet to retain credibility the High Seas Fleet had to be kept numerically threatening, which translated into the construction of seventeen dreadnought battleships and five battle cruisers before August 1914—a force big enough to fuel an arms race but too small to fight.

Dreadnought's bow wave was also felt across the Atlantic, where Sims and his scientific marksmen moved swiftly to acquire more all-big-gun battle-ships. Not before voices were raised in protest . . . in Congress, which remained skeptical, and also from the prophet Mahan. "We are at the beginning of a series to which there is no logical end, except the power of naval architects to increase size," he complained sourly.[16]

President Roosevelt counterattacked brilliantly, first asking Sims to write an article rebutting the technically challenged Mahan, and then concocting a gigantic publicity stunt. Gathering America's mixed-caliber battleships, he had them painted white and dispatched on an around-the-world tour. Having sent the white elephant fleet packing, Roosevelt turned to explaining the virtues of dreadnoughts to Congress, demanding four per year. The legislative body consistently cut his request by half; but in time Roosevelt and Sims had their way, with thirteen authorized by 1914. Meanwhile, Mahan's vast scheme of national aggrandizement was increasingly given only lip service. American dreadnoughts were built largely because other dreadnoughts were being built.

Elsewhere, the pattern was repeated. The Japanese, fresh from their victory at Tsushima, responded to the *Dreadnought* with their own *Satsuma* and kept building as fast as their limited industrial base allowed. The French, no longer much interested in sea power, found the trend irresistible, and by 1911 the first true Gallic dreadnought, *Jean Bart,* slid down the ways. In 1910 the Italians added the *Dante Alighieri* and kept building. The Russians, after losing nearly their entire fleet in 1905, were in no condition to resist dreadnought fever, laying down the *Pervozvannyi*-class as early as 1906, and then the *Gangut*-class in 1911. Even Austria-Hungary, with precious little coastline, built the 20,000-ton *Viribus Unitis.* Secondary powers without shipyards sufficient to fabricate dreadnoughts, like Turkey and Brazil, ordered them from others. Around the world it was springtime for dreadnoughts.

Nowhere was the enthusiasm for the ships greater than in their birth-place. With Jacky Fisher in charge the Royal Navy steered a course of naval expansion that left the rest wallowing in its wake. With the single-mindedness of all great fleet builders, Fisher orchestrated support over the protests of his most resourceful critics. Dreadnought fever climaxed in 1909 when Parliament authorized eight of the mammoth warships. Fisher's friend Winston Churchill wrote: "The Admiralty had demanded six ships; the economists offered four; and we finally compromised on eight."[17] Not every year proved so fecund, but by the fateful summer of

1914, the Royal Navy had accumulated twenty-one all-big-gun battleships and nine battle cruisers—thirty in all, or nearly three quarters of a million tons of dreadnoughts.

But instead of security, these vast expenditures of metal and effort brought only uneasiness. The ships were perceived as so valuable that it was felt the appropriately named Grand Fleet had to stay safely in home waters to thwart Germany's equally immobile but ironically mislabeled High Seas Fleet. Meaningful strategic objectives were increasingly sublimated into statistics. Gun size, broadside weight, turret armor thickness, and speed differentials of a few knots, these and simple numerical comparisons of fleets took on a life of their own. There were certainly war plans, but combat showed them to be window dressing. Dreadnoughts became ends in themselves, and the race to build them was a classic example of intraspecific competition expressed through weaponry. Not only were the ships large, loud, fast, armor-plated, and festooned with phallic naval rifles, but the actual competition was utterly symmetrical. Of over fifty dreadnoughts built before World War I, none differed in a critical respect any more than Brown Bess had differed from her Continental sisters.

Tactically, however, something fundamental had happened. The naval mainstream had led itself out on a technological limb. The character of the battleship had been decisively altered; it was no longer a brawler built to slug it out at point-blank range, but a weapon of precision whose destructive potential depended on the pinpoint accuracy of relatively few guns firing at ranges exceeding eight miles.

Behaviorally the transition was eased by the patterns of courage foreshadowed by the Duke of Parma's ill-fated lunch. Rather than swinging a saber in yardarm-to-yardarm donnybrooks, an admiral might feel suitably heroic on the bridge of a dreadnought, eyes locked on the horizon, oblivious to the towering shell splashes threatening to engulf him.

But in combat such spectacular misses proved far too frequent. The variables impinging on the precision of naval gunnery simply overwhelmed the relatively primitive optical and electromechanical fire control devices available at the time. When war came, ballistic variance, smoke, and the uneven lighting conditions in the North Sea confounded the battleships' ability to fight decisively, while subsurface weaponry steadily eroded even their freedom to find each other. Until then, however, dreadnoughts were magnificent in the shipyards and fleet reviews.

Meanwhile, the naval arms race inflamed the tensions already present in the pre-1914 European state system. Ironically, these tensions were largely based on a series of misperceptions of politico-military reality, and none was greater than the presumption of the dreadnought's potency. Paradoxically, the dreadnought's very shortcomings as an instrument of destruction provide the major point of comparison with our own nuclear deterrent. Both were far better not used than used.

Chapter 12

FALSE PINNACLE

In a self-congratulatory era, the revolution in weapons effectiveness seemed a crowning jewel in the West's diadem of progress. Few aphorisms cut closer to the bone than the words of Hillaire Belloc's Captain Blood, "Whatever happens, we have got: the Maxim gun, and they have not."[1] Suitably armed, Westerners were rewarded during the closing years of the nineteenth century with vast territorial gains in exactly those areas—Africa, the Middle East, and East Asia—that had successfully resisted conquest in the sixteenth century during the first great era of gunslinging imperialism. By 1900 Britain possessed an empire blanketing a fifth of the land on earth populated by 400 million souls, while France's holdings stood at six million square miles and over 50 million people. What could have been better proof of Western power and superiority? The imperialist states reveled in these acquisitions as much for what they symbolized as for the potential wealth that might be drained from them. Seen from this angle the white man's burden was light as a feather, an ultimate verification of strength.

Pointing the weapons revolution outward also conveniently eclipsed questions surrounding what might happen when vastly improved arms were used against similarly equipped Westerners. Instead, warfare—even in America, where the impact of the Civil War had largely worn off—was seen in an increasingly positive light. Weapons technology, especially in light of the three quick conflicts that turned Prussia into Germany, was interpreted almost universally as a factor in shortening wars. The prodigal success of imperialistic ventures also helped nurture the assumption that the new arms would favor the offense rather than the defense. There was some dissent among specialists, but hardly enough. So Europe floated along on a puffy cloud of optimism, temporarily unaware, like Wile E. Coyote, that it could never support the weight of reality and the bottom was a long way down.

The tools that made the second great explosion of Western imperialism possible were various; in fact, some of them weren't weapons at all. Sanitary engineering, especially an emphasis on clean water and sewage disposal, along with new classes of medicines, greatly increased the survivability and effectiveness of Western military forces in tropical climates, in some cases cutting the morbidity rate by an order of magnitude. Similarly, the development of water distillation equipment, canned meat, evaporated milk, and other preservative techniques allowed for longer periods between revictualization, thereby substantially extending the effective operational range of Western forces.

Communications also were greatly enhanced. Telegraphy was exploited, with landlines and submarine cables gradually connecting large portions of the globe with near instant messaging. At the turn of the twentieth century, Guglielmo Marconi pioneered wireless or radio communications, destined to be of even more military importance since it was not dependent upon fixed links. By 1901 Marconi had successfully transmitted radio signals across the Atlantic, and not coincidentally the British navy shortly after became his best customer.

Yet ultimately it was the sharp edge of European weaponry that proved most daunting; the intruders came equipped with weapons that made it difficult for native forces to retaliate without sustaining horrific losses. After one notable slaughter of 30,000 Sudanese at Omdurman in 1898, Winston Churchill neatly summarized the Western advantage: "It was a matter of machinery."[2]

As with the previous bout of expansion, new types of ships played a key role. Steam increased carrying capacity dramatically, and it allowed vessels to operate in the face of seasonal wind shifts and other inclement conditions that previously had limited sailors. Cruisers with their long ranges and big guns could descend without warning on major port cities and threaten bombardment; they were capable of reaching, targeting, and destroying virtually any urban facility. Heavily armed riverine craft could penetrate the interior by sailing upstream along navigable watercourses. From "showing the flag" to "gunboat diplomacy" to ultimately controlling the shoreline, steam-powered warships were nearly perfect vessels of intimidation.

Next came going ashore, which in terms of transport and logistics normally reduced the intruders to the same level as the natives—foot and pack animals. Among the first projects undertaken by Western colonists in the late nineteenth century was the construction of rail links. Generally, this took place during the post-conquest phase; but not always. Horatio Herbert (later Lord) Kitchener, a methodical soldier if there ever was one, approached Omdurman largely by building a railroad 200 miles through the desert. More typically though, invading Westerners marched or at least walked toward destiny, which meant traveling light. Though artillery played a significant role in several battles, in general it was simply too heavy to bring along in quantity. The new imperialism was built instead on small arms.

In 1891 a Mandinke warrior captured by the French was shown an artillery piece, which he dismissed as simply making a lot of noise; however, when his attention was directed to a Lebel rifle and he was asked if he knew what it was, he replied: "When this touches me, this death."[3] The new imperialism coincided with the introduction of small-bore, very-high-velocity, clip-loading rifles, and the invaders benefited greatly from these state-of-the-art small arms. Not only did their effective range extend beyond a thousand yards, but in expert hands they were capable of ten aimed shots a minute—ominous statistics for an adversary, especially one armed with a spear. In several engagements epitomized by the battle of Rorke's Drift, handfuls of Westerners felled literally thousands of highly motivated indigenous warriors. The degree to which the invaders craved and depended upon firepower was most starkly revealed in the story of the dumdum bullet, which also said a good deal about war in such circumstances.

In 1895 British colonial authorities discovered, or thought they discovered, that the standard issue .303 caliber Mark II high-velocity steel-jacketed bullet shot by the new Lee-Metford rifle was not doing enough damage when it hit native insurgents in the Chitral. Alarmed by stories of individual enemies surviving as many as six such wounds, Major General Gerald de Courcy Morton, the adjutant general in India, set his ordnance experts in the ammunition factory at Dum Dum working to develop a bullet that expanded on impact, like those used in big-game hunting. The resulting dumdum, and later the Mark IV variant, reliably inflicted large, jagged wounds, ideal for stopping native combatants, or for that matter, just about anything on foot, in their tracks. However, it also raised questions of legality, since Great Britain had signed the Saint Petersburg Declaration, banning exploding small arms projectiles.

The British defense of the dumdum could hardly have been more revealing. Their representatives at the International Peace Conference at The Hague in 1899 argued that enemies encountered in the colonies were simply not equivalent to the signatories of the Saint Petersburg Declaration. "Civilized man is much more susceptible to injury than savages. . . . The savage, like the tiger, is not so impressionable, and will go on fighting even when desperately wounded."[4] The conference agreed, and the dumdum bullet was viewed as entirely appropriate. When war came to South Africa the same year against the Boers—often depicted as primitive farmers, but still white farmers—British forces fought with fully jacketed ammunition.

The pattern held when general war came to Europe in 1914, and even well into the Cold War. The dumdum remained something of an exotic and did not see much combat until the so-called tumbling bullets of the American M-16 and the Soviet AK-74 were used, respectively, against the Vietnamese in the 1960s and the Afghans in the 1980s—both easily perceived as adversaries vastly different from the "civilized" troops fighting them.

Nor was this necessarily just a matter of racism. Colonial warfare was frequently a matter of hunting down the opposition, slaughtering them in great numbers, and cowing the larger populations through depredations even against women and children. Weapons that served these ends were welcomed. High-powered repeaters firing designer slugs were good; even better was anything that could mow down vast quantities of opponents.

After the failure of the Mitrailleuse in the Franco-Prussian War of 1870, the machine gun's adoption continued, but selectively. Capitalizing on the tradition of fast firing and good machine tools, Americans led the way, with all four of the major innovators—Gatling, Maxim, Browning, and Lewis—hailing from the United States. To a man they walked in the footsteps of Sam Colt, brash entrepreneurs ready to go anywhere to sell their wares. Europe was the obvious target; but all met with considerable frustration when dealing with local military authorities. As social historian John Ellis explained, "Military reactions to the machine gun . . . were rooted in the traditions of an anachronistic officer corps whose conceptions of combat still centered around the notions of hand-to-hand combat and individual heroism."[5] A gun that sprayed bullets like a garden hose did not just contradict such values, it promised to make a mockery of them.

The machine gun's cool reception at home was starkly contrasted by the enthusiasm with which European, particularly British, forces took them up in pursuit of colonies. A garden hose that shot was exactly what they wanted. Richard Jordan Gatling's eponymous machine gun, which he steadily improved through the 1870s and 1880s, soon could be found wreaking havoc in a variety of exotic places. The Gatling was heavy—basically the size of a small piece of field artillery—and expensive—costing in excess of $15,000 in today's money—and it required a hand crank to keep it shooting. Imperialists did not have long to wait for an alternative that made rapid-fire death available at the touch of a trigger.

The assistant commissioner of Bechuanaland plainly spoke for his imperialist colleagues when he wrote home in April 1890 asking anxiously: "When may I expect the Maxim gun?" When several became available, the British went into action. "And the white man came again," moaned one tribesman, "with his guns that spat bullets as the heavens sometimes spit hail, and who were the naked Matabele to stand up against such guns?"[6] The results were much the same elsewhere. At Omdurman just six Maxims provided most of the killing power. "It was not a battle, but an execution," commented one Western observer.[7] Similarly in the 1890s, during an encounter between representatives of the German East Africa Company and Hehe tribesman, an officer-surgeon and his assistant dragged two machine guns and plenty of ammunition into a mud hut with a clear field of fire, and killed around a thousand native combatants. Two men, two guns, a thousand dead.

Richard Jordan Gatling's
eponymous gun.

Lethal Maxims

The Maxim boys, Down-Easters from the state of Maine, did a lot for weapons. Brother Hudson got into the dynamite business, later moved to smokeless powder, and eventually invented Maximite, a particularly stable and energetic high explosive that found a home in torpedoes powered by his self-combusting creation, Motorite, and were detonated by special delayed-action fuses he also designed.

Yet it was his older sibling, Hiram, who proved the real inventive wizard of the family. Just twenty-six, he received his first patent for a hair curling iron, followed by hundreds more, including a locomotive headlight, a mousetrap, an automatic sprinkler system, and a variety of vacuum pumps and engine governors. Hired in 1879 as chief engineer of the U.S. Electrical Lighting Company, he exhibited an electric pressure regulator three years later at the Paris Exposition. Shortly after he had an epiphany: "In 1882 I was in Vienna, where I met an American Jew whom I had known in the States. He said: 'Hang your chemistry and electricity! If you want to make a pile of money, invent something that will enable these Europeans to cut each others' throats with greater facility.'"[8]

Propelled by these practical, if not exactly noble, sentiments, Hiram set about to bring the world a better machine gun. Within two years he succeeded. Maxim managed for the first time to harness the force of recoil in the service of the loading, firing, and ejection mechanisms. Once the first round was shot, the gun's operation became fully automatic, firing as long as the trigger remained depressed and the ammunition held out. The gun's craving for bullets was sated by a belt feed with rounds placed at regular intervals to match the cyclic rate of fire, a healthy 400 rounds per minute. Better yet, Hiram had packed all of this capability into a package only forty-four inches long and weighing just sixty pounds, with a price tag a fraction that of an equivalent Gatling gun. Unlike the Mitrailleuse and the Gatling, however, Maxim's gun shot bullets through a single tube, necessitating a characteristic bronze water jacket surrounding the barrel to cool it. While burst firing caused less stress, the gun was easily capable of ripping through several hundred rounds at a time—a feature Hiram used to great effect during demonstrations by spelling out prospective clients' initials on targets.

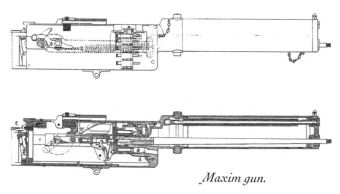

Maxim gun.

Despite such personal touches, Maxim found his European customers unwilling to mow each other down. All the major powers bought small quantities for experimentation, but acquisitions at home remained limited prior to 1914. Colonial targets were another matter. Here Hiram encountered no such sales resistance. Lethal Maxims were exactly what was desired.

If ever there was a weapon that gave vent to humankind's proclivity for predatory aggression it was the machine gun. Consider the case of one British soldier on a punitive expedition to Tibet in 1904, who in the middle of a battle was so sickened by the slaughter that he ceased firing his Maxim, only to be reminded by his commander that he should not regard his victims as anything but game, and that "the General's order was to make as big a bag as possible."[9] This and similar statements revealed a perception, however dim, that such weapons were inappropriate for use against humans. Only by expelling such victims from the ranks of *Homo sapiens,* a process made easier by their divergent physical characteristics, could such slaughter be rationalized. Of course, machine guns had nothing to do with hunting. They were meant solely to kill humans, black, yellow, and ultimately white.

Just as the South American Indians had lapsed into passivity in the face of the conquistador's arquebuses during the first wave of Western imperialism, so too did indigenous populations of Africa and Asia resign themselves to the power of dumdums and Maxims in the second. "War now be no war," lamented one resident of what is now northern Nigeria. "I savvy Maxim gun kill Fulani five hundred yards, eight hundred yards far away. . . . It no be blackman . . . fight, it be white man one-side war."[10] That would change soon enough. But at the time native peoples had to submit—open their ports to the white man's smoke-belching ships, accept his one-sided trade, arrogant officials, even his God. But beneath the mask of passivity grew a sullen rage and an awareness that, as Mao put it, "power grows out of the barrel of a gun."

Japan led the way. Tokugawa Ieyasu's extraordinary effort to undo weapons technology basically eliminated guns from Japan for upward of two centuries. The arrival of Commodore Matthew Perry's powerful squadron of American warships in 1853 came as a profound shock to the Tokugawa shogunate and its subjects. Perry confronted Japanese society with devices such as the telegraph, steam-powered conveyances, and, most compellingly, massively superior armaments, which he used to browbeat the shogunate into signing a commercial treaty soon to be followed by similar arrangements with other Western powers.

The Japanese were humiliated and fearful of further encroachments. Yet their reaction was quite different than other states facing subjugation. Japan had an earlier history of military innovation, epitomized by their initial embrace of the gun and the introduction of salvo fire routines by

Oda Nobunaga during the last half of the sixteenth century. Once again they moved purposefully to redress the military balance.

Since the beginning of the Tokugawa shogunate, military power had remained concentrated in the hands of the sword-wielding hereditary caste of samurai. Now they became the instrument of their own undoing. In Choshu a group of radical young samurai began reorganizing the local lord's army under the brilliant leadership of Takasugi Shinsaku. A student of the latest Western military practices and a fervent adherent to the *sonno joi* (revere the Emperor, expel the barbarian) movement, Takasugi began recruiting a force without regard to background and equipping them with the latest minié-ball-firing rifles. The Tokugawa perceived it as a peasant rebellion, and sent a samurai force armed only with bows, swords, and the tactics of old Japan. They were shot to pieces.

Out of the resulting ferment was born the so-called Meiji restoration, a coup d'etat, which replaced the Tokugawa regime with a group of younger samurai intent on ending most feudal prerogatives including their own, and substituting a modern political and economic system. At the heart of their agenda was a conscript army and a state-supported weapons industry. Between 1873 and 1877 recalcitrant samurai revolted four times to little effect, leaving their power broken forever and the Meiji resolutely embarked on a further expansion of their weapons program. The Meiji increased land arms expenditures 60 percent and naval appropriations 200 percent over the next decade. By 1903 the Japanese were spending a higher proportion of national income on armaments (10.3 percent) than any country in Europe. While other state-incubated heavy industries were gradually transferred into private hands, the arms industry remained tightly in the grasp of the state until World War II.

As Japan's military-industrial complex took shape, conscription began. Inspired by another student of Western military practice, Omura Masujiro, conscription was first written into the law in 1873 but only massively instituted after the last of the four samurai rebellions. It was at this point, argues E. H. Norman, historian of the Meiji period, that Japan's army made the transition from a force intended purely for home defense to one intended also for use abroad. Japan not only armed and built its force structure based on Western practice, it also embarked on a Western-style path of imperialism. Its focus was China, which at this point resembled a blue whale beset by orcas biting chunks from its outer perimeter. Through astute policy the Japanese avoided crossing the other predators, while still managing to fight a successful war against China in 1894–95. Five years later they even provided the largest element in the international force that suppressed the

Boxer Rebellion, the indigenous effort to rid China of foreigners. Nonetheless, a confrontation with one of the hungry Western powers was probably just a matter of time.

The clash came in 1904–5, and Japan's opponent was the least developed of the great European powers, Russia. The Czar's forces made a major effort, but Japan's stunning victories, particularly the annihilation of Admiral Rozhdestvenski's fleet in the Straits of Tsushima, marked a striking turn in the history of arms. Not since the fifteenth century, when the Turks successfully took up the gun, had a non-Western power shown such an affinity for the most modern weapons. With a fleet modeled on the Royal Navy's and a land force armed and trained according to German principles, Japan served notice that the West's key military advantage, weapons technology, was no longer an exclusive preserve. The Japanese continued to import large quantities of arms and copied the designs of others, but so did Western powers; indeed, imports had frequently been a major impetus for innovation. What mattered was that Japan had acquired and used state-of-the-art armaments to defeat a major European state.

Given their own racism, Westerners remained remarkably unperturbed. The victory certainly came as a surprise; but the Japanese continued to be viewed through a veil of condescension—a plucky race good at copying things, but incapable of true creativity and, therefore, not to be considered a major threat. Nor was there much concern that other non-Occidentals would follow in their footsteps.

Even more surprising was the "professional" interpretation of the Russo-Japanese War. The conflict was observed by some of the best soldiers in the world, including John Pershing, Ian Hamilton, and Max Hoffman, probably Germany's most brilliant operational planner in World War I. Russian successes with Maxim guns were duly noted, as was the Japanese alacrity in acquiring their own machine guns once exposed to what one observer labeled "the devil's tattoo." All recognized the increased lethality of high-explosive-based field artillery, the prominence of entrenchments, and the suicidal nature of frontal assaults. "A thick, unbroken mass of corpses covered the cold earth like a coverlet," noted one observer of the Japanese human wave attack on Port Arthur.[11] Yet dispatches were selectively interpreted in a manner that screened out the most prescient observations. The power of the defense and the prevalence of field fortifications were viewed as temporary aberrations. Everyone conceded that modern weapons were vastly more lethal; but the accepted view held that battles were won by aggressive action. Almost no one was willing to consider the consequences.

Nowhere were these beliefs more deeply entrenched than among the French, founded on the drubbing they took at the hands of the Germans in 1870 and the thinking of one of the war's victims. A willingness to take the offensive had become a matter of national honor.

The French were not alone. The officers of Europe's armies had convinced themselves that offensive warfare was very much in the offing. Many quietly conceded that casualties would be heavy; but this was just another reason to believe that the brewing conflict would be short.

Actually, it was no easy task to figure out how to employ the new tools of war, much less gauge their probable effects. All land forces periodically held maneuvers in which they tried to simulate the expected battlefield conditions. Together they amounted to little more than bad theater. How could it have been otherwise? True realism demanded unleashing artillery, high explosives, and machine guns to mow down advancing troops. Short of that it remained possible to believe massed infantry might move forward in the face of such fire, and that cavalry could exploit breakthroughs.

Certain lessons were learned. The problem of generating fire while moving forward surfaced, as did the need for better artillery support. The Germans, in particular, recognized the need for more machine guns. Rather remarkably, heavier-than-air flying machines proved useful for reconnaissance and artillery spotting. But that was as good as it got. No exercise was ever staged in which opposing forces became locked in a stalemate and fought it out from the shelter of trenches.

It was much the same at sea. Both sides' dreadnoughts lined up and rehearsed scenarios for the capital ship showdown expected when general war was declared. Reinhard Scheer, future German commander at Jutland, explained, "We were convinced the English would seek out and attack our fleet the minute it showed itself and wherever it was."[12] Meanwhile, on the other side of the North Sea, it was taken for granted by the British that the High Seas Fleet would quickly offer battle and be just as quickly smashed. One battle would settle the naval war as had Tsushima and Trafalgar, made more sudden and deadly by the enormous fighting power of the all-big-gun battleships. Both sides expected the other to make the first move. No one stopped to think what might happen if neither side budged.

It was not only soldiers and sailors who looked forward to war. Militarism had become infused in the body politic, not only in Germany but throughout the West. Citizens at all levels, spurred on by Darwinian

Philosopher of the Offense

The most important military-intellectual figure to emerge from the Franco-Prussian War was Colonel Ardant du Picq. He did so posthumously, killed, appropriately enough, by a Prussian artillery shell near Metz. Yet French officers were in no mood for irony. Instead they consumed his book, *Battle Studies,* with the credulity of true believers, since it appeared to give meaning to their humiliating defeat and the changes that threatened to overwhelm their profession.

They had simply lacked will and the right attitude in the face of superior weaponry. War, not military technology, was eternal. "Battles now more than ever, are battles of men," du Picq assured them. Courage trumped everything. "Attack is always, even on the defensive, an evidence of resolution and gives moral ascendancy." Forty-four years separated major encounters between Europe's two best armies and in that vacuum du Picq's ideas flourished.

Similarly, assertions such as "the improvement of firearms continues to diminish losses" and "rifled cannon and accurate rifles do not change cavalry tactics at all" appeared to hold water, particularly in a context that stretched Eurocentrism to Euro-exceptionalism.[13] Results from abroad were interpreted selectively, and the French were busy equipping all their cavalry divisions with lances in 1914.

In fact du Picq was neither blind nor an utter reactionary. Because he provided French soldiers a means of believing more in themselves than in their weapons, his words seemed to have eternal meaning and were passed faithfully through several generations of officers right up to Loizeau de Grandmaison and Ferdinand Foch, only to crumple against the Western Front.

slogans like "survival of the fittest," were attracted to military life and its culmination, armed conflict. The new urban-industrial order was widely perceived as enervating—given over to tedium, greed, and moral degradation. Combat, based on the example of Prussia's three quick wars, was by contrast seen as a short interlude of adventure, during which young men shed modern life's corrupting influences and tested themselves against the ancient code of the warrior. This view, "war as summer camp," increasingly held sway during the last years of the nineteenth century and culminated in the high spirits with which Europe marched off to battle in 1914.

Weapons were very much a part of war's good name. Not only were they viewed as triumphs of the Industrial Revolution, but the new imperial acquisitions provided tangible evidence of their worth . . . at least from a Western point of view. The trenches along the Western Front would shed a

very different light on the subject. Armaments proved a false pinnacle of
science, technology, and industrialism. Weapons development certainly
continued after World War I, but grimly and with none of the naive optimism that better armaments would improve war.

While the long era of Enlightenment gun control—having been based
largely on unspoken assumptions—was easily forgotten, some at least
reached the conclusion prior to 1914 that much more lethal armaments
might well kill many more people. Organized peace movements appeared
and expanded in both Europe and North America, particularly in the latter
half of the nineteenth century. While official responses were plainly
subdued, there was an effort to further codify warfare and to limit
weaponry, most notably at the disarmament conferences at The Hague in
1899 and 1907. As with the Saint Petersburg Declaration banning explosive
bullets, both conferences were called by Imperial Russia. The element of
self-interest was obvious. In 1899, the czarist regime was prompted by a
desire to avoid the expense of rearming with quick-firing artillery; in 1907
the Russians tried to limit weapons like mines, which had victimized them
in the recent war with the Japanese.

Still, the two conferences did reflect humanitarian concerns over
weapons effects and they were broadly attended—twenty-six nations
were represented in 1899. The records of the proceedings reveal a strong
bias against unconventional arms—"it is logical to prohibit new means
above all, when they have a barbarous character and partake of treachery."[14] Not only was the Czar's suggested ban on bombardment from
balloons affirmed at both gatherings, but conferees in 1899 agreed "to
abstain from the use of projectiles the sole object of which is the diffusion of asphyxiating or deleterious gases."[15] There were also proposals to
prohibit weapons that killed indiscriminately—machine guns, torpedo-
firing submarines, and seaborne mines—but these had only mixed
success, with the first two being postponed and only mines strictly regulated. Efforts at banning mainstream developments, such as exploding
shells and quick-firing artillery, were decisively rejected. Nor was there
much interest in freezing military budgets or limiting the number of
combat personnel.

The two conferences did produce a significant elaboration of the rules
of land and naval warfare, particularly the rights of neutrals. With no mechanisms for monitoring or enforcement, however, these legal edifices burned

up like so much parchment in the heat of World War I. Indeed, the entire effort proved a failure. Paper exercises could do little to damp the forces of nationalism and technology.

Both consciously and unconsciously Europe was girding for war. In public discourse this was evident in the popularity of the many combat fantasies that found their way into print after 1880. Most of this literature presented a childishly anesthetized view of warfare founded on naked chauvinism. Such books and articles occasionally did give vent to the effects of modern weaponry, but as a rule only the enemy got killed in large numbers.

There was another far more serious, if vastly outnumbered, genre of prewar literature. Some authors had grave doubts about the direction war and weapons development had taken, and several demonstrated remarkable insight into the impending catastrophe. Most famous was English journalist Norman Angell. His book *The Great Illusion* argued that in the modern world, war and conquest were incompatible with economic progress. The reading public took the book as assurance that war was not simply inadvisable but impossible, and *The Great Illusion* became an international bestseller. Though Angell did not actually make the extreme argument for which he is legendary, after the disaster of World War I his name was permanently associated with feckless optimism.

Better but far less well known was *The Future of War*, written in 1898 by I. S. Bloch, a rags-to-riches Polish banker. Bloch homed in on the potential impact of the new weaponry, and predicted with devastating accuracy the course of events along the Western Front. "War has become more and more a matter of mechanical arrangement. Modern battles will be decided, so far as they can be decided at all, by men lying in improvised ditches they have scooped out to protect themselves from the fire of a distant and invisible enemy. . . . War will become a kind of stalemate, in which neither army [will be] able to get at the other."[16]

Bloch also argued that the consequences of such prolonged conflicts would be measured in more than casualties and material loss. Modern war, he warned, had the capacity to rip apart the fabric of society. Bloch's vision of the naval future, while less comprehensive, was still penetrating. He understood the significance of the torpedo, particularly in combination with the submarine, and saw how both, if employed ruthlessly against seaborne commerce, might render great battle fleets irrelevant. Yet it remained for a more fanciful soul to elaborate.

One who did was also a novelist, but his specialty was science fiction. In early 1914, H. G. Wells, creator of *The War of the Worlds* and *The Invisible Man*, published a tale called *The World Set Free*. As literature it was not up to his

Danger!

Eighteen months before war began, A. Conan Doyle, the opium-smoking creator of Sherlock Holmes, wrote *Danger,* a war fantasy unlikely to warm British hearts. The story opens by describing an uneven dispute between England and Nordland, a small imaginary European state. Without much of a surface navy, Nordland's king is about to submit when he is reminded of his eight submarines by a trusted and particularly Anglophobic officer, John Sirius. "Ah, you would attack the English battleships with submarines?" asks the worried monarch. "Sire, I would never go near an English battleship."[17] Sirius instead proposes to wage an unlimited campaign of submarine warfare against merchant shipping, striking at the island kingdom's vital dependence on imported foodstuffs. The King approves the plan, and as soon as it is put into effect numerous transports are sunk without warning. "What do I care for the three-mile limit or international law?" growls Sirius. England is quickly pushed to the brink of starvation and forced to accept a humiliating peace, her great fleet idle.

The story was widely read, yet all but ignored by the Royal Navy. "I do not myself think any civilized nation will torpedo unarmed and defenseless merchant ships," argued Admiral Penrose Fitzgerald.[18] He was disastrously wrong; but his position was not necessarily absurd. *Danger* illustrated the fragility and interdependency of industrialized societies and the ease with which modern weapons could strike at their hearts. But few as yet caught the message.

usual standards; but as prophecy it was remarkable. Its subject was nuclear energy. The first scene depicts a scientist holding a phial of uranium oxide before an audience and forecasting "a source of power so potent that a man might carry in his hand the energy to light a city for a year." Yet peaceful uses do not prevail when nuclear energy is harnessed in 1953. "By the spring of 1959, from nearly two hundred centres . . . roared the atomic bombs. . . . Most of the capital cities of the world were burning; millions of people had already perished, and over great areas government was at an end."[19] At least in Wells's white-hot imagination the era of nuclear weapons had dawned, an ominous ball of fire that the work of Rutherford, Einstein, and Planck had made inevitable.

Weapons too powerful to use were destined to undermine war in a way never before contemplated; the tools would come to hold their masters hostage. Yet the Cold War was thirty years in the future and chemical energy–based warfare still reigned. So the West would render up to the gun a terrible and long overdue price—the penalty for its hubris.

Chapter 13

ACCIDENTAL
ARMAGEDDON

I t was a war of weapons. The bizarre course of World War I, perhaps
the greatest single disaster of the twentieth century, an event that cast a
decisive shadow over politics until the fall of the Berlin Wall seventy-
five years later, was essentially an accident of technology. When the collective pratfalls of Europe's great powers plunged her into general war, the
weapons available were precisely those destined to frustrate decisive military operations. At sea the resulting impasse left relatively few dead and
amounted to not much more than a collective professional embarrassment;
but on land the stalemate killed at least nine million soldiers, a price in
blood that left Europe prostrate and profoundly undermined the institution of war. Weaponry, and weaponry alone, brought about this disastrous
consequence. Armaments dictated the nature of the fighting, obliterated
strategy, and played upon the frustration and impotence of commanders,
leading them to send legion after legion to oblivion in hopeless assaults on
fixed positions that took on the aura of mass executions. From start to
finish, technology, heretofore emblematic of the West's presumed superiority over other societies, called the tune and beat the time in this catatonic
dance of death.

This is not a popular conclusion. For historians, blaming an event of
such magnitude on machinery plainly goes against their instincts. Instead,
they have concentrated on one or another aspects of a geopolitical system

that was unquestionably primed for war. Exaggerated nationalism, perceived differentials of power, the role of Europe's alliance structure, the glorification of warfare, the mechanistic nature of war plans, and the offensive orientation of everybody all received their due. None of these, however, explain the horrific nature of the fighting that ensued. Viewed retrospectively in the fiery backdrop of a mushroom cloud, the futility of the Great War was no aberration, it was a portent.

Nobody wanted mass carnage. There was considerable concern over the laws of war—a major factor in the Germans' ill-advised decision to halt the initial campaign of unrestricted submarine warfare. The brutality of the fighting increased somewhat during the course of the conflict; but there were no wholesale slaughters of prisoners or premeditated resort to the kinds of depredations that would come to characterize the Great War's sequel, World War II. Instead, after four years of nightmarish fighting, the terms imposed on Germany were relatively mild.

It began according to plan. Everybody had a plan, and everybody planned to take the offensive. The French charged toward Lorraine and the Ardennes, yet they were stopped dead in a hail of fire, thereby allowing the Germans to occupy a large portion of northeastern France and insuring that the remainder of the war would be fought on their soil. The Russians advanced into East Prussia until they were annihilated at Tannenberg, setting the tone for further military disasters that formed a direct path to the Revolution.

But the plan of plans, the most offensive of all, was German. Its creator, Count Alfred von Schlieffen had worked for fifteen years on the lightning-fast, forty-day knockout of the French, completed before the lumbering Russian bear could do much damage in the East. Schlieffen's plan, and its subsequent execution in 1914, involved massing the lion's share of Germany's ground forces in the northeast and then sweeping around the French in a gigantic wheeling movement. But there was a problem. For the muscular Teutonic right wing, getting to France involved driving through neutral Belgium, an invasion that combined a stunning victory with a disastrous consequence: England's declaration of war against Germany.

Right at the outset the Germans faced a daunting problem. In order for the First, Second, and Third Armies to secure their communications, the ancient gun-making center of Liège, surrounded by one of Europe's most formidable fortress complexes, had to be reduced and reduced quickly. Failure meant the whole invasion would bog down. The Germans had thought long and hard about Liège. Their plan entailed penetrating between the forts, taking the city, and then reducing the strongholds one by one. Initially

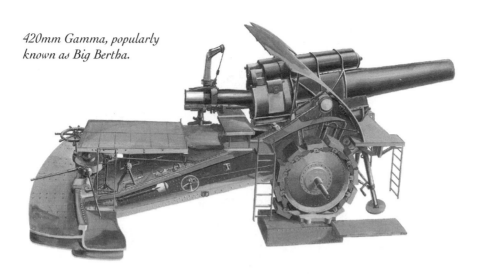

*420mm Gamma, popularly
known as Big Bertha.*

their attack produced little but piles of dead Germans, but then soon-to-
be-famous General Erich von Ludendorff rallied his troops and basically
bluffed his way into the city. The defenders of the forts remained optimistic, however, since their concrete and steel underground cocoons had
been designed against 210mm artillery fire, the largest anybody thought
they would ever face.

The Germans had bigger, far bigger. Since the 1890s Fritz Rausenberger,
ordnance director of the Kruppwerke, had been secretly working on a series
of progressively larger siege engines, culminating in the 420mm Gamma. It
was not a long-tubed gun, but a howitzer, whose high angle of fire was
calculated to send its two-ton shell plunging almost straight down. The first
of the great howitzers arrived a week after the attack had begun, dubbed
affectionately though not necessarily flatteringly after Krupp's wife, "Big
Bertha." She and her sisters—shortly joined by an Austrian cousin, the
slightly smaller 305mm "Schlanke Emma"—made short work of the Liège
forts, battering down the last in just four days. To the Germans and their
shocked adversaries, these great siege engines were novelties. For their high
trajectories were primarily effective against fixed positions, and everybody
still expected a war of movement. In fact Liège proved the howitzer's audi-
tion for a starring role in the long war to come.

Meanwhile, the road to France was now open and Schlieffen's wave of
field gray uniforms was ready to roll. But this time technology was on the
side of the Gauls. The French were now guided in their redeployment by an
aviator's report of confirmed enemy columns "gliding from west to east,"
the first significant use of the airplane in a war that brought it to promi-

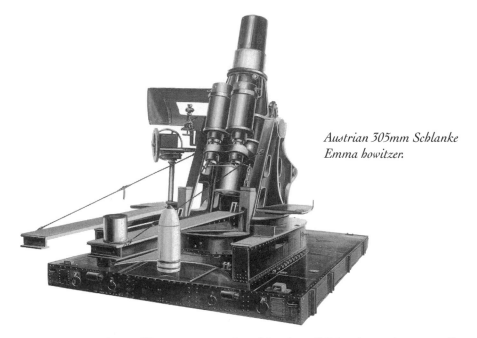

*Austrian 305mm Schlanke
Emma howitzer.*

nence as a major military system. Quickly, they fell back on their excellent rail net and internal lines of communication, transporting their troops by train and more famously by taxicab, to gain numerical superiority opposite the Germans just as the battle of the Marne began. The Germans had arrived on foot and by hoof. They reached the Marne tired and dusty, with their logistics stretched to the point of fraying.

An unprecedented array of firepower greeted them. The greatly improved range and rate of small arms fire was quickly evident; but at the Marne it was field artillery that saved the French. Long veiled, Mademoiselle Soixante-quinze at last had her day in the sun, mowing down German foot soldiers with frightful efficiency. So long as the targets remained out in the open, her low-trajectory 75mm shells had little trouble finding them. Like Jeanne d'Arc, however, this twentieth-century heroine found her dominance short-lived, ended by a change of venue that turned men into moles.

Defeated at the Marne, the Germans reeled backward. To the elated French the time was ripe to chase them to the border. "We need only legs now!" exulted General Foch. "The enemy is running away." That proved far from the case. On September 13 the pursuing French along with their British allies reached the southern edge of the Aisne valley. That morning an officer attached to the Moroccan Division headquarters overheard the commander on the telephone. "Barbed wire? Let them go around it, or cut it!"[1] The next sound he heard was a tremendous fusillade to the north, punctuated by the clatter of machine guns.

The Germans had dug in and turned on their pursuers, revealing the next lethal surprise in this war without precedent. Heretofore, the water-cooled machine gun's usefulness had been compromised by the fluid nature of battle, since it was too heavy to be carried forward rapidly by troops under fire. Now, however, carefully emplaced in field fortifications, it cut down charging British and French infantry like an invisible scythe. Repelled from the German trenches, the Allies were forced to dig in themselves. In this sector the resulting two parallel lines remained in almost exactly the same spot for four long years.

Realizing he could not punch through the Allies, the new German commander, Erich von Falkenhayn, tried to flank them by pushing reinforcements out of the north end of his trench lines—like toothpaste squeezed out of the tube and just about as militarily effective. Catching his intent, the Allies too stretched northward, trying to cap and turn the German defenses. There followed a month of mutual and futile flanking right up to the English Channel near Nieuport. By October 15 this so-called race to the sea left two parallel strips of field fortifications snaking around 470 miles to the Swiss border.

Next Falkenhayn turned to frontal assault, letting loose almost continuous attacks from the middle of October to the middle of November 1914 against outnumbered British and French divisions clinging to a narrow salient in front of Ypres. The fighting was desperate, but the only result was a grim combined total of around a quarter-million casualties. *Der Kindermord bei Ypern*—the so-called massacre of the innocents at Ypres—a hyped and much conflated legend, would be used by Adolf Hitler to pump martial ardor into the Third Reich. An ominous pattern of command also emerged, characterized by attacks prolonged far beyond the point of diminishing returns in terms of lives lost for territory gained.

Both sides were stuck in a ditch, or rather two ditches, and weapons would keep them there. Military participant and later historian J. F. C. Fuller noted that no attack or series of attacks from October 1914 to March 1918 succeeded in moving the front line even ten miles in either direction. The Western Front not only epitomized military frustration; but it was at the core of World War I's meaning. The Eastern Front remained intermittently mobile and saw perhaps even more fighting than the great bloodletting in France. But the Western Front had the best armies and the best weapons and produced only state-of-the-art futility. This became the central story of the war.

At the root of everything was a simple proposition: the flesh was weak, while weapons grew ever stronger. Artillery was the great killer this time, inflicting around 70 percent of the casualties. Because of their high arc of fire, howitzers became the heavy weapon of choice against trenches—the really big ones capable of reaching and obliterating even bunkers buried deep underground. Indirect fire techniques were greatly aided by spotters high above in barrage balloons and aircraft. Compounding increased accuracy was the sheer volume of fire, as new chemical propellants reduced tube pressure and extended barrel life, while precise recoil mechanisms allowed artillerists to load and shoot as fast as humanly possible.

Everyone had greatly underestimated ammunition requirements and soon ran short of projectiles. Factories on the home front proved up to the challenge, however, expanding munitions production dramatically and eventually more than satisfying the man-killing appetite of the Western Front's twin trench serpents. Shells themselves changed. It was assumed at the outset that shrapnel would be the standard antipersonnel munition. Yet the sheer power of high-explosive fill marginalized subprojectile rounds, frequently killing by concussion alone and fragmenting their steel casings so thoroughly and lethally that no other mechanism was required.

Nonetheless, the one weapon that best symbolized World War I is the machine gun. At 450 to 600 rounds per minute a single one of these bullet-spewing devices equaled the fire of forty to eighty riflemen. John Keegan in his groundbreaking book *The Face of Battle* noted that the machine gun was, after all, a machine, similar in some ways to a precision lathe. As such it had to be set up; but once this was done it operated virtually automatically, shooting down anything in its path with a reliability far exceeding that of a rifleman. It was, to employ a contemporary military euphemism, a "target-serving mechanism," and it made a mockery of the warrior ethic. Skill, strength, cunning, and bravery meant nothing in the face of a machine gun's stitching trace of bullets. Only percentages mattered. Human life was extinguished because it got in the way. This was a primary caveat of modern war, and nothing embodied it more starkly than a Maxim gun spitting bullets blindly across the killing fields. Tactically the machine gun was important though not critical. Symbolically it was everything; its staccato bark the inhuman voice of the military future. Troops hid in trenches, falling one by one to snipers and artillery, but still glad for the sanctuary against the day their officers drove them over the top. This was their home; this was their life, however temporary.

Weapons only made it worse with time. The Germans led the way. Since

Tourist in Hell

R obert Cowley, historian of what he called the "Unreal City," came up with a darkly whimsical notion. What if some connoisseur of misery had set about traveling the entire length of the trench lines, beginning at the North Sea on the Allied side, crossing over in Switzerland and retracing his steps up along the Teutonic counterpart? What would he have seen? What kind of place was it?

It was, first of all, a true city, the largest in the world at the time, with a combined population of up to six million. The task of construction was immense—the British alone issued 10,638,000 spades. Like any urban center there was constant need for medical attention and there were public health problems—particularly with rats, other vermin, and cadaver removal. There was also sort of an extended municipal park system—no-man's-land—green at first but gradually transformed by artillery into a vision of cratered desolation, reminding one French aviator of "the humid skin of a monstrous toad."

Reflecting inevitable urban diversity there were also many unusual features. Demographically speaking it was both grim and gay (homosexual, that is), since the entire population was male. Our Gulliver would have noted that the neighborhoods on the German side, as befit their craftsmanlike occupants, were much more elaborate, with cement- and steel-reinforced strong points, very deep dugouts, and even wood-paneled quarters for officers. He also would have found his route more complicated than he might have expected initially. Although the basic highway ran north to south, it was supplemented by a maze of parallel and orthogonal supporting and communications trenches to the rear. These provided both avenues of retreat and jumping-off points for counterattacks, together constituting a major factor in preventing breakthroughs. The French preferred a sawtooth pattern in trench design, while the British and Germans leaned toward crenellations, like the teeth on a Halloween pumpkin. Their purpose was the same, to prevent an intruder from being able to shoot down the length of a trench. Our tourist very probably would have concluded that construction on the Allied side looked temporary and jerry-built. Had he asked the local inhabitants, he would have learned that their commanders and the general staff, the city planners of record, wanted it that way, since it was believed anything more permanent would have compromised their troops' offensive spirit. This proved to be a two-edged sword. While any attack along the Western Front generally proved to be a bad idea, the Unreal City's life span was more on the order of Sodom and Gomorrah's than Rome's. In barely four years, all its inhabitants in any shape to do so left, not just voluntarily, but enthusiastically. Unlike for Lot's wife, looking back was not a problem. It had been the worst place on earth.

they were forced to fight a two-front war, the Kaiser's legions were chronically short of men. By the spring of 1915, they had come to grips with the static nature of combat and sought a war of attrition. Armaments, especially armaments optimized for combat in field fortifications, played a major role in this strategy. In addition to adding more howitzers (primarily the 150mm M1913 and its 210mm sister) and many more machine guns, they rapidly fielded a host of smaller, but still highly lethal, weapons. By March 1915, 15,000 improved Model 98 rifles equipped with telescopic sights had been issued to snipers. Unlike the Allies, German troops also began the war with effective grenades, mounted on throwing sticks—the so-called potato masher. For more distant targets, troops were soon issued portable grenade launchers. They also had *minenwerfer,* or "mine throwers," able to loft a 213-pound canister of high explosives that could level sixty-five feet of trench parapet.

The future, such as it was, belonged to much smaller trench mortars. Consisting of a base plate and a simple tube with a firing pin at the bottom, into which a projectile was dropped, these weapons—often called Stokes mortars after a particularly successful British design—required no machining and therefore proliferated quickly on both sides. Although viewed by higher authority as a cheap and effective means of harassment, the troops hated them. Their use invited quick retaliation, resulting in casualties on both sides with absolutely no tactical gain. For German commanders in particular, this was increasingly the point—simply to bleed the enemy, accepting massive casualties, but inflicting them faster.

The battlefield, archetypically the space between trenches known to all as no-man's-land, was admirably suited to accomplish this purpose. For the most part it was wide and difficult to cross, insuring slaughter. Barbed wire was strung in great quantities along the entire length of the trench line. Invented by Americans to keep cattle from wandering, it had the same effect on the battlefield. Troop movement was slowed and channeled, so men could be cut down in clumps. As the war dragged on, barbed wire was powerfully supplemented by the sheer pounding absorbed by the landscape, turning it into a sea of mud and craters that could only be crossed at a fatally slow pace.

Among the many problems of the PBIs (poor bloody infantry, a contemporary and ubiquitous acronym) was their fundamental lack of offensive power while crossing. Out there they were little more than targets. Planners sought to offset this disadvantage with an initial massive shelling of enemy trenches, followed, once the troops were on the move, by a "rolling" barrage out front aimed at blowing holes in barbed wire and generally creating a screen of dust and debris. Yet these barrages disoriented the attacking force, and proved difficult to coordinate. Defenders simply repositioned themselves within their trenches and put their machine guns to work.

Even when attacking troops were provided with machine guns of their own—typically air-cooled and therefore light enough for individual soldiers to carry across no-man's-land—it made little difference. All sides had them during 1915 (the Lewis and Hotchkiss were examples) but their setup time normally demanded that the forward trenches be reached before they could be used to effect. Instead, almost inevitably the initial charge was either turned back or subjected to a murderous series of counterattacks, sending the assaulting force back to their own trenches, sometimes followed by the defenders to renew the cycle. Back and forth they went, gaining little more than casualties.

Something was required that not only insured safe passage across no-man's-land, but precluded counterattacks . . . something, in other words, really devastating. In its pursuit both sides raised the specter of arms of mass destruction, exploring the outer limits of what actually constituted a weapon. The British fell back on sheer explosive force. Still clinging to their salient in coastal Belgium, they were intent upon seizing Hill 60 (the name derived from its sixty-meter, or 197-foot height above sea level) just north of the Messines Ridge, by blowing up a series of subterranean mines. In April 1915, during what became known as the Second Ypres, they succeeded in cutting it down a few notches, and then in 1917, as a prelude to the Third Ypres, they employed a much larger mine that left the much coveted mound more properly called Hill 42. In both cases, British tactical gains proved temporary; and in Second Ypres the Germans countered with an even nastier surprise.

Late in the afternoon on April 22, 1915, amazed aviators looked down to see a vast greenish yellow cloud rolling down from the German-held ridges toward Allied positions around Ypres. Special pioneer troops had released around 150 tons of compressed chlorine from thousands of storage cylinders secretly planted along their lines. Within minutes the cloud enveloped two French divisions, the 45th and the 87th. Those not immediately overcome ran away terrified, opening a four-mile gap in the front. By the end of the day 15,000 Allied soldiers were casualties, 5,000 of them dead. The incident marked the beginning of modern chemical warfare.

Seven months earlier, General Falkenhayn, frustrated over shortages of high-explosive shells and their inability to drive Allied troops from their trenches, gave personal instructions that a new type of munition be developed. The chemists thought they had just what he wanted. It was a tribute to the enthrallment of science to the German war effort that five future Nobel laureates (Walther Nernst, Fritz Haber, James Franck, Gustav Hertz, and Otto Hahn) eventually became involved in the chemical warfare program. When an actual chemical projectile was delayed, Fritz Haber stepped forward with an interim solution involving the release of bottled chlorine. Several senior commanders, including Crown Prince Rupprecht of Bavaria, expressed reservations, arguing that a gas attack was bound to provoke retaliation. Haber reassured them that the Allied chemical industries were incapable of producing gas in these quantities. So the attack was authorized.

Technologically a success, the use of chlorine on April 22 was a tactical failure. The German staff had lacked faith in the weapon, and had made no provisions to exploit the gap with decisive numbers of reserves. By the very next day Allied medical services had identified the gas as chlorine and provided improvised face pads to Canadian troops, enabling them to hold their positions when similarly hit.

Although chemical agents, subsequently launched by artillery, inflicted severe casualties—1.3 million from 66 million shells fired—never again was gas even as effective as at Ypres. The Allies managed to field workable chemical munitions within a year. Twice more the Germans upped the ante, filling their shells first with phosgene and then the insidious blister agent, mustard. Each time, to the Germans' surprise, the Allies responded in kind. Meanwhile, at least along the Western Front, improved gas masks were fielded fast enough to allow troops to bear, if not emerge unscathed, from such attacks.

The initial use of poison gas generated immediate and widespread vilification of Germany for violating the Hague convention. Once the Allies were able to respond symmetrically, however, protests were muted and the campaign took on a momentum of its own. It proceeded with little enthusiasm, though. Colonel Gerhardt Tappen, Falkenhayn's chief advisor on chemical warfare, was typical of his fellow officers in writing of its "unchivalrous nature," terming it "initially repugnant" to all concerned.[2] It was no more popular with Allied troops, even when they had the means to retaliate. Nobody who was gassed ever forgot it, in particular Corporal Adolf Hitler, whose memory of the experience was a major factor in Germany's abstention from chemical warfare during World War II.

Yet the continuing revulsion toward chemical weapons points to factors even more fundamental. The human fear of being poisoned or suffocated is very strong and basic. Building on this emotional bedrock was the negative impact of this class of arms culturally. Like the submarine, chemical weapons were virtually antithetical to the profile of preferred arms, particularly in the West. Rather than being loud, visually impressive, and confrontational, they were nearly invisible, silent, and capricious—so much so that a shift in wind might send them floating back to bedevil their users.

This combination insured that poison gas remained a cloud under a cloud, an object of general opprobrium and, ultimately, arms control. At the end of a war dominated by implements that killed randomly and senselessly, chemical weapons were singled out and their use made illegal by the 1925 Geneva Convention. (It was not until the Chemical Weapons Convention of 1992 that possession and stockpiling were actually banned.) During the Second World War, a conflict that saw the enthusiastic development and employment of virtually every form of weaponry including nuclear, the combatants nonetheless voluntarily refrained from using poison gas, and only poison gas. Even in an era of total warfare, some limits still remained.

"I don't know what is to be done—this isn't war," fretted Lord Kitchener, the same Kitchener who had presided over the slaughter of Africans at Omdurman by many of the same weapons that created and enforced the Western Front.[3] During 1916 conditions reached a nadir with the two greatest battles of the war, Verdun and the Somme. In the first Falkenhayn's overall intention was simply to kill, "bleeding them white," by attacking a point the French would fight to hold regardless of the cost—the great fortress complex Vauban had constructed as part of Louis XIV's Enlightenment version of Star Wars. Acting as the strategic magnet its builder intended, Verdun produced a battle that literally defied description, fought in surroundings torn apart by explosives, scorched by flamethrowers, and further fouled by phosgene—the latter two used massively for the first time. Soon it became virtually impossible to tell friend from foe, since all were the color of soil, and the soil itself was lethally infectious, made so through centuries of agricultural manuring. In the end, after 302 days of the fiercest fighting, only death remained real. Conservative estimates set casualties around 800,000, split about evenly between the two sides. Falkenhayn's plan drowned in a sea of German blood and he was sacked, while French infantrymen—the long-suffering poilus—remembered Verdun simply as "the hell." There were no winners.

Its counterpoint played out along the Somme, where what passed for conventional wisdom amongst the British leadership dictated that they double and triple their artillery, attackers, and reserves to achieve a breakthrough by still another frontal assault. Unfortunately, as John Keegan has convincingly demonstrated, the actual weight of explosives delivered

Fire!

During the early months of 1915 still another surprise leaped out of the German army's bag of dirty trench tricks. It scorched rather than choked. First French troops at Verdun, and then the British at Ypres, found themselves enveloped by outstretched tongues of fire, threatening on-the-spot cremation. Not since Byzantine dromons had first disgorged Greek fire had soldiers been so suddenly confronted by the horrific possibilities of pure flame.

With appropriate perversity, firemen were behind the flames. As far back as 1905 Leipzig fire chief Bernhard Reddemann had begun experimenting with military flame-throwing equipment, and soon after the outbreak of the war he and Berlin engineer Richard Fiedler managed to produce two effective nitrogen-propelled, heavy-petroleum-based flamethrowers—a large model and a more promising backpack version. Realizing that such devices demanded special skills, Reddemann organized what amounted to a private army, staffed primarily with those most knowledgeable and accustomed to things flammable—civilian firemen.

It was dangerous work, since Allied troops were quickly instructed to aim first at those carrying flame-throwing equipment. Nevertheless, by the end of 1915, Reddemann's band of firefighters had expanded from a single platoon to twelve full companies, and would keep on growing and improving.

Reddemann made tactical adjustments and equipment improvements. Automatic igniters were added and cleaner-burning petroleum replaced the "dirty oil" whose dense smoke clouds were obvious signatures for artillery spotters. The biggest change was the eclipse and then replacement of the large flamethrower by the backpack model.

Once they got in amongst the enemy's trenches these misanthropic Prometheans were at their most dangerous. A well-placed burst followed by a few grenades proved highly effective in breaking up machine gun nests, while the capacity of backpack flamethrowers to send jets around corners was particularly suited to the jagged nature of trench architecture. As the fighting dragged on flamethrowers were used with increasing frequency, and out of a total of 635 attacks over 80 percent were classified as successful.

Still, they were not decisive. In fact, their effect was mostly psychological. As one British officer wrote: "It looks like a big gas jet coming towards you, and your natural inclination is to jump back and get out of the way. . . . But in my experience the effective range of the flamethrower is very limited, and the man who manipulates it as often as not is shot. . . . The actual cases of burning by devil's fire have been very few."[4]

The Allied side feared them, but they also shunned them. Neither the English nor the French made more than a halfhearted effort to respond symmetrically while the fighting lasted, and flamethrowers were later listed among the weapons forbidden to postwar Germany under the Versailles Treaty. Flamethrowers did have a military future in the brutal Pacific Theater of the Second World War, where they were used to immolate suicidally inclined Japanese defenders lodged in caves and bunkers. Otherwise the flamethrower's primary impact was, like poison gas, simply to make war worse.

during the 1.5-million-shell preparatory barrage fell far short of the amount necessary to obliterate the German machine gunners sheltered deep below in bunkers built for just such a contingency. When the barrage ceased and the British Tommies went over the top, their executioners had plenty of time to emerge, set up their weapons, and begin one of the most horrific slaughters in the history of warfare. "I heard the 'patter, patter' of machine guns," remembered one member of the Third Tyneside Irish. "By the time I'd gone another ten yards there seemed to be only a few men left around me; by the time I had gone twenty yards, I seemed to be on my own. Then I was hit myself."[5] Kitchener was right, this wasn't war; it was human sacrifice and on a scale that would have done the Aztecs proud. By nightfall the British had suffered 57,000 casualties, the Germans 6,000.

The battle dragged on for another 140 days—the British continued the attacks and the Germans inevitably responded with counterattacks (one source counted 330) until they more than evened the score in losses. The French were involved also, and although they gained more territory and lost far fewer troops, they never came near a breakthrough. By mid-November the fighting had ceased. The butcher's bill was staggering—420,000 British casualties, 200,000 French, and 465,000 among the Germans. The lines had barely moved. More than any single event in the Great War, the Somme came to symbolize military folly.

French

British

German

Helmets used on the Western Front provided little protection against the storm of steel.

✖✖✖

The Great War at sea charted a similarly outlandish, if far less lethal, course. Here too the armaments generated an equivalently weird stalemate; but to a far greater extent the character of the two antagonists determined who prevailed.

The young German navy was best described as liberal, bourgeois, and technical. As the single truly national institution created under the German Empire, the navy was immensely popular with the middle class, and its flag officers included almost no nobles. Teutonic dreadnoughts and their auxiliaries were similarly solid products of a society that placed special emphasis on quality and durability, German U-boats in particular had capabilities and reliability truly remarkable when compared with the sputtering craft that passed for submersibles in other navies. Nonetheless, the Imperial German Navy was plagued by the most bourgeois of afflictions, an inferiority complex. Its officers never truly believed they could beat the English, and persisted in playing by rules dictated by their foes.

The British Royal Navy, despite a reputation built on three centuries of unparalleled success, was a deeply flawed institution. Its decision makers were almost fatally tied to a rigid set of outmoded assumptions. Not having fought for more than a century, it had rusted from within; a petrifaction that found its material representation in a nearly lockstep overreliance on the gun. The Germans treated mines and torpedoes with deadly seriousness; the British scoffed at their enemies' inexperience. Big guns in great ships won wars in decisive battles, and the Royal Navy steamed confidently forward guided by the vision of Trafalgar.

Yet the guns of August were barely heard on the bounding main. The decisive clash in the North Sea that everyone expected utterly failed to materialize. Kaiser Wilhelm, all too aware his dreadnoughts were outnumbered, was not about to let them accept battle as an underdog. Instead, he kept the increasingly oxymoronic High Seas Fleet at their docks in Wilhelmshaven, hoping the aggressive British, attempting a close blockade of the heavily mined waters nearby, might be decimated sufficiently to allow the Germans to risk battle.

The Royal Navy was not suicidal. After the destruction mines wrought in the Russo-Japanese War, it was clearly understood that they had made close blockade simply too dangerous. Consequently by 1911 the Royal Navy had evolved a new policy of "distant blockade." The Germans were to be confined to the North Sea by the British fleet poised far north, just off the tip of Scotland at Scapa Flow, ready to spring. But was it too distant to be effective?

In their two great wars against the English, Germany did not have a good record of communications security. They were adept enough at encoding; but their cipher keys kept turning up in the hands of the enemy. It happened twice. The latter instance had a major impact on the Second World War and is justly famous; the first is less well known.

On August 26, 1914, the new German cruiser *Magdeburg* lost its way off the Russian coast and ran aground. When all efforts to refloat her failed, *Magdeburg's* commanding officer, Richard Habenict, set about performing two essential tasks: blowing up his ship before the Russians arrived, and destroying the ship's naval codebooks. Both were bungled; the aft charge never detonated, and a Russian search party found an extra codebook stashed in Habenict's cabin.

On October 13 the son of the Russian ambassador to England handed the purloined treasure to the newly installed First Lord of the Admiralty, Winston Churchill. The future prime minister was intrigued, and the book quickly landed in the hands of a group of neophyte code breakers who set about applying it to a pile of intercepted transmissions. Within months they were able to read much of the German navy's wireless traffic, and in December they successfully predicted the Germans' battle cruiser raid on Hartlepool and Scarborough. The squadron's commander, Franz von Hipper, slipped the trap set for him; but on January 24 the British battle cruisers, tipped off again by the code breakers, were waiting for him off Dogger Bank. In the ensuing chase one of his ships, *Blucher,* was sunk, and Hipper and the Germans learned just how fatally vulnerable their lightly armed battle cruisers were to magazine-igniting flashbacks from shell hits. The British simply reverted to form.

Momentarily in high repute, the code breakers were given carte blanche by Jacky Fisher, brought out of retirement by Churchill to serve again as his First Sea Lord. Unfortunately, their immediate boss, the Director of Operations, Captain Thomas Jackson, was a much more typical British officer, who quickly threw a wet blanket on their work. (Cryptographic historian David Kahn reports that at one point, when the flow of deciphered intercepts temporarily ceased, Jackson called to express relief that he would no longer be bothered by such nonsense.) His attitude prevailed, and when the opportunity finally came in June 1916 at Jutland to catch and destroy the entire High Seas Fleet, Admiral-in-Chief John Jellicoe ignored key intercepts. There would be no Trafalgar, and the British would have to wait for another war to take full advantage of their code breakers.

Meanwhile, another German naval menace made its entrance into the conflict. Practically unnoticed, the untried submarine slipped into the waters that separated Germany from England. For almost a month nothing happened. Then disaster struck. Early on September 22 three 12,000-ton British armored cruisers, *Cressy, Hogue,* and *Aboukir,* were attacked by Otto Weddigen, captain of the German submarine U-9. Within an hour he had sunk them all, drowning 1,459 men. In terms of casualties it was the worst disaster to befall the Royal Navy in nearly three centuries.

There was no real answer to the submarine. In a mechanical approximation of revenge HMS *Dreadnought* managed to ram and sink U-9 six months later. But this was simply a matter of luck. In truth all the British could bring to bear against the submarine were ridiculous panaceas. Picket boats armed with blacksmith's hammers were sent out to smash periscopes. When this didn't work, the craft were given steel nets to catch them like cod. Later, seagulls were trained to perch on periscopes to make them easier to see, while sea lions from the London stage were recruited to seek out submerged hulls, which they shrewdly judged inedible and ignored. It was not until July 1915, with the formation of the Board on Research and Invention, that the Royal Navy made a wholehearted effort to generate effective countermeasures. Still it was nearly two years before hydrophones, the first practical acoustical detection system, in combination with depth charges were available in sufficient quantities to seriously menace submarines.

Meanwhile, the dreadnoughts had become beasts of prey. In early fall 1914 a host of submarine sightings (both real and imagined) drove the Grand Fleet out of Scapa Flow and eventually all the way to Loch Swilly on the northern coast of Ireland. Distant blockade was 300 miles more distant, but still there was no safety. On October 27 the new dreadnought *Audacious* struck a German mine and sank. The experience unnerved Jellicoe, who three days later wrote the Admiralty informing them that in any future fleet action, "If the enemy turned away from us, I should assume the intention was to lead us over mines and submarines, and decline to be drawn."[6] This is exactly what he did nineteen months later at the short-circuited Trafalgar, Jutland. Having found no respite in Irish waters, he led his wandering dreadnoughts back to Scapa. Soon enough German submarines ceased actively stalking them. These wolves of the sea found tamer, fatter flocks to plunder.

Conan Doyle's *Danger* was quickly forgotten by the English; but the book was called to the attention of Admiral Tirpitz and the Naval High Command, who kept it in the back of their minds during the frustrating first months of war. By November, given the accumulated disappointments on land and sea, the submarine seemed the only German weapon scoring notable successes. It was popular with the public, and newspapers indicated that a majority wanted it turned loose against merchant shipping without restraint.

In early February at the urging of his admirals, the Kaiser approved unrestricted submarine warfare, with only a civilian, Chancellor Theobald von Bethmann-Holweg, expressing reservations. All waters around Great

Britain were declared a war zone where merchant vessels, including neutrals, were subject to attack without warning. Initially, it seemed something of a bluff, since the Germans had only twenty-one submarines, including just nine of the superior diesel type. The initial results were modest. A mere thirty-four English transports displacing 104,000 tons—less than a quarter percent of the merchant fleet—were sunk before the end of April. Then on May 7 the famed 32,000-ton luxury liner *Lusitania,* laden with munitions and passengers, was sent to the bottom by U-20, claiming 1,198 lives, including 128 U.S. citizens.

America was outraged, particularly President Woodrow Wilson, who took special offense at the manner the ship was sunk. The British naturally joined the chorus of condemnation; given their lack of antisubmarine technology, there was little else they could do. Nevertheless, the broadside was surprisingly effective, especially considering that by August the U-boats' toll had risen to 183,000 tons a month. When the White Star liner *Arabic* was sunk on August 19 and Wilson threatened to break diplomatic relations, the Germans conceded his every point, and by the end of September had essentially abandoned the submarine campaign. With the exception of a single two-month stretch of sinkings in the spring of 1916, they did not resume unrestricted submarine warfare until February 1917.

This may have been a grievous error. The threat posed by the United States's entry into the war at this point was largely illusory. Even with the benefit of a considerable military preparedness campaign, it was more than a year after the American eventual declaration of war in April 1917 before U.S. troops began making a significant contribution in France. In 1915 it would have taken longer. The Germans should have known as much, since future Chancellor Franz von Papen, an energetic military attaché in Washington, was perfectly aware just how far the American military was from a war footing. It was obvious.

Accumulating enough submarines was plainly a problem for the Germans. Initial production was ragged, troubles were encountered with diesel design and fabrication, and inexperienced shipyards continuously fell behind schedule. Consequently, only eleven boats were delivered in 1914. From this point, however, production gained momentum with a total of fifty-two new submarines added in 1915, and a further sixty-one commissioned between January and August 1916. Meanwhile, attrition in this time frame remained very low due to the Royal Navy's dearth of antisubmarine weapons. Even including the later stages of the war when defenses were better, the exchange rate was 29.67 merchant ships, or 69,015 tons sunk, per U-boat lost.

Using these figures as a guide, it is possible to project what kind of losses might have been inflicted had unrestricted submarine warfare been pursued without pause. Due to the exigencies of production, the total of boats stabilized from September until the first part of 1916 at around fifty-eight, so it makes sense to keep the scores steady at approximately 250,000 tons per month. Since the next eight months roughly doubled the flotilla to 111, a steady increase up to 550,000 tons monthly is reasonable. (If these figures seem high, it should be remembered that actual sinkings peaked in April 1917 at 860,000 tons, achieved with 156 boats.) The net result from this hypothetical nineteen-month campaign would have been roughly 5.3 million of the 12 million tons of merchant bottoms with which the English entered the war. Meanwhile, the Admiralty would have had little chance of heading off a crisis, since it failed to keep anything approaching adequate statistics on shipping, and were blissfully unaware of the precipitous drop in replacement merchant vessel construction in 1915 and 1916. Meanwhile, submarine-induced congestion in ports is estimated to have reduced the annual carrying capacity of the affected ships by up to 20 percent.

All of this would have added up to major shortfalls in imports, especially foodstuffs, which could well have resulted in widespread hoarding and the perception of mass starvation. For the English people this crisis coming on top of a growing awareness of the huge casualties on the Somme and the disappointing news of the fleet action at Jutland might have been too much to bear. It is impossible to know if the British really might have dropped out, but had they quit, the French and Russians almost certainly would have collapsed. Suppose Imperial Germany had won all fronts, would there have been a Bolshevik revolution, a successful Adolf Hitler, or even a Second World War?

Speculation over outcomes aside, did Germany restrict its submarines purely out of fear of America? Perhaps the weapon had something to do with the decision. This becomes even more pertinent when the relatively light casualties inflicted by submarines (fewer than 13,000 for the entire war) are compared with the meat grinder on the Western Front. The submarine continued to be popular with the German public; but its operational characteristics raised questions among professional military men. Chief of the Naval Staff Georg von Müller referred to the submarine campaign as a "desperate measure," and even hard-bitten Admiral Tirpitz was forced to admit that "our appearance of uneasy conscience encouraged the English case that the campaign was immoral."7 So Germany gave the British the precious time to generate the means to meet the climactic but belated resumption of the unrestricted campaign of 1917.

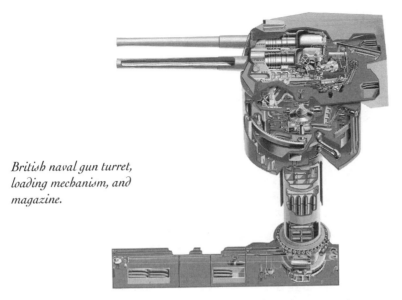

*British naval gun turret,
loading mechanism, and
magazine.*

Atop the waves the stalemate persisted. After the inconclusive action off Dogger Bank, the High Seas Fleet had not shown itself in the North Sea for over a year, and by the spring of 1916 many members of the Royal Navy were beginning to despair of ever seeing a Trafalgar redux. Yet events would soon put the two fleets on a collision course—the German armada had a new and more aggressive commander, Vice Admiral Reinhard Scheer, and renewed pressure to resume unrestricted submarine warfare forced the leadership to unleash the High Seas Fleet as the lesser of two evils.

Scheer's battle plan followed the basic attrition scenario with a new wrinkle: Hipper's battle cruisers were to be used as bait to lure units of the Grand Fleet first over a line of U-boats and then into the waiting arms of the entire High Seas Fleet. Warned if not exactly guided by their code breakers, the British knew the Germans were coming and planned an equivalent snare using their own battle cruisers commanded by the dashing David Beatty. Knowing their U-boats could remain on station just one more day, the Germans were pressed into action on May 31. The stage was finally set for the battle royal, and, as it transpired, each side would fall for the other's trap.

First it was the Royal Navy's turn. While investigating a suspicious steamer, Beatty's six battle cruisers spotted Hipper, who immediately turned his own five battle cruisers as if to flee. Beatty took the bait and soon both sides formed parallel lines and blasted away with their main guns at a range of eight miles, all the while racing toward Scheer's battle line. Within a period of thirty minutes, Beatty's *Lion* took a near-fatal hit in her Q turret,

The fate of HMS Invincible.

and both *Indefatigable* and the Royal Navy's champion gunnery ship *Queen Mary* were turned into fireballs and sunk with a near total loss of their crews. "There seems to be something wrong with our bloody ships today," Beatty was prompted to comment.[8]

There was. Unlike the Germans, who had learned their lesson at Dogger Bank and taken remedial action, the British had no idea how subject these ships were to catastrophic flashbacks from the turrets to the main magazines. Yet better protection against the contagion of flames would not have saved these flashy vessels. Products of Jacky Fisher's imagination, they were fantasy weapons that sacrificed everything for speed and big guns. Advertised by their creator as *Invincibles,* they proved little more than intricately wrought powder kegs.

Yet tradition still counted for something, and Beatty's squadron refused to panic. His scouts quickly recognized they were approaching the main body of the High Seas Fleet and signaled the danger in time to reverse course. Soon they were clear and leading the entire German fleet straight back into the jaws of Jellicoe, just fifty miles to the northwest. The predator was now prey.

John Jellicoe, master of one of the clumsiest weapons in history, was worried. Before he could fight, the Grand Fleet's twenty-four dreadnought battleships had to be deployed in a single file. Yet if done too soon the Germans might escape; if he waited too long, leaving his ships bunched up, the results might be fatal. Gradually, the rumble of heavy artillery became audible, and at last Beatty signaled that the Germans were approaching. One minute before they sailed into view the last British battleship joined a vast L-shaped line, which had been formed directly across the German course of approach. Not only was their T crossed, but the British blocked the way back to their home base.

In theory Scheer's fleet faced annihilation. But this was the North Sea, subject to all the imponderables of smoke and fear and confusion, and in the heat of battle the logic that spawned the dreadnought simply melted. After only nine salvos, the Germans disappeared. At first Jellicoe thought it was only due to "thickening of the mist," but soon it became apparent the Germans really were gone. Scheer had ordered a "battle turn to the right," a desperate maneuver unknown to the British. Rather than sailing around a common point to be pummeled, each German dreadnought turned along an individual axis, reversing course in less than four minutes and rapidly sailing clear of the tempest of English shells.

At this juncture, inexplicably, the balance between fight and flight again shifted in Reinhard Scheer's endocrine system, and he ordered his fleet back toward the British. "It would surprise the enemy, upset his plans . . . and facilitate a night escape," he hoped.[9] Instead, the High Seas Fleet ran straight into the middle of the British line, and, for the second time in one day, found its T crossed. "We were in a regular death trap," wrote one officer on the *Derfflinger.* Scheer ordered another "battle turn," but the sheer volume of the British fire caused the maneuver to be botched. In hopes of diverting the barrage, Scheer ordered his fastest dreadnoughts to "close with the enemy and ram"—what became known as the "death ride of the battle cruisers." As a last resort, Scheer also unleashed his destroyers, ordering all flotillas to attack with torpedoes. Amidst a hail of fire, they managed to launch thirty-one.

It was enough. Poised on the brink of victory and Nelson-like immortality, John Jellicoe blanched at the deadly fish. He had warned the Admiralty what he would do in case of such an attack and now he made good on his promise, ordering his entire line to turn away. When he returned, the Germans were gone for good. As darkness fell, Jellicoe ignored his code breakers, who transmitted Scheer's real course of retreat, and turned instead in the opposite direction. The High Seas Fleet slipped behind him and made its way to safety. The two fleets would never meet again in battle.

During the action the British lost an additional battle cruiser, *Invincible,* plus three of the ill-fated armored cruisers—a total loss of 110,000 tons of warships and over 6,100 killed—figures that compared unfavorably with the Germans' 62,000 tons sunk and fewer than 3,000 killed. Yet the Germans had run away, and the British still claimed command of the sea. In actual fact the stalemate simply had been reasserted.

With hindsight it was clear that Britannia no longer ruled the waves, and neither did the battleship. Twenty-one small ships firing torpedoes had chased the entire Grand Fleet off the field at a critical moment. The dreadnought was simply not a decisive weapon. The combination of smoke and confusion made accurate firing of the great guns, at the ranges torpedoes necessitated, impossible with available optics-based range finders. For the second time in Western military history, the basic naval unit reached the point of obsolescence. Just as the great battle of Lepanto marked the twilight of the galley, Jutland was, if not the end of the line for the ship-of-the-line, then at least a very important turning point. In his report on the battle to the Kaiser, Reinhard Scheer conceded that "only the resumption of unrestricted submarine warfare—even at the risk of American enmity—could bring a victorious conclusion of the war within a measurable time," while Jellicoe advised the First Lord of the Admiralty in October 1916 that the "ever-increasing menace of the submarine attack on trade is by far the most pressing question."[10] The war at sea had entered its final phase.

No combat arm suffered a more ironic fate during the Great War than the cavalry. It survived as a living relic of warfare, not discarded but held behind the lines by traditionalists like British general Douglas Haig hoping to exploit breakthroughs that never came. Useless and ignored, cavalrymen were also spared the pointless deaths that claimed so many others. But safety brought its own brand of anguish to the cast-off equestrians, forced to witness what must have seemed the irrevocable doom of their way of life and fighting. Remarkably, they were dead wrong.

Back in the fall of 1914, during the hectic days of the "race to the sea," a troop of French cavalry swept down on a flight of reconnaissance aircraft, parked helplessly in a field near the river Aisne. Ironically, these ungainly flying machines held the keys to the cavalry's future, or at least the future of a style of fighting for which cavalry had provided an outlet at least since the invention of the stirrup. High above the impersonal slaughter of the trenches, the classic paradigms of intraspecific aggression soon bloomed anew. Aircraft lent themselves to a number of combat functions—bombing, infantry support, reconnaissance, and artillery spotting. Nonetheless, the focus was on plane-versus-plane combat.

This brand of fighting grew out of early desultory encounters between observation aircraft, armed with bricks, steel darts, and a variety of small arms. Soon enough machine guns took to the air, but mounted in positions

that made them hard to aim—above the wings, or on pivots generally operated by observers. The key development came with the introduction of
devices—first simple deflection plates and then mechanical synchronizers—
that enabled machine guns to be shot safely through the whirling blades of a
propeller, allowing the pilot to aim at targets simply by pointing the plane.
The fighter aircraft was born. Within weeks pilots like Frenchman Roland
Garros and the ambitious Germans Max Immelmann and Oswald Boelcke
were stalking victims and engaged in swirling dogfights.

The Germans had the edge, in part because they flew the first great
fighter of the war, Tony Fokker's Eindecker—a monoplane with a broad,
shoulder-mounted wing, powered by an eighty-horsepower engine and
armed with a single Parabellum machine gun linked to a cam-operated
synchronizer. Immelmann and Boelcke were virtuosos of flight. Using the
Eindecker's superior maneuverability and firepower, they established tactical rules that are still studied to this day. So dominant did they become that
this period of the war was remembered as the "Fokker scourge." Yet it
remained a contest. Both Immelmann and Boelcke studiously kept track of
their aerial victories, as did other pilots, whose kills provided a heroic counterpoint to the impersonal death-dealing below.

Cavalrymen virtually stampeded into flying squadrons, aspiring to
become fighter pilots. Despite the rise of mass formations especially among
the Germans, nearly every important ace from Immelmann, to the brood-
ing "Winged Sword of France," Georges Guynemer, to the super-aggressive
Englishman Albert Ball, to the lethal American race driver turned pilot
Eddie Rickenbacker expressed a preference for fighting alone. Pilots deco
rated their planes with large initials or other highly visible insignia, the
most brazen example being the blood-red paint jobs of Manfred von
Richthofen's various fighters. Chivalric pretensions were also reflected in
the frequently voiced contention that death-dealing was secondary in air
combat. Georges Guynemer famously stopped a dogfight when fellow ace
Ernst Udet's guns jammed. Similarly, Rickenbacker later wrote, "I would
have been delighted to learn that . . . any pilot I had shot down had escaped
with his life."[11] Fighter aces became virtually the only heroes in a war that
had ground heroism into suicide.

Pilot skill and determination may have been a vital factor in individual
success, but overall the ebb and flow of air combat was largely determined
by which side had superior aircraft. Fighter development and deployment
took place at a breathtaking pace. By early 1916 the Eindecker was thoroughly outclassed by the elegant little French Nieuport 11 biplane, which in
turn was quickly supplanted by the highly streamlined and lethal Albatros

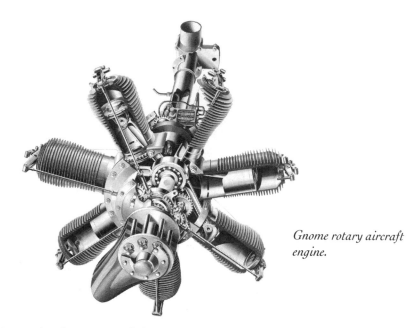

Gnome rotary aircraft engine.

D III—the first major fighter mounting twin machine guns. By the spring of 1917, however, its reign was cut short by a new generation of Allied planes, including the British Sopwith F.1 Camel, the Bristol F.2B, and the French Spad 13, all of which were eclipsed in early 1918 by the Fokker D.VII, probably the best all-round fighter of the war.

To a considerable degree the flight characteristics of World War I fighter aircraft were determined by motor type. At one end of the scale were what might be called "corkscrews," aircraft powered by rotary engines—air-cooled, composed of a radial bank of cylinders revolving around a fixed crankshaft at the same speed as the propeller. They were durable and extremely compact, though so limited in power potential as to prove a technological dead end. Appropriate airframes had to be as small and lightly constructed as possible. The result was extreme maneuverability, characteristics enhanced by biplane or even triplane configuration, and a generous dihedral (the wings' upward slant or angle). Also important was the torque effect of the rotary, rendering them somewhat sluggish in left turns, but lightning-quick in the opposite direction. Top speed and ability to withstand dives were not strong points, though. The Eindecker, the Nieuport, the Sopwith Camel, and the Fokker Dr. 1 triplane (Richthofen's favorite) were all corkscrews. They were unforgiving and hard to fly, but in the hands

Sopwith Camel.

of a seasoned pilot they were capable of carving a series of tight right spirals until they were in a completely dominant position.

At the other end of the spectrum was what might be termed the dagger, or stiletto, planes powered by engines basically identical to automotive designs. They were heavier, and water cooling rendered them significantly more vulnerable to bullets; but they were inherently powerful and economical in fuel consumption. Their speeds were higher and could be reinforced to withstand steep dives. Their pilots learned to avoid prolonged turning contests with more agile foes, and struck instead suddenly from on high, with the cold precision of a bird of prey. The Albatros, the Spad, the Bristol F.2B, and the Fokker D.VII were all stilettos.

The war ended with the daggers ascendant. Superior Allied engine technology allowed them to pack the last generation of fighters with enough horsepower to provide a clear edge in vital top speed. The Germans, on the other hand, excelled in airframe design, their last stiletto being the revolutionary Junkers D.I, an all-metal monoplane with totally internal bracing but lacking a power plant to do it justice. The rotary engine soon disappeared, but the basic conflict between speed and maneuverability was destined to continue well into the jet age.

Was aerial combat as honorable as legend portrays? Certainly not. Survivors fed on newcomers. "There was no sportsmanship; it was simply killing. You looked for the new boys," wrote Albert Ball, "the frightened ones with as little experience as possible."[12] Although there were memorable instances of long and skillfully conducted duels between equally matched opponents, the major aces characteristically preyed on clearly inferior planes and those flown by pilots betraying signs of inexperience. The average combat life span of fledgling fighter pilots ranged between three and six weeks. Those who managed to survive this initial period and picked up some fighting skills frequently lasted much longer. But in the end most of the renowned combat aces were killed, victims of nervous exhaustion, rickety planes, and a dearth of safety features—only the Germans had parachutes and not until the war was nearly over.

Nevertheless, the lionization of air combat was far from simple hypocrisy. The fighter aircraft was celebrated because it provided a rare syncretism of warlike preconceptions and high technology. Swift to the extreme, loud, and visually impressive, it was capable of absorbing all manner of innovation and still preserved its heroic essence. As such it remains today an object of fierce loyalty and affection, and not just among those who fly them. But if the fighter plane lived on as an icon of the compatibility between war and technology, it was a misleading one. Technology could be manipulated by cultural presumptions; yet its development was still driven by the laws of possibility. Hence military aircraft development also moved quickly in a far more ominous direction.

Early arms controllers quickly recognized the inhumane possibilities of overhead bombing from balloons, and banned it at both Hague conferences. Yet as later with poison gas, the Germans made short work of this prohibition, and by January 1915 were using their zeppelins to bomb English ports. By September this approach had escalated to squadron-sized night raids on London itself, and within a year the zeppelins had accounted for more than 400 dead and millions in property damage. Yet German airships were fundamentally flawed. Not only were they huge targets—most over 600 feet long—but they were filled with highly flammable hydrogen. A combination of British spotlights, antiaircraft guns, and incendiary-firing pursuit planes were soon turning the zeppelins into flying crematories with such regularity that attacks were halted in December.

The Germans logically switched to airplanes. The roots of the program reached back to autumn 1914 when one Major Wilhelm Siegert, a former balloon pilot, argued for an aircraft-based strategic bombardment program so effectively that specifications were issued for a large bomber, or G-type. A number of companies, including Gothaer Waggonfabrik A.G. Gotha, began work, and after two years of trying the firm finally produced a successful design, the G. IV Gotha, which was rushed into production.

The basic G. IV was a large, angular biplane with a wingspan of over seventy-seven feet and twin 260-horsepower water-cooled Mercedes engines linked to pusher propellers that drove the aircraft for more than 300 miles at a sedate eighty miles per hour. Yet the Gotha was no sitting duck. It was highly maneuverable, and even with a full fuel and bomb load could climb to between 18,000 and 21,000 feet—considerably higher than nearly all interceptors. Pursuit craft that engaged them at lower altitudes found G. IVs to be relatively

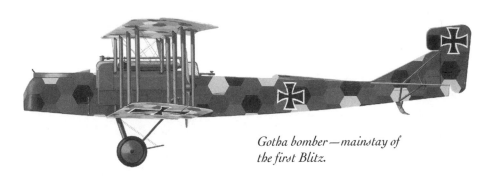

*Gotha bomber — mainstay of
the first Blitz.*

tough customers, armed with two electrically heated 7.92mm Parabellum machine guns, one shooting out a special tunnel in the rear of the fuselage.

Each Gotha carried 660 pounds of explosives, a fraction of the zeppelins' capacity, but Gothas came in waves and were far less vulnerable. There was at least a pretense of aiming. The G. IV was equipped with a Goerz bombsight employing a three-foot Zeiss vertical telescope, and racks accommodating bombs of either twenty-seven or 110 pounds, primarily useful against point targets.

When the Gothas were unleashed in late May 1917, as a complement to the renewed unrestricted submarine warfare program, it was clearly understood they were to be used against both military targets and civilians. Typical was the Gothas' first mass raid, which bad weather turned away from London. While the attack only lightly damaged military camps at Shorncliffe and Cheriton, it slaughtered sixty people, mostly women and children, in the Folkestone commercial district. The results were similar when Gothas succeeded in reaching the capital in June. The planes located and hit designated targets, but a 110-pound bomb plunged through the Upper North Street Schools killing or injuring sixty-four children. The bombing only grew more indiscriminate as British defenses improved and the early-warning net was streamlined to provide adequate reaction time. By the end of August, the bombers ceased daylight missions, and sought the cover of darkness.

So began what historian Raymond Fredette has called the first Blitz, a week-long string of random night bombings during the final days of September 1917. After four such attacks as many as 300,000 people were taking nightly refuge in the Underground, and numerous others hid in the tunnels beneath the Thames. For all anybody knew, the bombing might never end. In fact, the attack was a spasm, a supreme German effort. Virtually all Gothas flew, plus zeppelins, and several enormous unstandardized aircraft known as Giants. Typical of this small class was the R.39, with a wingspan of nearly 140 feet, four engines totaling almost 1,000 horsepower, a crew of nine, and a 4,000-pound bomb load—plainly a harbinger of strategic bombers. Yet only one Giant reached London.

The first Blitz was a dismal failure. Of ninety-two G. IV sorties, only fifty-five reached England and fewer than twenty found London. Thirteen Gothas—nearly one third the squadron—were destroyed that week. There were other raids, but the cost rose still higher. By 1918 the British had nearly completed an air defense system, which in every essential save radar was like the one that stunned the Luftwaffe in 1940. The Germans could have raised the toll themselves. They apparently never thought of poison gas bombs, and ignored the sinister magnesium-based Elektron incendiary bomb on the grounds that the war was already lost.

The memory of the Gothas nonetheless haunted the English. When war came again to the otherwise unprepared nation, there was not only air defense but civil defense and a heavy bomber force to wreak terrible vengeance.

Gridlock finally began to ease in early 1917, though the killing consumed another twenty-one months and millions more lives. Weapons were pivotal, yet the process was impelled in equal measure by desperation, opportunism, and exhaustion. Some military historians today point to tactical and technical solutions; survivors at the time saw mostly waste and tragedy.

On the first day of February 1917, the Germans resumed unrestricted submarine warfare, and by early April the United States joined the Allies as a full belligerent. Very soon William Sowden Sims, who had so much to do with scientific gunnery and the genesis of the dreadnought, found himself on the way to England as commander-designate of U.S. naval forces in Europe. It was bad enough that his ship struck a mine shortly before landing, but the next day John Jellicoe shared news of catastrophic losses to submarines, apparently climbing toward a million tons a month. Worse yet, Sims learned the Admiralty was against convoying transports with military vessels, fearful that any more attrition to the Grand Fleet's destroyer screen would leave it stranded in Scapa Flow unable to put to sea.

Sims realized that this situation, if allowed to continue, could knock England out of the war before the full weight of American participation was felt. He had followed the weird war at sea with a growing conviction that conventional naval weapons were incapable of winning it. Instead, he advised Washington that the American battle fleet should be held back and stripped of its auxiliaries for convoy duty, while an immediate emphasis be put on the construction of merchant bottoms and more destroyers.

The politicians were ready for a change. British Prime Minister David Lloyd George seized upon Sims's recommendations as support for his own faith in convoys, and by the end of April he forced the Admiralty to drop their objections.

No one solution was found, but a combination of new weapons, the convoy system, and fortuitous geography proved sufficient to blunt the belated German naval offensive. By this time submerged U-boats were no longer invulnerable or undetectable. In January 1916 the Admiralty had introduced the depth charge—basically an oil drum packed with 120 pounds of TNT set to explode at a predetermined depth. By August 1917 they were being lofted up to forty yards in tight patterns by howitzer-like devices mounted on the sterns of destroyers, guided by hydrophones, capable of picking up the propeller beat and general location of a submarine far below. Sinking a U-boat required exploding a depth charge within fourteen feet of the hull, no easy shot. Yet a sustained depth charge attack was an absolutely terrifying experience, frequently leaving a crew debilitated for days.

Whereas lone merchantmen frequently had been approached on the surface by U-boats and sunk with their deck guns, the convoy's depth-charge-slinging destroyers made submerged attack, frequent battery recharge, and torpedo expenditures imperative. A single U-boat was hard-pressed to claim more than one victim per convoy, and convoys presented predator submarines with either empty ocean or a surfeit of targets. Rather than employing mid-ocean scout submarines to direct groups, or wolf packs, to specific convoys, the Germans reacted sluggishly, clinging to lone wolf tactics.

Meanwhile, the British had effectively mined the English Channel, forcing submarines to waste fuel rounding the tip of Scotland. Soon this route was endangered by the American-developed antennae mine, which enabled the Allies to lay a three-dimensional mine field, 600 feet deep across 240 miles of open water between Scotland and Norway. By war's end U-boats were having major problems just getting on station.

Statistically, the corner had been turned back in July 1917, when monthly transport losses ceased exceeding Allied capacity to replace them. Yet the U-boat had not so much been defeated as mishandled. German leaders had not only allowed themselves to be talked into a self-defeating pause in the campaign; but in the end concentrated on the wrong targets. Not a single transport carrying American troops across the Atlantic was lost. Had only a few been sunk, the course of the war might have changed. Instead, France was soon awash with doughboys—fresh and all too ready to go over the top. Meanwhile, another new player had already entered no-man's-land.

Endgame arrived along the Western Front in 1917. With a supreme effort

Tanks Roll

The concept of an armored fighting vehicle had appeared intermittently in the Western military tradition—the wheeled battering rams of the Assyrians, the war elephants of the Hellenistic monarchs, the armor-plated combat carts of the Bohemian Hussite Jan Zizka, and in drawings in Leonardo's notebooks. Like the airplane and submarine, however, the tank's emergence had to await an appropriate prime mover. As World War I commenced, both sides were equipped with wheeled armored cars powered by internal combustion engines, and the British used theirs quite successfully until the trenches reached the sea, and the vehicles could neither flank nor traverse them. At this point the terrestrial branch of the military imagination got stuck.

Fortunately, the Royal Navy—still wed to the concept of armored vehicles at sea—took naturally to the concept on land and provided a home for further development. Winston Churchill, then First Lord of the Admiralty, played an important role as matchmaker and midwife. Inspired by dreadnought instigator Admiral Reginald Bacon's design for an artillery transporter employing an American caterpillar tread, along with the suggestions of Colonel E. D. Swinton, a shrewd staff observer in Flanders and coiner of the phrase no-man's-land, Churchill referred the matter to the irrepressible Jacky Fisher, who formed a "landships" committee. Chaired by battleship designer Eustace Tennyson-d'Eyncourt, the body moved deliberately to develop a practicable armored vehicle capable of crossing trenches. After several designs were considered and three prototypes built, the committee settled on the so-called Mother type. This vehicle, consisting of a lozenge-shaped hull with a track encircling the entire profile, was placed in production, but with a target output of only 150 units. Upon completion the new weapons were shipped in small lots to France disguised by crates and referred to deceptively as storage containers, or "tanks"—Swinton's term again.

At this point, nearly two years after the war began, the army took charge. The plan was to throw the new vehicles into battle on the Somme as soon as a small force could be assembled. Swinton demurred, warning that they "should not be used in driblets" but kept secret until sufficient numbers existed to launch "one great operation" coordinated with infantry.[13] Much later, the originator of Germany's panzer concept, Heinz Guderian, paraphrased these words when he directed his corps: "Don't dribble, pour!" But in 1916 Swinton was disregarded.

Instead, on September 15, the British, on orders from the Somme's chief architect, General Douglas Haig, let loose just forty-nine of the vehicles in a widely dispersed pattern. A substantial portion (the lucky ones) broke down before even reaching the jumping-off point. Some got stuck in no-man's-land or astride German trenches, while the remainder, cruising at less than one mile an hour, proved easy targets for artillery. The day ended in fiasco. Still, those few tanks able to get in among the Germans appeared to have had a terrifying effect. Haig was sufficiently impressed to order 1,000 more.

British Mark IV tank.

After accumulating tanks for over a year, the British attacked the Hindenburg Line at Cambrai in late November 1917, with 400—all of them. The two-day battle produced a six-mile advance, a gain unparalleled since 1914. But the tanks outran their infantry, which had been pinned down by machine guns, and losses of armored vehicles continued to be heavy. It was seven months before an equivalent success was achieved by massed British armor—August 8, 1918, the so-called black day of the German army—and in July 1918 the French managed a similar success using nearly 500 independently developed Renault light tanks.

The Germans were again slow to react. When they did, it was primarily with antitank weapons, since severe raw materials shortages made it impossible to compete with Allied tank production. By war's end they had fielded only twenty of the huge A7V "assault" tanks; but on April 24, 1918, one engaged a British Mark IV in the first tank-on-tank duel.

When the shooting stopped, the tank was hardly viewed as a decisive weapon. It remained frail, unreliable, and lacked endurance. But it plainly had a future. Operationally, tanks provided land combatants with potential answers to several key problems: how to fight out in the open; how to carry significant firepower on the attack; and how to dominate territory without necessarily occupying it. Psychologically, tanks had much to offer, being inherently large, armor-plated, visually impressive, and loud. Occasionally they even engaged in duels. Understandably, cavalrymen gravitated to armor units during the interwar years. Less romantic than fighter aircraft, tanks still constituted a kind of reconciliation between technology and the warrior ethos. World War II has been called a conflict of airplanes and tanks. This was no accident.

both sides gathered the resources for one final push, and in their desperation also unveiled several of the tools of war's future. First it was Germany's turn.

The Russians began to collapse in March 1917, when a combination of incompetence and war-weariness toppled the czarist regime. The process continued over the next twelve months, bringing the Bolsheviks to power in November 1917, and ending when they accepted extremely harsh terms at Brest-Litovsk on March 3, 1918—so harsh that the Germans had to leave forty-three divisions in the East to enforce it. Nevertheless, Germany was at last able to focus its combat efforts on the Western Front.

Erich von Ludendorff, now Quartermaster General and basically

Das Pariskanone

A roar, followed immediately by an explosion in the Place de la République, shattered the early-spring Paris morning. It was 7:20 on March 23, 1918—the beginning of the third day of Michael. Twenty minutes later there was an identical explosion in front of the Gare de l'Est Metro entrance. When the dust cleared, eight lay dead and thirteen others had been seriously injured. Worried eyes combed the skies for the German bombers that had staged nightly raids over the city. There was nothing—no aerial silhouettes, not even the drone of an aircraft engine. Yet the explosions continued, twenty-five that first day, killing sixteen people and wounding twenty-nine more.

The Paris Defense Service immediately began gathering metal shards and subjecting them to microscopic analysis. In less than three hours they reached a startling conclusion: Paris was being shelled. Yet the sixty-seven-mile distance to the nearest point along the German lines, combined with an additional ten-mile safety margin from counterbattery fire, meant the offending cannon had to have a range approaching eighty miles! Further study suggested a projectile approximately eight inches in diameter, with abnormally thick sides and rifling grooves cut into the steel body of the shell—all features corresponding to the gargantuan chamber pressures necessary to hurl a round this far. Careful scrutiny of local maps led the French experts to pinpoint a probable firing point in a stand of woods nearly seventy-five miles from Paris. By evening, orders went out to detach a battery of 305mm rail-mounted guns within range of the designated spot. The very next day the *Pariskanoniers* found themselves subject to a disconcerting if not lethal barrage. Amazingly, the French ballistics experts had been right on the mark!

The object of their study was mind-bending in its technical details. Often confused with Big Bertha, the *Pariskanone* was a far different and more specialized beast. Developed in two years at the Kruppwerke by Dr. Otto von Eberhard, the gun (or guns, since there were three in all, with seven barrels) was essentially a fifteen-inch naval rifle, tubed down to 8.26 inches, vastly reinforced, and lengthened to the height of a ten-story building—so long it was subject to drooping and had to be braced with a cable system resembling one half of a suspension bridge.

running the war, planned a vast offensive code-named Michael. He had two advantages. One was manpower—thirty-three divisions drawn from the East swelled the German strength in the West to nearly 175 divisions. Of these, forty-four "mobile" divisions and thirty "attack" divisions received special tactical training designed to overcome fixed positions—the second perceived edge. Training focused on the close coordination of concentrated and nonstop artillery firepower with rapidly moving teams of "storm troopers," men heavily equipped with automatic weapons and flamethrowers, instructed to infiltrate around strong points and to strike deep. Michael constituted a last gigantic throw of the dice, before the Americans had

What emerged was a veritable infernal engine. Fueled by 432 pounds of smokeless powder, the gun developed nine million horsepower as it accelerated its 264-pound projectile to 5,500 feet per second. The shell reached a maximum height of twenty-four miles in a minute and a half, and for at least fifty miles of its range the projectile traveled in a near vacuum—so high that calculation of its trajectory had to take into account the earth's curvature and rotation. The atmospherics within the gun were just as extreme, with the dynamic pressure on the breech lock reaching approximately 4,700 times normal barometric pressure.

Considering its near-mythic parameters, it is not surprising that the *Pariskanone* has been subject to considerable misinterpretation. While estimates of the number of rounds launched ranged as high as 10,000, Paris actually was shelled only 367 times, losing fewer than 260 of its citizens. The gun was conceived by the Germans as a strategic weapon, an integral part of Michael intended to bring the French capital to its knees. Instead, Parisians learned not to congregate in groups and accepted the casualties much like we accept traffic deaths. Today the great gun is most notable for its portentous similarity to the ballistic missiles that hold us hostage. Parisians at the time knew nothing of nuclear weapons. But they did know poison gas, and had the Germans filled the *Pariskanone*'s shells accordingly it might have been a true knife in the heart. The Germans spoke of the necessity for "frightfulness," but as with the submarine, they were not "frightful" enough. They had in their hands a true weapon of mass destruction, but they hesitated to follow the logic that within a lifetime would make the term "total war" an oxymoron.

Krupp Pariskanone.

accumulated in sufficient numbers to become decisive. Desperate times called for desperate measures.

The giant hammer blows of Michael produced results only at a terrible cost. The German assault of 1.4 million soldiers slammed into the British lines, driving them back almost forty miles in two days, causing 200,000 casualties and capturing 90,000. The English bent to the point of desperation, but they did not break. The trench lines stabilized, and on March 29 the British staged a counterattack. By April 5 this phase was over; Ludendorff had gained a substantial wedge of territory—now open to flank attack—but lost a quarter-million soldiers.

He next turned on the French. In May and early June the Germans cleaved a great salient and pushed it all the way to the Marne, only fifty-six miles from Paris. The French, like the British, flexed but held. On June 9 their intricate new defense-in-depth basically foiled German infiltration tactics. Two days later General Foch, now Allied Supreme Commander, staged a bloody counterattack at Château-Thierry and Belleau Wood supported by tanks and the U.S. 2nd and 3rd divisions. American Expeditionary Force Commander John J. Pershing took some delight in comparing his "superior" troops to the "tired Europeans." On July 19 Ludendorff was forced to order a general retreat. His dice had come up snake eyes.

Pershing had been exactly right. The effectiveness of German storm trooper tactics had been overrated. They helped but did not solve the problem of weapons-induced gridlock. Instead, the gains they achieved were largely a function of the general exhaustion of the combatants. British and French troops would not have retreated so far or so fast in 1915. Meanwhile, German soldiers betrayed similar signs of weariness. At key points (Albert and Moreuil, for example) they delayed the attack through widespread looting. There were instances when troops refused to move forward, and 20 percent of the reserves sent to the front failed to report. The German army had been bled white, its weapons worn out, its remaining soldiers undernourished. Now it faced a new nightmare: enthusiastic Americans.

The Americans arrived with uniforms, small arms, and not much else. Like Tennessee Williams's Blanche DuBois, they relied upon the kindness of strangers, in this case for planes and heavy weapons. This was misleading. America's huge and progressive industrial base was slowly turning itself into a war machine. Had the conflict lasted much longer it would have flooded Europe with high-quality arms. Sam Colt and the denizens of the Connecticut River valley were simply a prelude. Americans had a genius for making weapons. Their very innocence and optimism, along with their ingenuity, let them go where others might have hesitated, as the Great

War's sequel and the Cold War were destined to reveal. But even in 1918, as they struggled to duplicate the French 75's secret recoil mechanism, they were working on concepts of startling originality.

Colonel Henry "Hap" Arnold, a founding father and eventual head of what would become the U.S. Air Force, was looking for a place to hide. It was October 1918, and as assistant director of the Army's Aeronautical Division he was showing off a highly classified new combat aircraft to a grandstandful of officials assembled at a secluded airfield near Dayton, Ohio. Arnold watched as the little biplane soared into the air and then dove toward the crowd, who dove for cover. Fortunately, the plane crash-landed several hundred feet away. It had no pilot.

It was, in fact, the world's first guided (in this case misguided) missile, the direct ancestor of today's Tomahawk and Harpoon cruise missiles. The Bug, as it came to be called, was the pet project of General Motors's "Boss" Kettering, the man who revolutionized the automotive industry with the self-starter and fast-drying exterior paint in virtually any color. He was joined in his work by a design team that read like a Who's Who of innovation—Orville Wright, C. H. Wills (Henry Ford's engine designer), Nobel Prize–winning physicist Robert Millikan, and Elmer Sperry, whose pioneering work with gyroscopes was the basis of automated flight control.

Known officially as the Liberty Eagle, the Bug was basically a flying bomb—wings, tail, and fuselage, a four-cylinder two-cycle engine, and 180 pounds of high explosives. Since it didn't need to land, the Bug had no wheels and was launched from a sort of proto–shopping cart. It was designed to fly straight at its target, climbing to a predetermined altitude, governed by a highly sensitive barometer, and kept level by one of Sperry's gyros, explains aeronautical historian Dik Daso. Originally, only the wing's dihedral maintained directional stability, but after the ill-fated test lateral controls were added with good results. Thus guided, the Bug flew until a mechanical counter determined that sufficient engine revolutions had been achieved to reach the target, and its fuel was cut off.

The war was about to end. Fifty Bugs were hastily assembled and Hap Arnold was dispatched to Europe to convince John Pershing to use them before the fighting ceased. But Arnold was delayed and then got sick, and the Bug never flew against the Germans. Had they known about the project, many in the Air Corps would have considered this a blessing. Military aviation was barely out of its infancy and already the technologists had figured out a way to evict the pilots. The struggle continues to this day.

On the field of battle, John Pershing thought he had a tactical answer to gridlock. He called it open warfare, "irregular . . . formations, comparatively little regulation of space and time . . . and the greatest possible use of the infantry's own firepower to enable it to get forward."[14] All he really had was a slight update to the doctrine taught to him at West Point by Dennis Hart Mahan, Alfred Thayer's father, in the mid-1880s. Later Hunter Liggett, the brightest and also the fattest of Pershing's generals, admitted that no one had really "thought through" open warfare. In fact it amounted to frontal assault.

What Pershing really had, and what he won with, was fresh bodies. The concept of reserves, originally a tactical innovation, had now risen to the level of grand strategy. At Belleau Wood, Soissons, and later battles right up through the first stage of the Argonne offensive, American infantry attacked in what amounted to human waves. "The best way to take machine guns is to go and take 'em! Press forward!" advised General Charles Summerall.[15] The grizzled Germans mowed them down in great numbers, but the Americans kept coming, a desperate problem for an exhausted army on the edge of defeat.

Pershing was as stubborn and determined as his soldiers. Resisting enormous pressure from the Allies to parcel out his forces, he adamantly refused to let them fight in anything less than a division—big juggernauts of 28,000 men, twice the size of German and Allied equivalents. Tanks had their impact, the British and French rallied for one final push; but the relentlessness of the Americans must have been most discouraging to the dispirited Germans. In war timing is everything, and these fresh doughboys had arrived at just the right moment to pry the Germans out of their trenches.

On September 26 Ludendorff's staff found him on the floor foaming at the mouth; the next day he recommended an armistice. It was arranged on November 8, but the Allies insisted on a more dramatic date, even though it prolonged the killing. So it ended the way it was fought, accumulating useless casualties. On the eleventh day of the eleventh month of 1918 the shooting finally stopped.

Chapter 14

GRUDGE MATCH

All the participants in the Great War, including the Central Powers, had been moving toward an expanded interpretation of the rights and possibilities of the individual before hostilities broke out. Now they had to face the consequences of a war gone out of control, and then pursued with a single-minded obliviousness to the welfare of not just the combatants but large portions of the citizenry.

There were two reactions among the major powers. The first approach, most apparent among the liberal democracies, was a rejection of war and a growing belief that it was one of those institutions that was no longer appropriate or really useful in the new human condition. Unfortunately, the pacifism and efforts at arms control that emerged during the 1920s and 1930s were unrealistic, even dangerously so. Yet they were not necessarily fecklessly utopian. For example, although the 1925 Geneva Protocol banning poison gas was hedged by reservations, and the 1928 Treaty of Paris outlawing war was enforced by fiat and nothing else, they nonetheless registered a growing awareness, soon to be ratified by nuclear weapons, that certain arms were simply too powerful to use effectively.

Yet the immediate future was held hostage to an entirely different interpretation. Large segments of the developed sphere (essentially those who fared worst in the Great War) turned back to authoritarian institutions and values reflective of agricultural tyranny and applied them in an industrial setting, giving birth to modern totalitarianism. It was not entirely illogical. Given the apparent evolution of machine technology in the direction of economies of scale, centralized management, and the regimentation of huge labor pools, it made a certain sense to conclude that the pyramid-shaped hierarchies, not unlike the ones that once ruled the world's agricultural river valleys, could be superimposed on an environment dominated by machines.

The new tyrannies promised compatibility with military institutions in a time of great international instability and a widespread perception of scores to be settled. Rather than rejecting war, the totalitarian disposition reached the opposite conclusion. Aggressive warfare became the key means by which the new despotisms sought to fulfill their agendas. Agricultural tyranny's hunger for territory found an analogue in the presumed necessity to capture and physically control the resources necessary to run an industrial complex. There was also the same drive to shape demographics through force of arms, to move and sometimes eliminate whole populations. Implicit in all of this was the willingness to squander large armies in the pursuit of these objectives, a prospect that military technology was eminently capable of accomplishing.

Regardless of the price it had so recently exacted, the prospect of improved weapons continued to be embraced, and not just by despotism. Although the liberal democracies pursued peace not war, foreboding provoked by the totalitarians led the free states to seek better arms, if not single-mindedly, then at least consistently. Everyone was constrained, however—the Germans by the Versailles Treaty, the Russians by revolution and resulting social dislocation, the Americans, British, and French by the Great Depression and antiwar sentiment, and the Japanese by the small scale of their economy. Budgets were modest and not necessarily well balanced. But it was broadly appreciated that military operations were now truly three-dimensional, and that the air, and to a lesser extent, submarine environments offered important opportunities for advancement. So programs were formed and resources found. Research and development is relatively cheap (seldom more than 20 percent of total expenditures) and characteristically takes a long time. This is also true of the process of figuring out how best to employ new weapons, known in the military as doctrinal development. Both were a good fit with the 1920s and early 1930s—long enough to digest the military problems posed by the Great War and to address technological challenges methodically. So, when rearmament accelerated after 1937 there was plenty to mass-produce.

Not everything worked. But, with the exception of the atomic bomb, World War II was decided by weapons developed during the interwar period. They proved extraordinarily lethal, particularly to civilians, and they were frequently used in a way calculated to maximize casualties. Together they represented a fulfillment of chemical energy's military possibilities, a consummation of Leonardo's visionary notebooks, which brought human warfare to the outer edges of what was tolerable. Ironically, the major players from the Great War directed their innovative strategies not necessarily at the areas most responsible for victory or defeat, but those that had the greatest psychological impact. The two were not necessarily incompatible; but it was also a sign that the central venue of warfare was shifting from the battlefield to the realm of the imagination.

The British response was particularly illustrative. English land forces had been savaged by machine guns and artillery inflicting over a million casualties, while submarines had posed a real threat of driving the country out of the war; yet it was the small-scale and largely ineffective German strategic bombing campaign that haunted the nation's nightmares. "The bombers will always get through," asserted conservative politician Stanley Baldwin, epitomizing his countrymen's deepest fears. The submarine problem was swept aside by a proto-sonar known as ASDIC kept so secret that only a small number of the Royal Navy's officers knew it was normally incapable of detecting submerged assailants. Similarly, the British had invented both the tank and the aircraft carrier during the Great War, but entered its sequel with obsolete equipment and doctrine in both. Nevertheless, when the time came the Royal Air Force was ready to begin bombing others, and most importantly to stop others from bombing them.

On the face of it, British strategic bombing looked weak. The Germans not only entered World War II with the world's largest fleet of bombers, but had fielded the Knickebein blind-bombing system, employing crossed radio direction signals to enable them to attack at night with some degree of accuracy. But the real strengths of the English program were less obvious—a near fanatical commitment to strategic bombing and a broad developmental pipeline that would generate aircraft until satisfactory models were found. The Royal Air Force had been granted independence from the other services due to public anger over the German raids on London, and its leader during much of the 1920s, Sir Hugh Trenchard, along with the acolytes who followed him, Charles Portal and Arthur "Bomber" Harris, never forgot. They pursued the strategic bombing mission with the zeal of true believers, albeit without as much funding as they wished.

A considerable portion of the bomber force entering the war was composed of single-engine, light-bomb-load Vickers Wellesleys and Fairey Battles, planes that combat proved to be little more than flying coffins.

*British Vickers
Wellington bomber.*

Next in line, however, were two full generations of two- and then four-engined bombers with much greater strategic reach and ordnance capacity. Among the two-engine platforms—the Hampden, the Whitley, and the Vickers Wellington—only the last proved a true winner, with an airframe whose geodetic construction allowed it to absorb tremendous punishment, leading to a production run of over 11,000 beginning in 1937 and spanning the entire war. The results were similar with the follow-on generation of four-engined bombers—the Short Stirling proved to be something of a sitting duck, the Handley Page Halifax a useful workhorse, but the Avro Lancaster was a thoroughbred. The key to success lay in a remarkable power plant. Both the Halifax and Lancaster had been planned originally as two-engine airframes, slated to use the Rolls-Royce Vulture, still in development. When it proved underpowered, the planes were redesigned to accept four Rolls-Royce Merlins, the most versatile and famous aircraft engine in the war, with 150,000 built in all. The performance of each plane improved dramatically, especially the Lancaster, which flew at a top speed of nearly 300 miles per hour, had a range of 1,700 miles, and carried up to 18,000 pounds of bombs—nine and eighteen times the capacity of the Vickers Wellesley and Fairey Battle respectively. This was a true instrument of mass destruction destined to wreak near biblical havoc on Germany's cities. The Luftwaffe had nothing like it, and would not get one for lack of sufficient engine technology. Only the Americans had an equivalent, the Boeing B-17. For all the success of mainstream British bomber acquisition, there was an alternative.

Just as Vitorio Cuniberti had helped give birth to the dreadnought with a vision of an invincible ship, air theorists of the interwar period, inspired by another Italian dreamer, Guilio Douhet, conceived of strategic bombing in similarly idealized terms. Defense against bombers was pointless they thought, largely because they would fly so fast and high and come in such great numbers. By implication the bombers did not require protection in

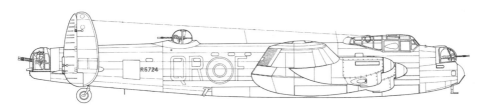

*British Lancaster bomber—a true instrument
of mass destruction.*

the form of fighter escorts or self-defense weapons. This point of view plainly slowed development of true long-range fighters, but military men instinctively shied away from bombers incapable of defending themselves. Thus the Halifax went to war carrying nine machine guns, the Lancaster had ten, and the American B-17 was festooned with thirteen.

Geoffrey de Havilland was a dissenter. The independent aircraft designer watched the 1938 Munich crisis with growing dismay, realizing that Britain would need good airplanes that could be produced quickly. He decided to take matters into his own hands. Without government support, he developed the most versatile military aircraft in the war and also the most iconoclastic. He named it, appropriately, Mosquito.

Spurred by memories of the rickety canvas-covered crates of the Great War, aviators assumed that duraluminum construction was a giant and irreversible leap forward in aeronautical engineering. Not de Havilland: he built his Mosquito out of wood, plywood, and balsa sandwiches held together with a new generation of adhesives that imparted extraordinary strength-to-weight ratios. The result was an extremely light airframe, yet one strong enough to accommodate two Rolls-Royce Merlins. The initial flights were everything he expected and more. Not only did the plane have the maneuverability of a fighter, but its performance was staggering: nearly 400 miles per hour top speed, a range of 2,000 miles, and a ceiling of 35,000 feet, fully 15,000 feet higher than the other British bombers. Its ordnance load was only one ton, but cheapness and ease of production promised a veritable swarm of Mosquitoes—four or five could be built for the cost of one Lancaster. One thing the Mosquito did not have was self-defense guns. As a bomber the plane depended entirely on its speed, ceiling, and small size to foil interception.

Not surprisingly the Air Ministry rejected de Havilland's initial proposal. When the prototype Mosquito's performance could not be ignored, they sought to load it down with machine guns. Bomber Harris, in

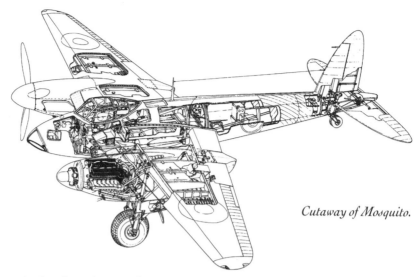

Cutaway of Mosquito.

particular, hated it. Only RAF director of production Sir Wilfred Freeman, a future champion of the Whittle jet engine, saved the plane. It soon earned a legendary reputation as a fighter, fighter-bomber, and reconnaissance aircraft. But, as historian Chris Van Aller points out, it was never really let loose as a strategic bomber.

If planes could laugh, the Mosquito had the last one. Its production run stretched all the way up to 1950, long enough to watch defensive armament beginning to disappear from the new generation of Cold War jet-powered strategic bombers. Meanwhile, the Mosquito's wood construction gave it a tiny radar cross section—a stealth bomber a half-century before its time.

In 1938, radar cross sections would have meant nothing to anybody in England, except those directly involved in building an air defense system that saved the country from horrible devastation. Fortunately, there were some in the RAF who believed the bombers wouldn't always get through, and were willing to stake everything on it. One was Air Marshal Hugh Dowding, chief of the "few" to whom so much was owed by so many.

Upon taking command of the RAF's research and development, Dowding issued specifications for the first British combat monoplanes, guidance that would lead to the Hawker Hurricane and the Supermarine Spitfire, two of the most successful fighter aircraft in history. Again the key was the trusty 1,030 horsepower generated by the liquid-cooled Rolls-Royce V-12 Merlin. The chunky 320-mile-per-hour Hurricane came first in 1935. Armed with eight machine guns and later four 20mm cannon, it was the designated stiletto destined to carve to shreds Nazi bomber

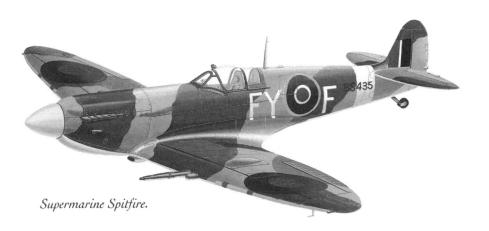

Supermarine Spitfire.

formations over England. The corkscrew followed two years later, the sleek and beautiful 350-mile-per-hour Spitfire, whose slim fuselage and artfully proportioned wings allowed it to turn inside of the skid prone Messerschmitt Bf 109. Together they formed a one-two combination that eventually cleared the skies over Britain. But they never would have found enough Germans to shoot down had not Dowding's vision of radar extended beyond that of the naked eye.

When the Nazis came, the British were not only looking, they were listening. Since the introduction of telegraphy, the German military had been fascinated with the possibilities of instantaneous communications. They were quick to adopt wireless transmission, and rather than radio silence they relied on encryption to protect the security of a characteristi-cally vast flow of message traffic. The loss of the *Magdeburg*'s codebooks led to only minor damage in the Great War, but the British insured that the penalties were much greater in its sequel.

During the 1920s a German firm developed the Enigma machine, an apparently unbreakable means of secure communications. Linked to a radio transmitter, Enigma employed a series of rotors and connectors to turn a typed message into a stream of encrypted gibberish, which could then be sent to an Enigma receiver that, using the same settings, trans-formed it back into plain text. Enigma was portable and apparently tamperproof, and the German army, navy, and eventually the Luftwaffe seized upon it as the perfect complement to their very chatty communica-tions doctrine.

The portability should have worried them. In the early 1930s Polish

Radar

Ironically, it was the Germans who laid the groundwork. As far back as 1904 Christian Hulsmeyer claimed his "telemobiloscope" could transmit radio waves and receive their reflections off a moving object. During and after World War I, however, official Germany focused on radio not radar.

But at lower levels, locational beam research continued. In 1933 Rudolf Künhold suggested that centimetric wavelengths might be able to pinpoint surface ships and possibly aircraft. Two years later Wilhelm Runge was able to measure signals reflected from planes at fifty kilometers even at night and through fog. When he reported his achievement to Luftwaffe procurement chief and former fighter ace Ernst Udet, the latter remarked: "If you introduce that thing you'll take all the fun out of flying!"[1] The Germans planned to be on the offensive, and the most obvious implications for radar was air defense—so they made little of it.

The British wanted protection and were far more flexible. "The Germans would not have been surprised to hear our radar pulses," Winston Churchill later wrote. "What would have surprised them, however, was the extent to which we had turned our discoveries to practical effect and woven all into our general air defense system."[2] Dowding's role was critical. In 1935 he helped set up a small committee under Henry Tizard to explore scientific concepts that might be turned into weapons. When the committee asked Robert Watson Watt to prepare a paper on death rays, he had little good to say about them but did report an incident over Geoffrey de Havilland's aerodrome that seemed to indicate that passing planes were reflecting beams in a way that could be measured. This led to a groundbreaking paper in early 1935, "Detection and Location of Aircraft by Radio Methods."

Tizard's committee moved swiftly, conducting experiments on pulsed signals that could be projected on cathode ray screens calibrated for distance, and developing a plan for a double line of transmitters built over a long front. But it was Dowding who provided the £10,000 out of his limited budget so the work could continue and the necessary equipment developed. By the time of the Munich crisis, five long-range radar stations were in place. A month later Dowding's Fighter Command opened their filter room at Bentley Priory, where all incoming data was received and analyzed so that aircraft assignments could be made rapidly and economically. At war's outbreak there were eighteen stations, all linked by landlines to the center, plus thirty low-level installations. Structurally, it was the same system that had been in place in 1918; but now it operated at the speed of light and looked into the sky for hundreds of miles.

intelligence purloined an Enigma machine without the Germans realizing it, and with the help of several mathematicians were soon reading their traffic. In August 1939, shortly before their country was overrun by the

Wehrmacht, the Poles presented the English with an Enigma machine and a full analysis of the German system.

The British task was daunting, nonetheless. Because Enigma procedures called for rotor settings to be changed every day, anyone wanting clear text had to obtain lists of those adjustments. This was key to the Germans' misplaced confidence. They anticipated that their enemies might obtain an Enigma machine and even lists of the settings for a specific time period; but they continued to believe the decoding had to stop once the lists were updated.

This equation grievously underestimated the determination of the British—the brilliance of their cryptographers sequestered at the Bletchley Park communications analysis center, their future success at obtaining enciphering tables, their ability to mask their knowledge, and the very magnitude of their effort. The Germans were also betrayed by the laxness of their own procedures and the carelessness of operators, especially the Luftwaffe's operators, who persisted in using cribs and broadcasting the same message headings on successive days at the same time. They would pay dearly and not even know it.

On the ground the Germans were focused on far less subtle means of war making; but like the British their future plans were deeply influenced by impressions made by a weapon that had only an ancillary effect on the decision of 1918. Tanks plainly did not chase the Germans out of France and out of the Great War, but they certainly caught their attention. Germany remained a land power and mobile armor promised a means of taking the offensive and sustaining it—not just through the killing zone but deep into enemy territory. Though sorely constrained at home by the Versailles Treaty, the contracted and reconstituted Reichswehr found ways of learning from others. Consequently, when the British carried out a series of innovative field experiments with armor during the late 1920s and then ignored them, the Germans were paying close attention and carefully evaluated the results.

But no state was more helpful than the new Soviet Union that had sprung up out of the wreckage of Imperial Russia. By 1922 the once and future enemies had secretly joined forces to explore new ways of killing each other's soldiers. By 1928 the joint tank school at Kazan, the chemical warfare school at Volsk, and the aviation center at Lipetsk were all running with an efficiency that left future German War Minister Blomberg extremely impressed. Until Adolf Hitler ended the relationship in 1933, great progress was made in training Soviet and German officers in integrated battlefield concepts combining mobile artillery, mechanized infantry,

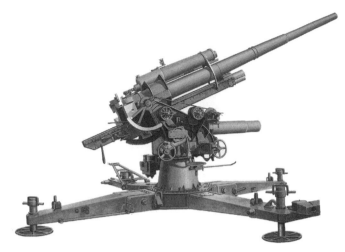

German flak 88mm antiaircraft gun — one of the best and most versatile weapons in World War II, also used with great success as a tank and antitank gun.

and tanks—the essence of panzer warfare. As the Germans left Russia, the conceptual framework for the coming land war was set, only the tools remaining to be gathered.

The German military had no intention of fully complying with the weapons provisions of the Versailles Treaty. A secret Krupp design bureau for tanks and other ground force weapons was set up in Berlin, while its artillery, antiaircraft guns, and light tanks were produced under license by Bofors in Sweden. The international character of Europe's defense industries served the Reichswehr's covert strategy particularly well in the sphere of advanced weapons, all forbidden to Germany under the treaty. Conceived and designed surreptitiously in Germany, they could be built and tested elsewhere. Thus U-boats were constructed in Holland, torpedoes were developed and produced in Spain, and virtually the entire aircraft factory of Tony Fokker was secretly spirited out of Germany. By 1933 an arsenal of prototypes was ready.

Hitler brought them into being. As a veteran of four years in the trenches he knew what he wanted. While he would have little feel for truly advanced arms such as radar and the possibilities of nuclear energy, he had a real grasp of what was needed to avoid the kind of stalemate that had ruined the German army along the Western Front. Although he played little role in the actual development of mobile armored concepts, he reacted immediately upon seeing his first demonstration of Heinz Guderian's panzers. "That's what I need! That's what I want to have!"

Hitler believed from the beginning that if he did not win quickly, his prospective enemies would eventually outproduce him. The general quality of the weapons was good and in some cases—such as the remarkably versatile

88mm antiaircraft gun—excellent. But until near the end of World War II, little else would follow. By this time the Allies, especially the Americans, were generating a profusion of new models, and producing them in staggering quantities. Germany's prewar land arms program was primarily bent on forging the ten panzer divisions available in 1940, the steel tip of what remained a mass-conscript infantry force not fundamentally different from the one that marched into France in 1914. Similarly antiquated was the supply system, its prime mover between rail heads and the front lines being the horse.

The air arm followed the same pattern. German air doctrine was excellent. The Luftwaffe brilliantly exploited the possibilities of close air support, becoming a sort of vertical artillery for the army primarily by exploiting the inherent accuracy of dive-bombing. Largely as a result of experience in the Spanish Civil War, German air units also learned to use Ju 52s to transport bombs, fuel, and spare parts forward, leapfrogging from airfield to airfield to gain a true measure of strategic mobility. Yet aerial resupply was (and remains to this day) a limited, even precious resource inevitably exceeded by the mass operations of World War II, forcing the Luftwaffe, figuratively and even literally, to "get a horse."

Its air force relied on four basic types: the twin-engined He 111 level bomber with an ordnance capacity of just four tons; the Bf 109 fighter, fast and relatively nimble but prone to crash on landing; the Ju 52 transport, and the Ju 87 Stuka dive-bomber, a sinister-looking aircraft whose close coordination with tanks came to epitomize blitzkrieg tactics. By casting aside several superior prototypes and concentrating on these easy-to-produce airframes, Luftwaffe head Hermann Göring and the German aircraft industry managed to build up a combat-ready force of around 2,000 planes in 1940. In reality, the production base was not very efficient, and when

German Bf 109 fighter.

Germany's aircraft were exceeded in performance, it had trouble generating and manufacturing better models.

The Soviets, meanwhile, set their sights on the long range. If nothing else they believed in planning and redundancy, and cast these concepts into steel, concrete, and smokestacks by the thousands. At great sacrifice they constructed an enormous and robust military-industrial complex during the first, second, and third Five-Year Plans, which brought forth, just when most needed, a flood of high-quality land arms.

Ironically, the chief overseer, Joseph Stalin, just as relentlessly loboto-mized the Red Army at nearly the same critical juncture. In 1937 he began a blood purge that all but destroyed the Soviet officer corps, zeroing in on the cadre clustered around M. Tuchachevsky, the leading advocate of mechanized warfare and what was termed "deep battle"—rapier thrusts into the enemy's rear designed as much to disrupt as destroy. Armored formations were dissolved, and the army reconfigured to fight linear battles planned and led by commanders largely without experience and wary of concepts that had gotten so many of their brothers shot. So they fell prey to the Wehrmacht's version of deep battle. Millions were killed and their country almost wrecked, but miraculously they recovered, saved as much as anything by good weapons and the ability to produce them. One in particular stood out.

On July 8, 1941, the seventeenth day of Operation Barbarossa, the German invasion of Russia, triumphant panzer units scattered in the corn-fields around Senna were confronted with an armor-plated apparition of their eventual doom. Out of the stalks rumbled a squat tank with severely sloping sides, a massive turret and gun, and extremely wide treads for trac-tion, which proceeded to cut a nine-mile swath of destruction through the Germans before they managed to destroy it. The vehicle served notice that the supposedly primitive Russians had managed to build what German Field Marshal Paul von Kleist called "the finest tank in the world."

The origins of the T-34 stretched back to the early 1930s, when Red Army planners supported a prototype tank designed by the American J. Walter Christie, its innovative suspension siring a series of Soviet BTs, or "Fast Tanks." Industrial legend has it that Christie's designs embodied most of the advances that later made the T-34 such a superior machine, but Soviet engineers vastly improved everything.

The real progenitor of the T-34 was the revolutionary A-20 experimen-tal tank put together in 1938 by the talented team at the Kharkov Design Bureau. Critically, the A-20's engine bay held the first truly practical tank diesel, the aluminum alloy, twelve-cylinder V-2, the basis for a line of tank

Soviet T-34 tank — "Hammer of the Proletariat."

engines extending right up to the late Cold War's T-72. While the V-2's power and economy were certainly useful, the fact that diesel fuel was far less flammable than gasoline constituted a key advantage in armored combat with the Germans, who for all their experience with diesels never managed to field a tank during World War II with anything but a gasoline engine.

There was great pressure to produce the A-20 as it was; but many Soviet soldiers returning from the Spanish Civil War testified that its gun and armor were not heavy enough. With Stalin's concurrence, the engineers proceeded with a major redesign that included tracks made particularly wide to deal with Russia's endemic mud and dust, a high-velocity 76mm gun, and hull armor fully forty-five millimeters thick. Manufacturing such a tank was no simple matter. At thirty tons it had the thickest armor any Soviet tank had ever carried, and coordination with component producers—particularly the makers of the V-2 diesel—presented complications. Nonetheless, by June 1940 the first mass-produced T-34 rolled off the assembly lines. It was destined to be replicated 50,000 times—a tidal wave of tanks that overwhelmed the Wehrmacht.

Germany had nothing to match it. Their doctrinal emphasis on speed caused them to concentrate on far lighter machines, and as late as 1939 only 10 percent of the army's tanks exceeded ten tons. Even the larger Panzer IIIs (30mm armor and 37mm gun) and Panzer IVs (seventeen tons, 20mm armor, and a 75mm low-velocity gun) were completely outclassed by the T-34.

In November a group of German engineers visited the front to examine captured examples of the enemy wonder tank, and were begged by officers

German Tiger tank; note 88mm gun.

to copy and quickly mass-produce a Teutonic version. They did no such thing; their technological vanity precluded it. Instead, they generated a series of complex designs that eventually bore fruit in the lumbering fifty-six-ton heavyweight Tiger, the temperamental middleweight Panther, and the monster tank destroyer Ferdinand—namesake of its creator, Dr. Porsche. A combined total of barely 7,000 were produced. Their fighting qualities were excellent—both Tiger and Ferdinand mounted the ubiquitous and deadly 88—but symbolically, the notion of fighting in Russia with Porsches spoke for itself.

Among totalitarians, the Soviets had the right idea. Communism grew out of an interpretation of the industrial age as it appeared in the 1920s and 1930s. It was about heavy industry, assembly lines, and above all mass production. The endless ranks of T-34s that clanked forward to meet the Wehrmacht were truly representative of the environment, and in a war of factories, this is why they won.

Far to the west the French nation remained traumatized and thought only of the defensive. Twice within a half century huge German infantry armies had marched into France and ground ahead toward Paris. The image of the lumbering historical Hun hypnotized the French and caused them to build, as had the Imperial Chinese, a great wall. And like the Middle Kingdom in the face of an equally implacable threat, the French could look to a tradition of massive deterrent fortifications epitomized by the strategic necklace Sébastien de Vauban fashioned for Louis XIV. Psychologically at least the Enlightenment master builder provided ample inspiration for its twentieth-century counterpart, the Maginot Line.

A massive undertaking, it grew to include literally hundreds of separate but supporting fortifications concentrated on the northeast frontier but

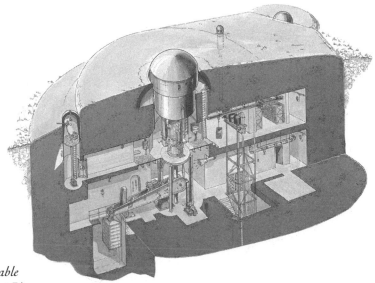

*Detail of retractable
turret in Maginot Line.*

extending all the way down along the border with Italy. Inspired by wounded war hero and politician André Maginot ("I'm like my leg; I don't bend"), the bill by 1935 was already seven billion francs and climbing.

Technically it was impressive. Based largely on the hero of Verdun Marshal Petain's belief that only subterranean defenses could survive large-caliber artillery, the major forts consisted of cupolas, turrets, and block-houses all protruding only slightly above ground and linked to each other and a central underground fortress by tunnels, many of them featuring electric trains referred to as the Metro. Much of the Maginot Line's firepower came from its 152 electrically powered turrets. Weighing up to sixty tons apiece, they were both retractable and rotated—putting to rest the myth that the big guns they held fired in only one direction. Beneath the main trunks of the fortresses the French constructed small underground cities of barracks, ventilation and power generation machinery, ammunition maga-zines, along with fuel and water storage tanks sufficient to keep the major fortifications going for months. To provide continuity, the French filled the gaps with a maze of smaller fortresses, casemates, bunkers, and observation posts placed so as to channel an attack toward the major strong points.

Seen purely as a weapon, the Maginot Line was not such a bad idea. The element of mobility introduced by armored vehicles and tactical air support was not yet unstoppable. When lines were properly constructed and defended, such as in Italy or along the Rhine in 1944, they proved exceed-ingly difficult to punch through, even for heavily mechanized forces. Operationally, however, a cogent criticism of the Maginot Line was that there was not enough of it.

The French had left uncovered what Clausewitz called "the pit of the French stomach," the ancient invasion routes across the border with Belgium. In part, this was a military issue: the high water table in parts of Belgium precluded underground fortification, and the territory of the Ardennes forest was fatally presumed rugged enough to provide France time to mass its forces there in case of an attack. But there were deeper reasons. After October 1936, when Leopold III revoked the Franco-Belgium Treaty, France was free to extend the Maginot Line to the sea along its own territory. But Frenchmen proved unwilling to bear the cost and chose instead to live with their vulnerability. Complete or incomplete though, the Maginot Line betrayed France's lack of political will. As a strategic signaling mechanism it invited attack and promised no retaliation. They would regret it.

As putative opponents in what became World War II's Pacific annex, America and Japan had been eying each other pugnaciously since the conclusion of the Russo-Japanese War in 1905. So palpable was the collision course by 1925 that Hector C. Bywater—then the world's leading naval journalist—was able to write a futurist novel, *The Great Pacific War,* that outlined with remarkable accuracy the subsequent course of events in the Pacific Theater—the Japanese surprise attack on Pearl Harbor, their subsequent assaults on Guam and the Philippines, the American response of amphibious island-hopping, the gradual weakening of the Asian power. His sole mistake was to predict a decisive role for battleships, particularly at the war-ending "Battle of Yap." Bywater was considered a prophet by both the American and Japanese navies, who continued to believe in battleships.

No sooner had the Great War ended than the victorious naval powers immersed themselves in a great and nonsensical race to build battleships; a total of thirty-six being under construction by the spring of 1921. No contestant was apparently more enthusiastic than the United States, which had embarked on a massive expansion of its battle line despite its complete uselessness during the world war. Beneath the surface, however, the impact of more progressive forces was felt. Most notably, Admiral Sims had turned on the great ships. As president of the Naval War College, he encouraged a series of war games which indicated that the aircraft carrier was destined to become the capital ship of the future. It was an excellent choice. Unlike the lethal submarine, which contradicted virtually every premise of naval orthodoxy, the carrier, though still unproven in combat, was inherently a large surface ship that promised to bring the romance of flying to the fleet.

Obligingly, General Billy Mitchell, former head of U.S. air operations in Europe and easily the loudest American advocate of airpower during the interwar period, managed to sink the surrendered German battleship *Ostfriesland* with a squadron of his bombers during the summer of 1921. The widely reported spectacle of a waspish flight of planes sending a dreadnought to watery oblivion was not missed by the new secretary of state, Charles Evans Hughes, who shortly after placed a radical agenda of disarmament before the Washington Naval Conference. The resulting treaty suspended battleship construction for a decade; set a fixed 5:5:3 dreadnought tonnage ratio among the Americans, British, and Japanese, and encouraged aircraft carrier construction with very liberal allotments for all three powers. (So liberal that the United States and Japan were both allowed to convert two nearly completed battle cruisers into flattops, giving them each a pair of veritable supercarriers at the dawn of naval aviation.) At the other end of the scale, the extremely unpopular submarine was the target of a serious attempt at prohibition, but it came to nothing, with no agreement reached even on its limitation.

Their courses bounded by diplomats, the American and Japanese militaries were now empowered to chart new ways to equip themselves for a real war. While the mainstream in both navies remained focused on gunships, small elements gained sufficient initiative to explore the cutting edge. One was the imaginative band of American naval aviators, who, under the leadership of Rear Admiral William Moffett, managed to consistently improve carrier aviation until they had fashioned a formidable weapon. Working through the navy's omnibus Bureau of Aeronautics, aviators cultivated close ties with both Congress and the aircraft industry to obtain the resources and equipment they needed.

As a foundation they fostered the development of radial engines (successors to the rotary with one and later two banks of fixed cylinders arranged in a circular configuration around a rotating crank shaft), whose inherently high power-to-weight ratios and air-cooled survivability made them naturals for naval missions where engine damage often spelled loss at sea. Radials powered a series of Curtiss, Boeing, and later Grumman carrier aircraft that paced U.S. fighter and even bomber development, and set the stage for a number of lethal World War II navy fighters like the Grumman Hellcat and the Vought Corsair.

American naval aviators were busy learning how to use their new tools.

The need for mass fly-offs prompted vital work on catapults and arrester gear along with deck-park techniques, which markedly increased the number of aircraft that could be usefully based on U.S. carriers. In the air they not only pioneered the lethal technique of dive-bombing, but used Marine pilots to test it in combat against the Nicaraguan rebels of Augusto Sandino during the late 1920s. Even more significant was a series of fleet exercises beginning in 1923, which bore fruit in far-sighted concepts of how carriers might be employed. Despite being tightly bound to the battle line to provide reconnaissance and air defense, aviators longed for independent operations.

Their chance came during an exercise in 1929. In Fleet Problem IX, the first which included both *Lexington* and *Saratoga* (the two big American carriers converted under the Washington Treaty) along with 260 aircraft, the aviators made an indelible impression. At a critical juncture the *Saratoga* was cut loose from the battle line to stage a surprise raid on the Panama Canal, which culminated in sixty-nine aircraft making mock bombing runs against two key lock complexes. The notional devastation marked the birth of the carrier task force. Although the carrier's tactical reach, already measured in hundreds of miles, was an important consideration in the vastness of the Pacific, until Pearl Harbor the battleship remained, in the words of the naval mainstream, "the backbone of the fleet." Meanwhile, flattops remained mostly leashed to the battleline as defensive watchdogs. When the time came, however, teeth were sufficiently honed that the transition to independent offensive carrier operations took place quickly and effectively.

Conditions were much the same in the Japanese navy. A youthful service, it had been raised from infancy on the writings of the prophet Mahan and the luster of the Royal Navy's dreadnoughts. Consequently, until shortly before Pearl Harbor, no key figure in the naval establishment expected carriers to operate on their own.

Nonetheless, as with the Americans, the sheer size of the Pacific and corresponding requirements for reconnaissance made carrier aviation attractive. The Japanese invested heavily in this arm, putting their two proto-supercarriers, *Akagi* and *Kaga,* to good use exploring the possibilities, as well as serving as the basis for a construction program. By December 1941 they held a ten-to-eight numerical advantage over the U.S. Navy in this crucial class.

Their real forte, however, was pilots and planes. Although few in the West suspected it, the carrier pilots who took off toward Pearl Harbor were the best in the world. Hardened by a Spartan and comprehensive training program, they averaged more than 300 hours flying time before they joined the fleet—far more than their American counterparts. To complement their skills, the Japanese quietly developed a series of increasingly more capable carrier-based aircraft optimized for a wide range of missions. Particularly impressive were their torpedo planes—the long-range Type-95 with a reach of almost 1,300 miles, and the tactical Nakajima B5N2 (Kate). Yet the most memorable member of the team bore the most insignificant designation.

Zero

Mitsubishi A6M Zero — who knew?

The Imperial Navy was one tough customer. It wanted a fighter nobody in Japan thought they could build. In May 1937 naval aviation had issued requirements stipulating an aircraft with extreme maneuverability, a climb time of 3.5 minutes to 9,800 feet, a top speed of 310 miles per hour, a maximum endurance of eight hours or two hours of combat, and armament including twin machine guns and two 20mm cannon plus a small bomb capacity.

Only two companies, Nakajima and Mitsubishi, responded positively and after a few months Nakajima gave up. Jiro Horikoshi, Mitsubishi's chief designer, struggled to remain optimistic as he encountered one obstacle after another. Finally, on April 1, 1939, the product of his labors was wheeled out of its hangar and readied for its first big test. The A6M1 hardly even looked warlike. Small, not very streamlined, topped with an outsized bubble cockpit fashioned out of multiple panes, it gave the impression of a generic toy airplane. But when test pilot Katsuzo Shima strapped himself in and took off, it flew like an angel—racing to altitude, pirouetting and turning in tiny circles (as befit the image, it was far less good at diving)—all the while sipping rather than gulping aviation gas. In short order, it exceeded every specification save top speed.

Jiro Horikoshi responded by installing a more powerful engine, the NK1C Sakae radial with fourteen cylinders producing 950 horsepower. Tests on December 28, 1939, dazzled the navy and especially Pearl Harbor architect Admiral Yamamoto Isoroku, who insisted that it be pushed into mass production. When early models achieved ninety-nine confirmed kills against Chinese aircraft, the Japanese military was ecstatic. The reaction in the West, much to designer Horikoshi's amazement, was equivalent to the plane's popular nickname Reisen, or Zero. Only when bullets from its guns began tearing into Grumman Wildcats and out-of-breath Curtiss P-40s did American pilots know for sure that the threat from the little Japanese airplane added up to a great deal more than nothing.

Meanwhile, both sides geared up for island-hopping. The Japanese quickly became aware of the need for effective naval gunfire support and the rapid ship-to-shore movement of troops. They discovered that shells designed to penetrate armor plate were ill-adapted to destroying field fortifications and worked on more effective rounds and firing techniques. They also explored the possibilities of amphibious craft capable of transporting not just soldiers but sufficient ammunition and heavy weapons. By 1930 Japanese assault forces had acquired two basic landing boats—the fifty foot, hundred-troop Type A with bow ramp and a smaller thirty-troop Type B. These worked well enough but logisticians were uncomfortable with using their lightly modified transports to bring everything to the staging area. By 1935 the 8,000-ton *Shinshu-maru* was launched and accepted, an all-purpose amphibious assault carrier that presaged today's LHA and LHD classes rather than the more specialized types deployed by the Allies. *Shinshu-maru* carried landing craft in a special well deck that could be flooded, allowing them to depart via an open stern gate; it had the capacity to discharge vehicles from a parking garage directly onto a pier; it could transport and unload aircraft, and it even had two catapults. British and American observers watching her working off Shanghai realized *Shinshu-maru* marked an important development. But other priorities and the limited Japanese production base forced curtailment of the program. It was not the only time Japan's ambitious military reach exceeded its industrial grasp.

The Americans' situation was essentially the reverse. While the United States had plenty of factory space, American amphibious developments proved less imaginative than those of Japan and even the British, who pioneered the LST. They did come up with two outstanding pieces of equipment, however. Knowing they would inevitably encounter coral reefs in the Central Pacific, the Marines focused on an amphibious tractor, and pursued it with what historian Allan Millet calls "the determination of the quest for the Holy Grail." Beginning with an original design by tank-man J. Walter Christie, the path stretched from 1923 to 1940, but ultimately paid major dividends in terms of lives saved by the 18,000 American-built amphibious tractors that bulldozed their way through World War II.

The other was a matter of patriotism and free enterprise. The navy's Bureau of Construction and Repair had so much trouble during the 1930s developing anything close to an adequate landing craft that participating Marines were driven to the edge of despair. Into the breach, or rather up on the beach, raced Louisiana bayou boat builder Andrew Higgins, who offered one of his Eurekas for testing. Not only did it have a superior power to weight ratio, but its center of gravity was well aft giving it a very shallow draft forward. With the power to motor through a shore break and then move troops all the way to the beach, it was universally praised by the Marines and navy during testing in 1939. But the picture was completed when Higgins, after seeing a photograph of a Japanese Type A, added a bow ramp. His was the best and most admired landing craft in World War II, and Higgins produced them by the thousands right in New Orleans in eight plants, where he hired whites and blacks, males and females, and paid them all the same. By this time Eureka was long forgotten; everybody called it the Higgins boat.

Affection was not a term associated with submarines. Had they been any less effective during the Great War, the international naval establishment would have gleefully cast them aside. But in the huge Pacific a general absence of choke points and the boat's inherent endurance made them impossible to ignore. Instead, domestication became the preferred solution, and the concept of the "fleet submarine" soon surfaced in both the U.S. and Japanese navies. The submarine, like the carrier, was to be tethered to the battle fleet as a defender employed almost exclusively against military targets.

The American experience with the V-class fleet submarine was instructive. With a top speed of eighteen knots on the surface (the highest attainable with contemporary diesel technology) it lagged badly behind the battle line's twenty-one. Moreover, its large size insured it was slow and difficult to construct. Only six were completed by 1930, and these proved slow divers, unwieldy and unstable—characteristics exaggerated by the navy's insistence on mounting guns of up to six inches on the deck. Not surprisingly, their performance with the fleet was uniformly dismal and the concept was quietly dropped.

The Japanese were more persistent. Despite serious technical difficulties, in 1932 the Imperial Navy laid down the first of the Fleet Type 6s displacing fully 2,000 tons. In line with their intended role as scouts and skirmishers for the battle fleet, technicians managed to graft first observational aircraft and then midget attack submarines onto the commodious Type 6s. None of this worked very well, but the Type 6 remained at the heart of the Japanese navy's submarine plans, which never wavered from the intent to decimate naval assets rather than commerce.

Back in the United States during the early 1930s a group known as the Submarine Officers Conference began making recommendations for a general-purpose boat of something over 1,000 tons, mounting a large number of torpedo tubes and having a range of up to 12,000 miles. Just what this boat might do was not stipulated. But the submariners knew what they wanted, and they got it.

In particular, engine performance was improved dramatically. By matching specifications with those needed for railway locomotives, the navy was
able to stimulate competition among American firms to produce compact,
high-speed diesels suitable for submarines. These were then employed in a
series of "long-range patrol boats" of increasing displacement culminating
with the ten-tube, 1,475-ton *Tambor* class. With four diesels, these boats
could recharge their batteries while cruising and power air-conditioning
units, which greatly improved crew performance during the long cruises
necessary in South Pacific waters. They were speedy, fast divers, and agile—
perfect for quick kills of targets of opportunity. The psychological impact of
World War I weaponry had once again asserted itself; the country that had
most vilified unrestricted submarine warfare had produced the perfect
commerce raider.

There was one problem: it shot duds. The Bureau of Ordnance had
developed the Mark 14 torpedo, a presumably advanced model with both
magnetic and contact exploders. Yet because of the expense, the bureau
carried out the Mark 14's testing with light dummy warheads, and consequently failed to realize that they ran much deeper—frequently too
deep—when loaded down with explosives. To make matters worse, bureau
engineers remained equally oblivious to the fact that neither the magnetic
nor the contact detonators actually detonated. When the British reported
problems in 1940, the bureau compounded its sins by refusing live tests on
the Mark 14, and therefore failed to correct the problems until 1943.
Meanwhile, only the Japanese really knew what was happening, its crew-
men having reported the unnerving sound of these fish without teeth
bouncing down the length of their ships as they steamed forward.

American sailors were spared similar experiences. Torpedoes, according
to naval historian Samuel Elliot Morison, were the prime technical achievement of the Imperial Japanese Navy, exemplified by the oxygen-fueled
Type 93 with a speed of nearly fifty knots, a range of 12.5 miles, and a large
warhead that exploded reliably when it hit something. Combined with the
formidable Japanese capacity for night operations such torpedoes spelled
the end for many an American combat vessel.

Of course it was a rare military device that worked exactly as advertised,
and in fact the Japanese and Americans each dwelled under far more conse-
quential delusions as to the effectiveness of key members of their arsenals.
If Japanese seaborne operations came to grief in part because they were
based too heavily on naval tradition and tradition's favorite weapon, the
ship-of-the-line, then Americans were guilty of looking too far into a

*Boeing B-17
Flying Fortress.*

future when enemy assets could be discretely targeted and precisely destroyed from above.

As far back as 1926, Army Air Corps documents were talking in terms of "vital parts," the loss of which would cause an economy to grind to a halt. During the 1930s these concepts were refined at the Air Corps's Tactical School into a focused conception of air war pursued at high altitude, during daytime, with precision bombing that would take apart an industrial economy's infrastructure piece by piece. Although the doctrine would be practiced most rigorously and stubbornly by the U.S. Eighth Air Force over Germany, the major justification for building a strategic bomber force during the mid-1930s was the Pacific. It was thought that strategic bombers could fill the gaps in plans to defend the Philippines left by the inadvisability of stationing the battle fleet so far forward.

By August 1935 the Pacific planners' panacea was flying in the form of Boeing Model X299—a low-winged bomber powered by four Pratt & Whitney R-1690-E Hornet radials, with a top speed of over 250 miles per hour and a maximum range of 3,100 miles, an aircraft destined to win fame as the B-17. A reporter watching it take off for the first time and noting its external armament exclaimed: "Why it's a flying fortress." The name stuck, but he was wrong. As rugged and well armed as the B-17 was, there were severe limits to any airplane's ability to absorb punishment before plunging.

The Americans were convinced that a flight of bombers could fight their way through contested skies during daylight without prohibitive losses. During the mid-1930s bomber speeds and ceilings matched or exceeded those of available American fighters, but strategic bombing advocates failed to realize that this was a temporary condition destined to disappear with the next generation of interceptors. Consequently, the Americans joined the British in overestimating the effectiveness of bomber self-defenses and underestimating the need for long-range escort fighters—misjudgments that would cost the lives of thousands of young aviators.

Not to mention that the difficulty of finding, targeting, and then accurately striking key targets was underestimated by all sides prior to World War II. The Americans were especially overconfident. In part this stemmed from the Air Corps's single-mindedness; but it also reflected their faith in a single device, the Norden bombsight.

This time the tools were ready. When world war reignited, jumping and rebounding across continents and oceans until most of the inhabited globe was engulfed, the implements necessary to reach out and strike across this vast domain were sufficiently mature and available to allow effective fighting. Not only were the weapons chosen mainly the right ones, but this time most command staffs had thought through in fairly realistic terms how new types of arms could be used to effect, and at least tried to train accordingly.

Some failed in combat, others worked in ways not anticipated; inevitably adjustments had to be made. Hence new weapons were developed as the conflict raged, and while their deployment was (with one exception) not decisive, they did make a measurable difference in the results of several very important campaigns. For the most part, this proved the prerogative of the victors. Thus the French, swept away by a flood of Germans, had no opportunity to right the strategic wrongs embodied in the Maginot Line. The Axis powers were also severely handicapped in this regard—Japan by the small size of its military-industrial complex and Germany by its feckless organization and incomplete mobilization until near the end of the war. The Allies as a whole were far more successful.

The British were particularly innovative in the face of a dire menace. Having been thrown off the Continent in June 1940, the island nation's very survival was almost immediately threatened by its old nemesis, German submarines, now provided with easy access to the Atlantic by bases along the western coast of France and working in groups, the aptly named wolf packs, to counter the surfeit of targets embodied in a convoy. Although the standard U-boats were too small—around half the tonnage of the *Tambors*—and few in number—Admiral Dönitz, the master planner of the campaign, controlled only twenty-nine boats in July 1940—the British shortage of escorts and neglect of antisubmarine warfare (ASW) enabled them to inflict heavy damage very rapidly.

America's Most Famous Secret Weapon

Publicity photos of B-17 noses almost inevitably bore the explanatory blurb: "Note the jacket draped over the top-secret Norden bombsight." Using it, American fliers were reputedly able "to put a bomb in a pickle barrel" from three miles up. Pickle manufacturers the world over may have shuddered, but there was limited cause for alarm. The curse of the Norden bombsight was that although it worked, it was unable to achieve anything like this level of accuracy. In part this was a technical matter, but it also stemmed from the kind of behavior inspired by secret weapons.

Carl Norden, a consulting engineer from New York City, had been working on optical bombsights since the early 1920s; but he developed his first really advanced unit for the pioneering naval aviators in 1930 to allow them to accurately target maneuvering ships from high altitudes. The key to the Norden sight's superiority was its ability to compensate for relative speed, bomb ballistics, drift, and deflection; it simply correlated more variables than other devices.

Army observers were so impressed by early navy tests that they contracted with Norden for twenty-five, which similarly excelled in their own April 1933 trials. Air Corps procurement agents immediately campaigned to obtain as many of the bombsights as possible, only to have their own service declare it too secret to mass-produce! Although the Air Corps reversed this decision within six months, the navy continued to guard its Nordens like a mother hen. As to future development, the Air Corps's new chief, the very determined General Hap Arnold, was informed that his service could use any Norden bombsight data so long as it did not pertain to "technical descriptions of features, parts, or methods employed." In effect, the bombsight was also branded too secret to improve! Not surprisingly, the United States entered the war with a serious shortage of bombsights, and the ones available needed further development.

Despite all the security, word of the Norden's existence leaked, and soon enough the legend of the by now nearly sacred bombsight's miraculous accuracy grew to towering proportions. Not only did this inflate the official image of what might be accomplished with the Norden, but it drew spies like a magnet. The FBI later learned that German agents managed to obtain vital blueprints for the sight as early as 1932. Significantly, Germany never chose to copy the Norden. Having realized that electronics, not optics, was the future over cloudy Europe, they developed the Knickbein radio-based system instead.

Nordens at war, struggling to hit the right part of the right factory, were not up to the task without help. They had to be operated in conjunction with automatic flight control equipment, which locked the aircraft's attitude, altitude, and speed and effectively turned B-17s into sitting ducks as they approached their targets. Too frequently the Norden simply had nothing to aim at, since targets were obscured by a mask of impenetrable clouds and smoke. Hence the Norden's optics were overshadowed by the radar-based BTO (bomb through overcast) system, which could penetrate cloud cover but remained less precise in delineating targets. Today the Norden remains secure in an obscurity more complete than was ever possible when it was America's most famous secret weapon.

In October 1942 the English lost around 350,000 tons of merchant shipping, setting off a frantic research and development program in the Royal Navy and Air Force. ASDIC's performance was belatedly improved both in terms of target location and range (hydrophones aboard submarines still could hear the beat of surface ships' propellers ten times further out, but with little direction resolution) and became the basic means of underwater detection. To find surfaced submarines, the British placed radar units on ships and patrol planes. Due to their speed and endurance, aircraft proved extremely important in hunting submarines, especially after the British equipped them with aerial torpedoes. ASW vessels, meanwhile, were fitted with improved depth charges in the form of hedgehogs, which lofted projectiles forward so as not to interfere with ASDIC. Better high-frequency radio direction finders also played a key role in locating the sources of what amounted to a torrent of messages between Dönitz and his submariners to orchestrate the search for convoys.

The admiral and his minions, however, did not realize that the British knew what was being said. Of all the contributions made by the code breakers, none was more crucial than their role in the war against the U-boats. Through captured keys and cribs the British broke into Dönitz's entire operational scheme—how long U-boats stayed at sea; how wolf packs were put together, and how they derived their immediate intentions—and were thus able to divert convoys until losses dropped to a bearable figure of around 100,000 tons a month. So successful were they that in early 1942 the U-boats switched hunting grounds to the waters around the Eastern United States, where, in an operation labeled Drumbeat, they feasted on coastal traffic not yet organized into convoys. Around the same time the Germans added a fourth rotor to their Enigma machines, shutting Bletchley Park out for an entire year. The combination of slaughter in American waters (the U.S. Navy, like the English, had neglected ASW through most of the interwar period), and no readable traffic, once again plunged the battle of the Atlantic into crisis, culminating in the grim winter of 1942–43 when accumulated losses, especially of crucial oil tankers, reached truly dangerous proportions.

The British and their American allies persevered. In December Bletchley Park conquered Enigma's fourth rotor and broke back into U-boat message traffic. By the spring of 1943 the cumulative effect of sonar, radar, air surveillance, and a variety of antisubmarine weapons—including the American Fido, an aerially delivered torpedo with an acoustical head that searched for and then homed in on the propeller beat of submerged submarines—began to take a devastating toll. Between May and October

1943, 135 U-boats were sunk. At this point the battle of the Atlantic was won, and the German U-boat—though not the submarine—defeated.

Meanwhile, from a naval perspective at least, the war against the U-boat might have been won significantly sooner if only air planners had been less stingy with their strategic bombers, leaving a mid-Atlantic gap in coverage and a welcome zone of shelter for harried German crews. The truth was that Bomber Command and the U.S. Eighth Air Force needed every plane they could lay their hands on just to make up losses.

After the collapse in France, the British were left only with bombers as a means of striking directly at Germany. Strategic niceties melted in the face of enemy air defenses, and daylight precision attacks gave way to nighttime "precision attacks," which, in turn, became nighttime carpet bombing of German cities. When, in August 1941, Winston Churchill's scientific adviser, Lord Cherwell, ordered a formal investigation of accuracy, the resulting Butt Report revealed that only a third of attacking aircraft hit within five miles of their aim points (an area encompassing seventy-five square miles). The reaction was twofold: it spurred better technology; and it caused Cherwell to invent "dehousing." After analyzing the results of the German Blitz against London, he concluded that what most bothered victims was losing their homes—"People seem to mind it more than having their friends and even relatives killed." Extrapolating this to Germany, where 22 million were concentrated in just fifty-eight cities, he estimated that a bombing campaign lasting eighteen months could dehouse a third of the country, and "break the spirit of the people."[3]

By February 1942, Churchill had found just the man to carry it out—Bomber Harris, now Air Marshal. He was suspicious of technology, and he didn't like the idea of economic choke points. His single concept was ravaging German cities. At this point the radio navigation aid Gee and significant numbers of Avro Lancasters were coming on line, and the results were awesome and also appalling. Before the campaign was over, the British, with the support of the Americans, killed or wounded over a million German civilians and destroyed 20 percent of all the dwellings in the country. Particularly devastating was the emphasis on incendiary ordnance, which typically overwhelmed German firefighters and allowed the destruction to spread unchecked.

The raids came like drumrolls at an execution. On March 28 Bomber Command burned the wooden city of Lübeck, dehousing over 15,000. Worse awaited the equally flammable Rostock at the end of April 1942. A month later came a 1,000-plane attack on Cologne, incinerating the city's center. The spring of the next year found Harris's bombers dealing out terrible punishment to the industrial Ruhr valley. But by this time German air defenses were better and the cost had grown very heavy—872 bombers lost in this season alone. In July, however, Bomber Command introduced Window—strips of aluminum foil that flooded German radars with false returns, rendering them helpless. For the next six months British bombers ran wild. On July 25, 722 of them lit the world's first firestorm in Hamburg, featuring 1,000-degree temperatures and 300-miles-per-hour blasts of superheated air incinerating everything flammable in their path. Similarly destructive raids followed in the fall on Mannheim, Frankfurt, Hanover, and Kassel, where a second firestorm was lit and still smoldered after seven days.

Then Harris overreached. "We can wreck Berlin from end to end," he had told Churchill. "It will cost us between 400–500 aircraft. It will cost Germany the war."[4] He was wrong on both counts. Berlin was located deep in Germany, demanding that British bombers fly long distances in the face of enemy air defenses, almost all of it without fighter escorts. To make matters worse, the Germans' new airborne radar, the SN2, defeated Window by operating at a longer wavelength. The raids stretched from November 1943 to March 1944, but when they were called off Germany remained very much in the war. The British had lost over 1,100 airplanes, almost all big four-engined bombers. Night bombing without long-range fighters had taken on suicidal overtones.

The American experience was no better. The Yanks arrived in England with Air War Plan D/1, a detailed guide to dismantling the German war economy. After a string of delays, they put it into effect with a series of unescorted daylight B-17 raids in the spring of 1943. Their primary objectives were all strategic—fighter engine and airframe manufacturers, submarine yards, ball bearing factories. But they seriously underestimated German air defenses. On April 17 a B-17 raid on the Focke Wulfe facility near Bremen resulted in 40 percent losses among the attackers. In June the Eighth Air Force staged two more raids beyond escort fighter range, and again took heavy losses. During July, American bombers arrived in the daylight over targets at Hamburg, Hanover, Kassel, Kiel, and Warnemunde, and lost another eighty-seven of their number, along with nearly a thousand crew members.

It only got worse. On August 17 the Eighth Air Force staged a dual raid on the Messerschmitt assembly plant and the infamous Schweinfurt ball bearing factory. It proved to be an epic bloodletting. In one day the Eighth Air Force lost over 10 percent of its bombers and 17 percent of its crews.

Equally bad, the apparent destruction of these choke points didn't seem to be choking the German economy. Ball bearings continued to roll off assembly lines, new submarines put to sea, and fighter losses were replaced. (Postwar analysis demonstrated that mangling factory sheds was far easier than pulverizing the machine tools below.)

Throughout the fall bomber losses oscillated, sometimes reaching as high as nearly 25 percent. On October 13 the situation reached its nadir when the B-17s returned to Schweinfurt. German flak and fighters shot them to pieces—fifty-nine bombers went down, leaving 594 men missing in enemy territory. Nearly every plane that returned limped home with damage and casualties. One survivor of this "Black Thursday" gave vent to the feelings of his comrades: "Jesus Christ, give us fighters!"[5]

By late 1941 two large fighter aircraft, the twin-engined P-38 and the radial-powered P-47, were flying and in the early stages of production. Both proved formidable combatants with good inherent range, but neither could make it all the way to Germany and back with the bombers—even using drop tanks in the case of the P-47.

Ironically, it was the English who stumbled over the solution—an orphan American plane they didn't even want at the outset. In April 1940 the British Purchasing Committee, desperate for fighters, approached North American Aviation to license production of the mediocre Curtiss P-40 for them. "Dutch" Kindelberger, the company's president, countered by offering to build a much better plane around the P-40's wheezy Allison engine.

The NA-73X (soon to be P-51) prototype first flew on October 26, and immediately demonstrated great maneuverability, an uncanny ability in dives, and a top speed forty miles per hour higher than the P-40's—with unexpected endurance. The key was the plane's superior aerodynamics; Kindelberger's designers had included an advanced laminar flow wing and a drag-reducing radiator inlet system well back in the fuselage.

The RAF snapped up 300, the first of which arrived in Britain on May 1, 1941, and was designated Mustang Mk 1. The plane flew well, but was relegated to tactical reconnaissance due to the chronically poor high-altitude performance of its Allison V-1710. Then in September 1942 the British engineered a Cinderella-like transformation; they pulled the Allison and replaced it with a Merlin V-12, which developed more horsepower at 25,000 feet than the American motor did on takeoff. Once again the Rolls-Royce engine made the difference; it not only pushed the Mustang's top speed to an eye-catching 440 miles per hour, but its fuel economy

North American P-51 Mustang.

combined with the Mustang's efficient airframe extended the plane's range significantly.

Back in the United States, tests with Packard-produced Merlins were similarly successful, and by the spring of 1943 experiments with drop tanks achieved what the bombers so desperately wanted—a plane they could take with them to thrash their Luftwaffe tormentors. So the Mustangs sallied forth to become the best fighter aircraft in World War II, shredding the ranks of Bf 109s, twin-engined Bf 110s, and Focke Wulfe 190s until the only German pilots left were half-trained rookies.

Soon too they were out of gas. As the skies were cleared over Germany, Allied planners made a conceptual breakthrough in their understanding of choke points. In a war powered by internal combustion, the petroleum industry and the rail net were the trump cards. Although many of Hitler's key operational decisions had to do with acquiring or protecting petroleum resources, historians Williamson Murray and Allan Millett make the point that, with few exceptions, the Allies had as yet made no concerted effort to strike at them. This changed dramatically in May 1944 when the U.S. Eighth and Fifteenth Air Forces began a full-court press against the German synthetic oil industry and Romanian refineries. Meanwhile, the success of Allied bombers in destroying the rail links in back of the Normandy beachheads, and its impact on the defenders' ability to reinforce and resupply, encouraged a similar effort in Germany itself. Before they were finished, virtually every gas tank in the Wehrmacht and Luftwaffe was empty—planes could not fly, tanks could not move, there were no trains to transport troops and supplies, no coal was delivered to factories—the Nazi war machine sputtered and stopped. The end was near.

Adolf Hitler, in the meantime, had retreated into a last redoubt of science fiction, where secret weapons meant salvation. Yet personal experience and his limited grasp of science and technology led him to undermine new initiatives as soon or even sooner than the Third Reich's technologists created them.

Back in 1936 Dr. Gerhard Schrader, searching for new insecticides, discovered tabun and sarin, fantastically deadly chemical agents that blocked acetylcholinesterase, the substance that allows muscles to relax. A factory was built at Dyhernfurth, near the Polish border, and 12,000 tons of nerve agent were stockpiled there. A lethal dose amounted to just a few grams. Yet Hitler, the World War I gas victim, never pressed for its use. He certainly allowed innocent Jewish civilians to be gassed in profusion with more conventional agents; but insecticide for soldiers had little appeal even for Hitler. Meanwhile, the Allies had no equivalent and only discovered the existence of German nerve agents at war's end.

Meanwhile, he had insisted that the Me 262—a brilliantly conceived jet-powered interceptor with a fuel consumption almost as high as its 540-mile-per-hour top speed—be turned into a bomber, wasting over a year of precious development time. Similarly, Hitler and his armaments chief, Albert Speer, overlooked the potential of the Waterfall—the world's first surface-to-air missile, which passed over fifty tests and proved briefly devastating in combat—in favor of the V-1 and V-2 offensive missiles, the vaunted "weapons of revenge."

Messerschmitt Me 262.

German V-1.

The V-1, consisting of little more than a pulsed jet engine, fuselage, wings, and a warhead, was at least cheap to build and burned low-grade fuel oil. The V-2 was a thirteen-ton, forty-six-foot-long rocket that consumed alcohol like a drunk and delivered a warhead no bigger than that of its simple sister. While Allied bombers during certain months of 1944 were averaging 3,000 tons a day of bombs dropped on Germany, the trade-off in explosives from an optimal thirty-launch day of V-2s was twenty-four tons. Nonetheless, Hitler told Speer that the V-2 was "the decisive weapon of the war." It was given equal priority with tank production, and vast quantities of the potato crop were diverted from the mouths of an increasingly hungry Germany to make its alcohol fuel. After imposing an economic burden on Germany equivalent to that of the Manhattan Project, the V-1 and V-2 inflicted around 30,000 British casualties and had no appreciable effect on the war's outcome—though they might have killed a great many more had they been tipped with nerve gas.

In a last desperate burst of creativity, the Third Reich's technologists forged virtually a complete prototype arsenal for the coming Cold War and a new era of armaments. Besides developing and mass-producing the first tactical ballistic missile, German rocket scientists at war's end were groping toward models with true intercontinental ranges, and they had schemes on the drawing board for launching modified V-2s from submarine-towed canisters. And just as the V-1 served as the primary link between Boss Kettering's Bug and the modern cruise missile, so did the Waterfall mark the primary path of future air and missile defense.

Among aircraft, the Me 262 was far from the only German plane without propellers. Few realize that Heinkel built a jet fighter equipped with an ejector seat, the He 162, and that Messerschmitt also fielded a rocket-powered interceptor, the 163B Komet. (Its fuel, "T-stoff," was so corrosive that, when some leaked into the cockpit, it dissolved the pilot!) There were

1. Explosive warhead
2. Guidance systems
3. Gyros
4. Helium (to prevent oxygen and
   alcohol from exploding)
5. Oxygen tank
6. Refrigeration tanks
7. Alcohol tank
8. Oxygen feed pipe
9. Alcohol feed pipe
10. Hydrogen peroxide tank
11. Steam generator
12. Pump (driven by turbine)
13. Steam turbine
14. Steam outlet pipe
15. Fuel injectors
16. Igniter
17. Fuel coolant (alcohol, which is
    heated and mixed with oxygen)
18. Combustion chamber

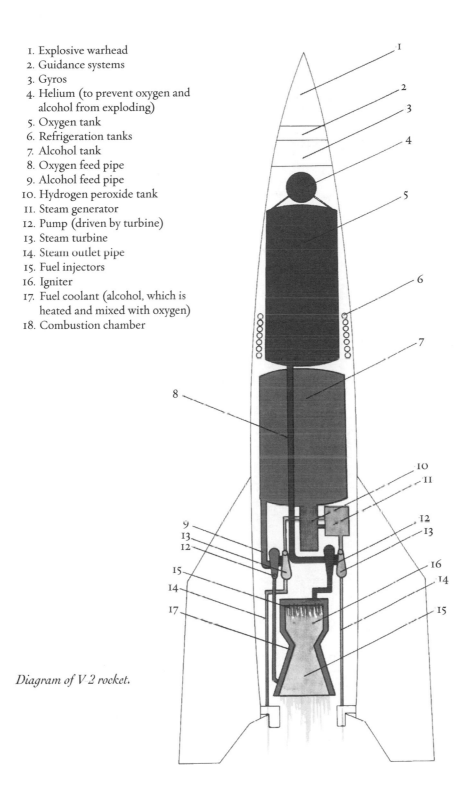

Diagram of V 2 rocket.

also jet bombers—the two-engine Arado Ar 234B introduced in 1944 and the four-engine Junkers Ju 287V, the first operational aircraft featuring a forward-swept wing. Heinkel was working on a similar wing for the He 162 along with a combat plane with a vertical takeoff and landing capability, while Focke Wulfe was involved in developments that included a jet-powered flying wing that bears a startling resemblance to the contemporary American B-2.

Pioneering efforts were also devoted toward improving ordnance accuracy. A team under Dr. Herbert Wagner at the Henschel firm brought the radio-guided Hs 117 Butterfly air-to-air missile to an advanced stage of development before Germany's surrender. Other Henschel engineers were working steadily toward true precision-guided munitions, culminating in the television-controlled Hs 293D bomb and the Hs 294 aerial-guided torpedo.

Seen in this light, postwar weapons development takes on a somewhat different cast. Something besides Hitler and desperation was pushing German technologists. Advanced weapons concepts were emerging from virtually every quarter. In other periods, just before basic transitions in weaponry, there have been clear signals that a major change was about to take place. Thus shortly before the proliferation of firearms the English had begun using longbows in gun-like fashion, carving up battlefields like Crecy and Agincourt into killing zones. German technologists on the cusp of nuclear weapons' inception were doing something analogous.

The thing they missed, of course, was the bomb itself. In the late 1930s Germany had led the world in nuclear physics. But many of the best scientists had been Jews or profoundly anti-Nazi, so they fled. Consequently, the German bomb program went nowhere and was scrapped in the fall of 1942. Later Hitler occasionally referred to the principles upon which it was based as "Jewish physics." In a twisted way he was right. Largely because of their departure, the bomb was developed elsewhere and the world was spared the specter of nuclear-armed Nazis.

The path was narrow and circuitous, but it led to America. In December 1938, when two German physicists, Otto Hahn and Fritz Strassman, managed to split a uranium atom by bombarding it with neutrons, they remained unsure what they had done and consulted former colleague Lise Meitner, a Jew who had fled to Sweden. Examining their results, she quickly realized that not only had they broken apart the uranium atom, but that a small amount of its matter had been transformed directly into energy. Using Albert Einstein's equation $E=mc^2$, she calculated that one pound of uranium fully fissioned was equivalent to the energy derived from burning seven million pounds of coal. Meitner immediately informed Nobel laureate Niels Bohr, who relayed the information to Enrico Fermi at a Washington meeting of theoretical physicists in January 1939. Fermi, a refuge from Mussolini's Italy, quickly realized that a sufficient amount of uranium (a critical mass) could generate enough atom-splitting neutrons to sustain fission through a chain reaction. By early March, Leo Szilard, a Hungarian expatriate, demonstrated that Fermi was right by bombarding uranium to produce more neutrons. An atomic bomb was possible, and these expatriates from Fascist Europe were fearful Hitler was building one. The alarmed scientists convinced Albert Einstein, the world's most famous living scientist and also a Jewish refugee from Germany, to sign a letter warning President Roosevelt.

FDR was sufficiently concerned to order further inquiries. This led to the formation of the Uranium Committee, a small research grant to Fermi, and eventually the patronage of Dr. Vannevar Bush, head of the Office of Scientific Research and Development, which had the backing and resources of the War Department. Further work demonstrated feasibility and on the day before Pearl Harbor the President ordered that everything possible be done to construct an atomic bomb. By the summer of 1942 the Manhattan Project was in existence, though it remained essentially a small-scale research program until December 2, when Fermi and his team engineered the first nuclear chain reaction by allowing their "pile" underneath the University of Chicago football stadium to go "critical." Fortunately for the rest of Chicago Fermi's calculations were correct and the experiment was terminated quietly; but it demonstrated conclusively that uranium 235 could produce plutonium and that both could be harnessed to generate a controllable explosion.

Almost immediately the Manhattan Project itself went critical in the form of two massive facilities at Oak Ridge, Tennessee, and Hanford, Washington, to produce fissionable material, along with a self-contained "science city" at Los Alamos, New Mexico, where an unprecedented talent pool of theoreticians and engineers gathered to design and build both a uranium- and a plutonium-based bomb.

As early as November 1944, American officials were aware that Germany had no viable nuclear program, and its surrender in May 1945 made this question irrelevant. Nevertheless, work not only continued but accelerated, until on July 16 the Los Alamos team set off plutonium device Fat Man, lighting the desert and producing an explosion the equivalent of 17,000 tons of TNT. Plainly the bomb had been brought into existence by a fear of Nazi Germany, but it was built to be used, and now there was only one enemy left.

Kamikaze!

"**N**othing [that] happened during the war was a surprise—absolutely nothing except kamikaze tactics," wrote Fleet Admiral Chester Nimitz. As the events of September 11, 2001, have shown, he was not the last to be caught unaware. Deliberate suicide in battle was the reductio ad absurdum of the warrior ethic, the final refuge of the militarily challenged.

Yet the Japanese had little else to fall back on as the war in the Pacific wound down. As the most industrially constrained of World War II's major combatants, they were forced to fight to the finish with the same weapons they had at the beginning—by now completely outclassed. Kamikazes represented strategic jujitsu, making something out of nothing.

The concept is sometimes attributed to Vice Admiral Takijiro Onishi. Actually, it began extemporaneously during the Leyte Gulf campaign, when aviation commanders concluded that their young ill-trained pilots were better used by simply diving their aircraft into American warships, rather than die in futile attempts to drop bombs and launch torpedoes. The attacks, which began on October 25, 1944, had a stunning impact, severely damaging seven American carriers and inflicting heavy casualties on numerous other vessels by the first of the year. Four months later during the Okinawa invasion, the U.S. Navy lost sixty-four ships to kamikazes, with another sixty damaged. Soon the self-destructive aviators were joined by suicidal speedboats and manned flying bombs modeled on German V-1s; even the huge dreadnought *Yamato* was sent on a kamikaze mission. Although many of the 4,000 pilots who gave their lives did so reluctantly, it seemed to the defenders that they were being assailed by maniacs. Already deeply angered by Pearl Harbor and subsequent Japanese brutality, to Americans kamikaze tactics were the last straw, one that swept away any inclination toward restraint. Suicide bombers, then as now, played upon our deepest fears. There is no more terrifying enemy than one determined to die: against him there is no deterrence.

With certain exceptions such as the Dresden raid, Americans generally abstained from mass bombings of civilian targets during the European campaign. This forbearance did not extend to the Pacific Theater, however. A month prior to Pearl Harbor, George Marshall had instructed aides to develop contingency plans for "general incendiary attacks to burn up the wood and paper structures of the densely populated Japanese cities."[6] Three years later, with the arrival of the very-long-range B-29 heavy bomber, the M-47 and M-69 napalm bombs, and General Curtis LeMay to command the Twentieth Air Force, these plans came to fruition. On the night of March 9, 1945, 334 B-29s armed only with incendiaries attacked Tokyo at low levels, destroying 267,000 buildings and killing 83,000

B-29 Enola Gay opens the nuclear era.

people, most of them burned alive. Japanese air defenses against such attacks were almost nonexistent, nor would they improve. By June, over 40 percent of Japan's six most important industrial cities had been gutted and millions dehoused.

Coming on top of this, the atomic bomb simply constituted a labor-saving device. Certainly, there was concern among the scientists and an awareness of unprecedented power on the part of those who made the decisions. There was also terrible anger, a wrath that had been ignited by the manner in which the Japanese had fought. But in the end nuclear weapons were employed because they promised to shorten the war and avoid the horrific American casualties that would have resulted from an invasion of Japan. The two weapons—the uranium Little Boy dropped on August 6 by the B-29 *Enola Gay* on the industrial city of Hiroshima, and the plutonium device used against Nagasaki—almost instantly killed at least 180,000 people and effectively ended Japan's resistance.

Yet there was little sense that something fundamental had changed. Less than a month earlier, the bomb's architect, J. Robert Oppenheimer, as he watched its first test, remembered lines from the Bhagavad Gita: "Now I am become death, destroyer of worlds; waiting the hour that ripens to their doom."[7] Today these words are considered prescient, but not then. Most wandered into the nuclear age just as Charles VIII had sauntered into Italy, believing the rules of the previous era still applied.

They didn't, and this was probably a good thing. Around 60 million people died as a direct result of World War II, roughly 40 million of them civilians. Many were simply stabbed, shot, or blown up by fairly conventional means. But new weapons, particularly military aircraft, took a terrible toll. Of all the arms conceived of by Leonardo, only chemical weapons had been handled with some forbearance, and that excludes their terrible use in the Nazi death camps. The era of the gun had been a time of death and instability from its inception, and at its end the tools made possible by chemical energy were taken up with terrible and vindictive enthusiasm. No event so completely ripped off the smiling mask of civility to reveal in its full magnitude the homicidal capacity of ordinary people, or at least males. Now weapons technology promised the capacity for complete destruction. Civilization's prospects did not look good.

Chapter 15

COLD WAR—
INFERNO OF ARMS

It was the strangest of wars a conflict waged nearly without death, but always promising mega-death; a showdown of strategies twisted like pretzels by swords grown omnipotent; a duel of defense budgets so huge that they became the real instruments of destruction; a clash of paradoxes so profound that at its end some declared history finished. We see now, a decade later, that history lives on. So does war. But it survives more like a rat than a lion seeking dark places, probing for weakness, dodging strength, and drinking the blood of innocents. This is the legacy of the Cold War.

It was from beginning to end about weapons—nuclear weapons so powerful that unleashing them implied suicide. Both sides had them, and each promised retaliation far in excess of any conceivable reward for aggression by the other. This was the essence of deterrence. Only the paths of madness and mishap remained open. But if deterrence blocked the main avenue to war at its highest level, each side had different explanations of why it did so and how best to build the barriers. In a nutshell they chose more, we chose better, and the ebb and flow between these two strategies mapped the topography of the great struggle between the United States and the Soviet Union.

The Soviets appear to have taken Flavius Vegetius Renatus literally, building arms that in their quantities and qualities reflected his very ancient

and very Roman advice: "Let he who desires peace, prepare for war."
Consequently, Soviet weapons were designed and deployed in ways that
would make them as useful as possible in what were perceived as realistic
wartime scenarios. Even when this proved beyond their grasp they sought
to create this impression. Unfortunately, such an approach was prone to
misinterpretation, since it left no clear criteria to differentiate preparing for
war from intending war.

The American slant reflected much more clearly what came to be
understood as the primary message of the nuclear age; that general war was
bound to be an unmitigated disaster. Therefore, the central thread of U.S.
armaments policy—though it was never clearly articulated—was continual
innovation to make war appear an increasingly unsavory and ultimately
impossible proposition. The overall impression to be gleaned from American arms was intimidation, and only secondarily an ability to perform
successfully under wartime conditions. But like the Soviet strategy, the
American path was subject to misunderstanding, especially by a very pessimistic adversary with a penchant for ideological preconceptions.

These approaches were not limited to strategic weaponry; they extended
throughout the arsenals of both parties. In part this is why they were so
mutually threatening. Americans consistently viewed the masses of simple,
robust Eastern bloc weapons as symbols of the inexorable nature of the
Soviet threat and its potential for aggression; while the Soviet leadership
apparently perceived the across-the-boards sophistication of American
arms as evidence of unpredictability and a desire to destabilize the international environment through technological surprise. On two occasions this
climate of suspicion nearly led to tragic consequences.

America emerged from the Second World War in excellent condition to
generate new weaponry. Not only had its massive armaments sector
emerged uniquely unscathed, but the conflict had nurtured a number of key
technologies and industries—aerospace, electronic control, and computers—that would prove critical to the evolution of the weapons that dominated the Cold War. The U.S. military-industrial complex, motivated by
the promise of substantial profits, would function as an incubator for new
weapons concepts, percolating them up through the Pentagon to provide
decision makers with a wealth of choices. This bottom-up phenomenon
would endow America with a major advantage in the forthcoming arms
competition.

It was an advantage that would be almost compulsively pursued. Again and again, from beginning to end, Americans sought state-of-the-art advantages in both performance and versatility. Eventually these gains came at an outlandish cost and in the face of inevitably dwindling numbers; but this only served to demonstrate the centrality of the urge for weapons that intimidated first and served as tools of warfare second. Tokens of destructiveness were required for the American vision of deterrence.

At heart it was a bluff, but a bluff that had to be real enough to be taken absolutely seriously by the other side. It was a difficult and dangerous business, especially in the immediate postwar environment. The withdrawal and subsequent demobilization of U.S. ground forces left unchecked an apparently huge Soviet army poised at the gates of Western Europe.[1] Yet the psychological advantage imparted by sole possession of nuclear weapons initially was deemed sufficient to deter. Despite having only seven bombs on hand in 1947—a number reinforced only by Sandstone test series data indicating many more could be built—the Truman administration decided in the summer of 1948 to rely on nuclear weapons and long-range bombers, not massive conventional forces, to dissuade the Soviets from moving west. It was only when pressed that the United States shifted the emphasis from symbols to tangibles.

That pressure was almost immediately applied when it was discovered that, contrary to all expectations, the USSR had exploded its own nuclear device in August 1949. This information, when combined with the knowledge that the Soviets were already mass-producing replicas of an American B-29 that had crash-landed in Vladivostok in 1944, caused great consternation among the Americans.

The initial jolt was compounded by the unexpected effectiveness of Soviet weapons used against U.S. forces in Korea. Particularly unsettling to American technological pride was the performance of the MiG-15 jet fighter; which, although it was shot down regularly by superior U.S. pilots, still proved faster, flew higher, and in the right hands could be a formidable opponent. By 1955 Soviet conventional arms, with their frequent upgrades and massive production runs, had plainly gained the respect of Western military analysts.

Together, the USSR's atomic bomb and the Korean experience constituted the first in a series of weapons-related shocks that would periodically propel the Americans into an almost frenzied reappraisal of the military balance. While the U.S. response did include fielding more systems, its core message was to be found in a series of startling performance advances.

Mikoyan-Gurevich MiG-15.

Three types in particular—the atomic submarine, the Mach 2 fighter, and the high-tech armored vehicle—not only epitomized this approach, but made a lasting impression on the Soviets.

By the late 1940s the USSR was embarked on a program to build a massive flotilla of conventionally powered submersibles, based on captured German models, which would ultimately grow to 450 boats. Rather than responding with more boats of their own or concentrating on antisubmarine warfare, the Americans leaped ahead of previous submarine designs with the nuclear-powered *Nautilus,* which went to sea in 1955. The boat's revolutionary power source, combined with parallel advances in carbon dioxide scrubbers and oxygen generation, enabled *Nautilus* on its first voyage to travel submerged approximately ninety times longer than any previous submersible. No longer tied to the surface, it became virtually undetectable, leaving the Soviets with a fleet of nearly new but completely obsolete boats, and the necessity of building their own nuclear submarine if they wanted to keep up. While the nuclear submarine's inconspicuousness and endurance was most advantageous when combined with the navy's ballistic missile program, the original *Nautilus* had no such objective and was based solely on the potential for improving performance.

This was also the case with the creation of the "Century Series" (F-100, F-104, F-4D/E, F-105, and F-106) of U.S. fighter aircraft during the 1950s and early 1960s. Although the frequent usefulness of sheer top speed in World War II air combat along with the unexpected swiftness of

Lockeed F-104 Starfighter.

the MiG-15 were influential, it was primarily technological possibility that drove the development of the fighter planes destined to contest the skies over Vietnam. Not only were these planes intended to exceed supersonic speeds, but the availability of thrust-enhancing afterburners, more sophisticated aerodynamics, and variable air inlet ducts allowed speeds to be pushed all the way past Mach 2. In the eyes of the American military and the aerospace community what was possible became necessary, therefore a Mach 2 capability became obligatory for members of the Century Series. Yet when these fuel-gulping super-stilettos actually had to fight in Vietnam, the record would show that only a few minutes of combat time ever took place at even Mach 1.4, much less at anything above. Meanwhile, jet-powered corkscrews, in the form of the MiG-15's follow-on, the MiG-17, more than held their own in low-speed-turning fights. Still, the Century Series led the way in an international stampede that not only filled the runways of Russia with Mach 2 chimeras like the MiG-21 and the Su-15, but also left both the British and French with equally speedy but only moderately useful aircraft like the English Electric Lightning and the Dassault Mirage III. But submarines and fighter aircraft were at least high-tech items to begin with; tanks provided a real challenge for state-of-the-artists.

Super Sheridan

In the world of early Cold War weapons the pace of armored vehicles development had been relatively glacial—basically a sequence of bigger guns, thicker armor, more powerful diesels, and the gradual introduction of electromechanical fire control. Yet even in this placid quarter the U.S. Army managed to conceive and develop a tank calculated to befuddle the opposition. Begun in 1959 and officially designated an "armored reconnaissance assault vehicle," the Sheridan light tank was designed to include no fewer than eleven cutting-edge advances. Not only was it intended to be amphibious and air-droppable, but it would mount the M-81 152mm main gun built to fire a ballistic round with a combustible cartridge case and also launch the 17,000-part Shillelagh antitank guided missile.

Almost none of these features worked under combat conditions. When sixty-four Sheridans (minus the Shillelaghs) were shipped to Vietnam, in short order they accumulated sixteen major equipment breakdowns, 123 circuit failures, forty-one mission misfires, 140 ammunition ruptures, twenty-five engine replacements, and persistent malfunctioning of the M-81 main gun. Nor did the Shillelagh-equipped models work any better; the missile generated so much carbon monoxide that the army recommended no more than four firings a day. Numerous fixes were applied to the Sheridan, most aimed at simplification, but eventually its controversial gun/missile was withdrawn from service. Nonetheless, the Soviets were impressed, and embarked on a program that almost two decades later bore fruit in the form of a tank with a missile-firing gun of its own.

*The Sheridan light tank —
the U.S. Army's vision of
high tech.*

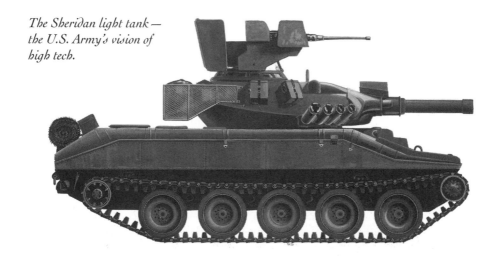

*Boeing B-52 Stratofortress — the first American intercontinental bomber, this Methuselah
of the air is still flying combat missions in 2002.*

Even if the United States was a genius among arms racers, it was a
neurotic one, plagued by self-doubts and fears of being overtaken. And
when the Soviets sought to compensate for their disadvantages by playing
upon these emotions, the United States overreacted quantitatively and at
the highest levels of the strategic competition.

In response to the Soviet bomb, production of American nuclear
weapons accelerated, and by 1950 over 300 atomic bombs were available to
be delivered by a fleet of nearly 400 giant prop-driven B-36s and over 1,600
medium-range jet-powered B-47s. The program intensified under the
Eisenhower administration's doctrine of massive retaliation, with the "new
look" air force being expanded from 110 to 137 wings, including the long-
lived B-52, America's first intercontinental jet bomber, and the short-lived
B-58 Hustler supersonic bomber—basically a flying fuel tank whose statistical capabilities obscured any particular usefulness as a delivery vehicle.

As bombers proliferated, the first successful U.S. test of a thermonuclear
device in November 1952 not only raised the ante of explosive power by an
order of magnitude, but also opened the way to much more economical use
of fissile material and still more bombs. By the time the Eisenhower administration left office, America's nuclear arsenal was nearing 30,000 mega-

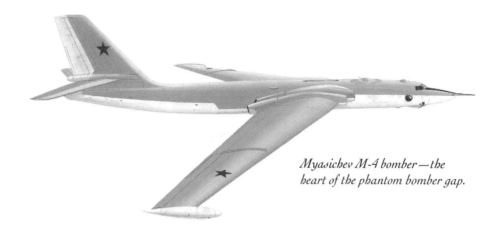

*Myasichev M-4 bomber—the
heart of the phantom bomber gap.*

tons, or the equivalent of around 20,000 pounds of TNT for every man, woman, and child on earth—deterrence deluxe.

Meanwhile, the Soviets were not only missing the point, they actually seemed to be thriving. For a time it appeared that the American military-technical lead was not just shrinking, it was being reversed. In August 1953, less than a year after the United States did, the USSR staged its first thermonuclear test, Joe 4. Although Joe was actually a "fusion-boosted" type and had a much lower yield, it was advertised as a usable bomb, as opposed to the truly thermonuclear American Mike, which was purely a test device weighing some twenty-one tons. In July 1955 Americans attending the Moscow air show were distressed to observe "wave after wave" of newly minted Soviet intercontinental bombers flying over them, prompting a U.S. Air Force spokesman to announce that the USSR would soon achieve a substantial lead in this category.

More serious than this alleged "bomber gap" was the impending "missile gap." In August 1957 the Soviets announced the successful full-range testing of an intercontinental ballistic missile (ICBM) well ahead of the air force's Atlas program. Although this event largely escaped the notice of the American public, in October the launching of Sputnik, history's first artificial satellite, had a riveting effect. To drive the message home, a month later the USSR orbited a dog-carrying satellite, whose 1,200-pound payload validated Nikita Khrushchev's claim that the USSR had developed a working ICBM.

America was left reeling. Although a growing body of technologically derived intelligence was coming to cast doubt on the extent of Soviet bomber and missile deployments, information available to the public portrayed the United States as clearly losing the arms race. Given this impression, there was little question that Congress would further accelerate strategic programs that were already bringing several revolutionary weapons to fruition. Eisenhower, with access to technical intelligence, knew better and urged caution; but eventually he too bowed to a combination of public and military-industrial pressure. The desired force levels of the Atlas and Titan liquid-fueled ICBMs were roughly tripled, while the follow-on solid-fueled Minuteman program was expanded and accelerated. Planned deployments of Polaris intermediate-range ballistic missiles on board nuclear submarines were also raised, in this case to over 300 on thirty-four platforms. Atlas and Titan represented transitional technologies and never reached these numbers. But Minuteman and Polaris—the former to be buried in increasingly blast-resistant silos across the American heartland, and the latter made virtually invulnerable by the vastness of the oceans, the endurance of nuclear submarines, and the limitations of antisubmarine warfare—came to constitute particularly lasting components of the emerging structure of deterrence. This was due in part to their chemically stable solid fuel configuration, a technology the Soviets would not master for nearly two decades. But these positive signs were hard to detect in the general gloom.

The newly elected President Kennedy chose to disregard now definitive intelligence that the Soviet "missile gap" was nonexistent, and continued the strategic buildup. By mid-1964 total mega-tonnage had doubled, and the number of warheads increased by 150 percent. Kennedy also undertook a very substantial expansion of conventional forces, setting about to double the number of active ships in the fleet and increase active army divisions from eleven to sixteen. U.S.-Soviet relations were at a very acrimonious state at this point, and Kennedy, having run and won largely on the "missile gap," was in a difficult position to admit that his campaign rhetoric was false. Even so, the series of Soviet weapons surprises had built such political momentum that the numerical search for security would have been difficult to restrain in any case.

Paradoxically, the net effect was to drive the destructive potential of the nuclear arsenal well beyond any rational scenario for using it, a condition first reflected by President Eisenhower's vacillation between hardheaded planning and utter dismay at the implications of what came to be known as "overkill." This judgment was ratified by Robert McNamara's well-documented horror at the full-bore Single Integrated Operational Plan (SIOP) upon taking office as Kennedy's secretary of defense. While McNamara and later the Nixon, Carter, and Reagan administrations would undertake fruitless searches for graduated response schemes that held out a

U.S Minuteman III ICBM.

reasonable prospect of avoiding an all-out exchange, the initial U.S. reliance on nuclear weapons and the expansion of the arsenal are best interpreted on a symbolic level. Ultimately, the result—if not exactly the aim—was to cow the Soviets, not to kill them.

Consequently, it was only in the aftermath of staring down the USSR in the Cuban Missile Crisis that sufficient national confidence was restored to accept the point that further expansion was irrational. From this juncture, further American growth in destructiveness was more a matter of technology than politics. Yet the fact remains that the numerical levels of American ICBM and SLBM (submarine-launched ballistic missiles) deployments during the Cold War were arrived at more by fear than planning. This is especially significant, since the other player in the grand game of nuclear poker was also involved in bluffing, only much more consciously so.

The USSR opened the arms race with a mixed hand of strengths and weaknesses. The economy as a whole remained deeply wounded after World War II; but the military-industrial sector was in better shape, since much of it had been evacuated east of the Urals in the face of the Nazi invasion. Of at least equal significance, the entire economic structure was optimized and would remain optimized to develop and produce weapons. While the overall quality of Russian devices and production technology might lag far behind that of the United States, the weapons community was consistently assured the best of what was available. Nevertheless, the strengths of the system were plainly vested in areas most relevant to less sophisticated items of military hardware, particularly land arms.

Even so, the USSR was not without a significant capacity to generate more esoteric military devices. For one thing, Russia was blessed with a strong cadre of theoretical scientists, particularly mathematicians and physicists. To back them up, history handed the USSR a wild card in the form of thousands of German scientists and engineers brought back from the project teams that fostered such a profusion of advanced prototypes near the end of the war. In practically every category of Cold War weaponry, their corporate memory provided the basis for Soviet advances. And what was unavailable from the Germans could frequently be pirated from the West. The Soviets were not only well organized to conduct scientific and technical espionage and the covert importation of embargoed devices, they grew extremely adept at copying, or reverse engineering, the take. From beginning to end "borrowed cards" were a stable factor in allowing the Soviets to compensate for their own technical shortcomings. Thus the MiG-15 owed its high top speed and overall performance to progressively upgraded versions of the Rolls-Royce Nene engine, just as Soviet sonobuoys of the 1970s were found to have IBM-based circuitry, and the lookdown/shootdown radars of the 1980s generation Soviet fighters acccurately reflected documentation from the U.S. F-18.

Yet even with these compensatory factors, the USSR's materials, device, and manufacturing technology lagged badly, forcing a low-risk, simplicity-oriented design approach generating weapons capable of being produced by Soviet industry, but lacking in flexibility, versatility, and growth potential. In short, the Russians were cast in the role of pluggers. But they were very organized pluggers, and this itself constituted a significant advantage. The nature of the Soviet state not only enabled its leaders to mask its weaknesses and personally direct limited resources, but it also allowed them to enlist other elements of national power in the coherent pursuit of a strategy that maximized the nation's threat potential.

By nurturing certain technologies, the Soviets were able to graft an esoteric facade on their substantial base of conventional military power. The USSR's approach to military technology reversed the U.S. pattern, with relatively little spontaneous innovation actually taking place until external events provoked high-level political intervention to prioritize development in a desired direction. This top-down strategy was inherently parsimonious; but its net effect was to develop the technology base in a narrow and irregular fashion that worked against fortuitous synergy. They could do what they could do, but little else.

Yet state control of information successfully disguised the true nature of the cards in the Soviet technical hand, while it took calculated advantage of the almost instant media feedback and information receptivity of the West. Not only were the Soviets able to create the illusion of a massive bomber force in the 1955 air show by flying the same planes over the field again and again; but it did not immediately occur to American observers that this was a ruse. Yet the faked "bomber gap" was nearly child's play compared to the Soviet campaign that led to an American perception of a "missile gap."

By exploiting the presumed link between progress in space and military rocketry, Nikita Khrushchev managed to erect a high-tech version of Minister Potemkin's fabled fake villages. Backed up by a drumroll of space shots, he proclaimed the safety of the socialist revolution behind a ring of invincible missiles—a conception of deterrence not greatly different from that of

Space Bluff

Nikita Khrushchev had almost nothing to do with the launching of the original Sputnik, which had been virtually bulldozed into orbit by the charismatic godfather of Soviet rocketry, Sergei Korolev. Yet the wily peasant-turned-statesman was nothing if not opportunistic. "We never thought you would launch a sputnik before the Americans. But you did it," Khrushchev told Korolev at the Kremlin. "Now please launch something new in space for the next anniversary of our revolution."[2]

When Korolev obligingly sent Laika the dog into orbit, he had unwittingly stepped on a space launch treadmill that would leave him gasping to provide Khrushchev with a string of new firsts, all carefully timed to further trump the Americans. In 1959 he managed to hit the moon with Luna 2, which scattered 150 hammer and sickle badges across the lunar surface. Throughout 1960 and the early months of 1961 Korolev's team raced to beat the widely heralded American astronauts into space, succeeding on April 12 when they orbited cosmonaut Yuri Gagarin, and then returned him safely to earth. This proved the high point. As the massive American space program began to take advantage of superior resources and technology, the Soviet equivalent took on a circus atmosphere.

Korolev wanted longer flights to test human reactions to weightlessness, but for each one Khrushchev made him pay with a first. Thus a three-day flight in 1962 was accompanied by the near simultaneous launching of a second Vostok capsule into near orbit. Korolev got a week-long flight the next year, but was required also to launch Valentina Tereshkova, textile factory worker and Young Communist League activist, who became the first woman in space. As very large American booster rockets came on-line capped with bigger two- and then three-man capsules of the Gemini and Apollo series, the Soviets were forced into a sleight of hand. The Vostok capsule became Voskhod, basically through the installing of three seats and packing their occupants like sardines. Voskhod flew only twice, the second being highlighted by a nearly fatal spacewalk, the last of the Russian firsts. The Soviet space bluff had reached the end of the line.

the Americans, except there was little behind it. The very success of his bluff not only provoked a fearful spasm of American missile deployments; but fear of being called eventually drove Khrushchev into a desperate ploy.

On the eve of the Cuban Missile Crisis the United States had already fielded nearly 300 ICBMs, half of which were solid-fueled Minuteman Is. Meanwhile the Soviets lagged badly with a force around sixty SS-6s and 7s, all of which lacked blast-resistant silos and inertial guidance. Worse yet, in October 1961 Deputy Secretary of Defense Roswell Gilpatric had announced, on the basis of satellite observation, that the United States was well aware how small and vulnerable the Soviet force really was.

Although recent information indicates that Khrushchev and the Soviet leadership were motivated to deploy eighty-four medium- and intermediate-range missile in Cuba in part to protect that country's communist revolution, it is hard to believe that strategic calculations did not play a role. Missiles in Cuba would have not only helped balance the scales quantitatively, but their short flying times would have left their targets very little time to organize a retaliatory strike.

Yet the gambit failed, and in doing so assured a basic change in Soviet strategy. By the fall of 1964, when Khrushchev fell from power, "Potemkin deterrence" must have looked positively disastrous in Soviet eyes. That year U.S. ICBM deployments, sustained by bureaucratic momentum, hit an all-time high with over 400 Minuteman Is added. Given this performance, the air force's insistence on a force of at least 3,000 must have seemed entirely credible. In fact, a few days after the election of Lyndon Johnson, Robert McNamara decided to stabilize American ICBMs at 1,054 and submarine-launched missiles at 656. Yet by this time they faced a new Soviet leadership. "You Americans will never be able to do this to us again," Soviet Deputy Foreign Minister Vasily Kuznetsov told a high-ranking U.S. diplomat shortly after the Cuban Missile Crisis.[3]

A new stage in the arms race had begun, with both sides reverting to form—the Soviets building and the Americans improving. Between 1966 and 1969 the USSR's ICBM deployments surged by approximately 300 new silo-based launchers per year, and by the latter date surpassed the American land-based force for the first time. The missiles represented in this expansion, the third generation of Soviet ICBMs (SS-9, SS-11, and the unsuccessful solid-fueled SS-13), were inertially guided, and not nearly so vulnerable to nuclear weapons–induced electromagnetic pulses as their radio-beam-riding predecessors. Not only was the new generation more robust, but the very large, high-yield warhead of the SS-9 left certain Americans believing that it was fashioned to include some ability to destroy missiles in their silos, implying an interest in preemption. Though the SS-9 was not a significant threat to our Minuteman force, around 1964 the Soviets began development of a fourth generation of ICBMs (SS-17, SS-18, SS-19), which eventually came to constitute a much more credible hard-target kill capability. Although it appears that the high throw weights of these missiles, particularly the SS-18, were initially intended to compensate for expected inaccuracies, when guidance improvements promised greater precision the Soviets wasted little time filling the fourth generation's commodious nose cones with silo-threatening multiple independently targeted reentry vehicles (MIRV). If Americans had pioneered MIRV essentially as an engineering exercise, the Soviets responded with characteristic and misplaced confidence that technology could be made to serve strategic ends. "While rejecting nuclear war and waging a struggle to avert it," senior foreign affairs official V. V. Zagladin noted, "we nonetheless proceeded from the possibility of winning it."[4]

If the MIRV-laden ICBM force symbolized Soviet-style deterrence at the highest tier of potential conflict, the army represented it at ground level. Beginning in the mid-1960s the USSR began relentlessly modernizing their already armor-heavy ground forces, creating a veritable land armada. At the center was the tank, which after the heroics of the now venerable T-34 had become an icon of Soviet military power. Using this model as a jumping-off point, designers at Kharkov and Niznij Tagil generated a series of increasingly more heavily armed and armored successors, including the T-54, T-55, T-62, T-64, T-72, and finally the T-80—the last three with 125mm autoloading main guns, and the B models of the T-64 and T-72 vested with the long sought capacity to shoot an antitank missile through these tubes. These tanks were not only incrementally improved, but fielded in numbers far greater than the NATO and U.S. equivalents facing them. To screen the fleet the Russians gradually added a family of auxiliaries. Beginning around 1967 Soviet industry began churning out thousands of the BMP infantry combat vehicles, dedicated to carrying foot soldiers under armor, but also mounting 73mm guns—enough firepower to accompany and in certain instances support the tanks. The same themes of protection and mobility were also reflected in fire support, when the Soviets supplemented and replaced their towed artillery with a range of self-propelled guns, howitzers, and even mortars, beginning in the early 1970s.

Soviet concern over ground attack planes and helicopters, the natural enemies of armored columns, was represented by the acquisition of a number of highly mobile gun- and missile-based antiaircraft systems (ZSU-23-4 and SA-4, 6, 8, and 9), the combat ranges of which were intended to provide concentric zones of coverage above advancing formations. Finally, to round out the force, Soviet armor was joined by an array of mobile bridging, mine-clearing, and logistical systems dedicated to insuring that neither natural obstacles, tactical impediments, nor supply shortages would slow the progress of Soviet ground forces. Lumped together, this armored juggernaut numbered around 120,000 vehicles by 1980.

By all appearances this was a blatantly offensive instrument tailored to punch through NATO defenses and conduct "deep battle" operations aimed at overrunning Western Europe in the shortest possible time. As Marshal Akhromeyev conceded somewhat lamely: "The USSR did not intend to attack anybody. . . . But the methods of troop training it used were of the attack style."[5]

To knit the web of Soviet-style deterrence still more seamlessly, the USSR conducted major expansions of its naval forces, air forces, and a very ominous modernization of the missiles in its intermediate nuclear forces (INF).

At sea the Soviet navy not only grew quantitatively, but its qualitative nature more than hinted at its objectives. Of particular concern was the steady growth of the submarine fleet. It was not only a matter of the USSR producing equivalents to American ballistic missile submarines; the coastal location of so many U.S. strategic assets raised the specter of these Russian vessels being used to conduct lightning-fast depressed-trajectory attacks intended to rob national command and control of the warning time to orchestrate a concerted response. Meanwhile, the steady expansion of the USSR's nuclear attack submarine flotilla constituted a growing menace to U.S. sea lines of communication, particularly in the North Atlantic. These boats were noisy and subject to detection, but over a quarter were equipped with cruise missiles enabling them to attack surface ships from much further away than equivalent torpedo-only U.S. units, which did not begin to receive the antiship Harpoon system until 1977.

The USSR's growing surface navy also placed heavy reliance on cruise missiles. In addition, the practice of keeping weapons loaded, along with a lack of ships' sustainability, convinced many in the West that preemption had to be central to the naval mission. But if Russian naval combatants were equipped for what U.S. naval officers sardonically referred to as "a short exciting life," they were not alone.

Although the Soviets retained their respect for American combat aircraft and pilots, they were unwilling to concede control of the skies over Europe. Instead, they conducted a very costly expansion of tactical aviation, which eventually resulted in a nearly two-to-one Eastern bloc advantage over NATO by the end of the 1970s. Also suggestive were changes in aircraft capabilities. Previous Soviet fighter-bombers had been transparent conversions of interceptors, chronically lacking in range and payload. The new generation of attack aircraft, epitomized by the MiG-27 and Su-24, were carefully tailored to the offensive mission. Meanwhile, the organization and subordination of Soviet Frontal Aviation made it apparent that these new aerial assets were meant to serve the larger purposes represented by the offensively configured ground forces. This theme was further driven home by the introduction—once again in very large numbers—of the Mi-24 attack helicopter, which its designer, Marat Tushenko, referred to as "our aerial BMP."

Yet nothing was more indicative of the Soviet passion for employing new weaponry as props for their elaborate strategic edifice than the deployment of the SS-20 nuclear-tipped intermediate-range ballistic missile, which began in 1977. Advertised as simply a replacement for the obsolescent SS-4 and SS-5, the solid-fueled, triple-MIRVed, mobile SS-20 was so much more capable than its predecessors that it promised a significant shift in the East-West power balance. With planned deployments of 441 launchers and 1,323 warheads, the SS-20 was poised to drive a wedge between the United States and its allies. By directly threatening them with discrete attacks, which would not necessarily demand an American strategic response, the new missile called into question the whole concept of extended deterrence. It was an ingenious ploy, representing just the psychological message the Soviet brand of deterrence was meant to convey.

It was also a grave error, since it left the USSR open for a devastating counterdeployment. Yet the SS-20s symbolized an even more fundamental problem. Soviet power did not just threaten Western Europe and the United States. The USSR's quarrel with China resulted in a huge military buildup opposite that nation, while intransigence toward Japan's territorial claims accompanied by threatening military gestures (SS-20s again) left that country, reborn as an economic giant, fearful and quietly hostile. Because deterrence à la russe caused competing states to be categorized as potential enemies and required that plans be made accordingly, it acted as a self-fulfilling prophecy and called forth countervailing military power.

The effect on the USSR itself was no more salutary. Security, seen in this light, became virtually an open-ended proposition. So, more than a generation after the last shots were fired in World War II, the Soviet Union remained an armed camp. There were huge opportunity costs associated with conducting the arms competition with an economy configured according to ideological considerations. Analogous to running in cement track shoes, the USSR stayed in the race only at the price of increasing exhaustion and diminishing blood supply to everything not vital to maintaining its lumbering gait. Paradoxically, the central perversity lay in the fact that the military-industrial sector could be made to operate, if not exactly efficiently in Western terms, then at least predictably through administrative fiat and selective prioritization. The net effect was to create a privileged and parasitic enclave.

Structurally, the military economy, like the rest of the economy, remained divided into a series of massive hierarchically organized ministries, a box-on-top-of-box organizational model, adopted during the 1930s from the Ford Motor Company, but conceptually not that different from the bureaucracies that had run ancient agricultural tyrannies. Inevitably, the flow of work remained stubbornly vertical, while horizontal integration, whether it involved multiministry products or the proliferation of new technology, remained difficult.

As long as the Cold War was being waged, several compensatory factors operated to prevent complete stagnation. Unlike the civil product environment, Soviet weapons acquisition had to exist in an inherently competitive arena, where continued American innovation acted as a goad to at least some creativity. And although Soviet weapons developers remained unable to duplicate Western product standards, from the late 1940s to the mid-1970s they had considerable success approximating them through clever systems engineering and packaging, narrow performance and mission envelopes, and a design philosophy that emphasized incremental steps forward, the use of common components, avoidance of exotic materials, simple electronics, and a low level of product finish except when functionally necessary. Just how far this could be taken became apparent to the American defense establishment, when a notable example, the Foxbat, literally dropped out of the sky.

Jumbo production runs were the key means by which the Soviets compensated for qualitative shortcomings. The net result was an enormous stockpile of modest, but continually modernized, arms—a profusion of tanks, artillery, fighter aircraft, missiles, and submarines—a veritable cornucopia of military adequacy. Yet the cost was extraordinary. At the time, U.S. analysts pegged it at somewhere between 12 and 17 percent of GDP, but they greatly overestimated the productivity of the Soviet economy as a

Foxbat

On September 6, 1976, Japanese air controllers were intently following a single aircraft, which had streaked out of Soviet airspace and entered their own. Nearly out of fuel, it managed to find a suitable airfield and landed safely. Local authorities made an effort to cover the intruder with a tarp, but it proved impossible to disguise with its huge ramp-type air intakes and twin vertical stabilizers. At the end of the runway sat a MiG-25, the Soviet Union's fastest and most advanced fighter, spirited away by one Lieutenant Victor Belenko.

Within days an engineering team from America was pouring over the plane, taking it apart piece by piece, examining every detail. Anatomically Foxbat (its NATO code name) was full of surprises. It was not only big but extraordinarily heavy, almost 80,000 pounds with a full load of fuel. It didn't take long to figure out why. The Mach 3 aircraft was not fabricated from lightweight but hard-to-work titanium or other high-tech materials designed to sustain the high temperatures generated at these speeds; instead it was welded together by hand out of stainless steel. This hot rod paperweight was pushed through the air by two huge engines, not advanced turbofans, but simple Tumanski R-15BD-300 afterburning turbojets, both with nearly 25,000 pounds of thrust and a matching appetite for jet fuel. Not surprisingly the combat radius was limited to 186 miles.

Further examination revealed more elaborations on the theme. The instrumentation and layout of the cockpit was virtually identical with those of the MiG 21 and 23, as was the ejection seat. The radar, using vacuum tubes instead of integrated circuits, could not spot targets lower than 1,600 feet, but was so powerful it could burn through jammer signals from hostile aircraft.

By the time the exploitation was over and the MiG-25 put back together and returned to the Soviets, it was apparent that the plane was designed to do one thing well. That was to

whole. It really ran around 40 percent, a crushing burden with little relief in sight.

Yet there was another supremely ironic penalty for the militarization of the Soviet Union; it choked the broad-based technological creativity that became increasingly important as weapons grew even more sophisticated. This began to register during the late 1970s, when Soviet weapons development was having an increasingly hard time even approximating the new creations of the Americans. The third generation of Russian jet fighters, bought at a huge cost and headlined by the MiG-23, remained caught in the statistical trap of previous Mach 2 aircraft, chronically unable to convert speed and acceleration into tactical utility and certainly no match for the

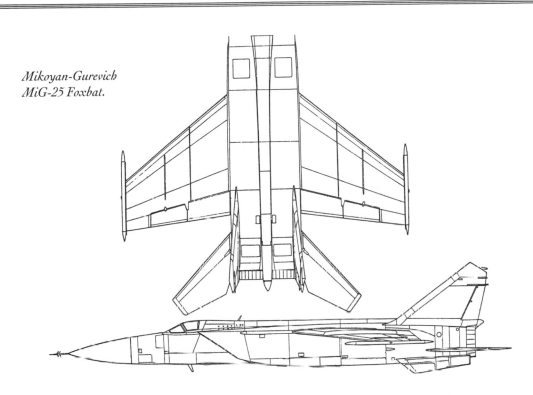

*Mikoyan-Gurevich
MiG-25 Foxbat.*

intercept a bomber flying in excess of Mach 2.5 at 75,000 feet. There was only one aircraft that fit this description, the U.S. B-70 Valkyrie, which had been canceled during flight testing in 1966 and never deployed. True to form, Soviet aerospace ground out over 600 Foxbats anyway.

computer-generated agility and firepower of the coming wave of new American planes like the F-16 and F-15. No matter how heavy the main gun or thick the armor of the Red Army's latest armored vehicles, weight constraints necessarily left them vulnerable to top attack—precisely the direction intended by a coming host of American high-tech tank killers including the A-10 "flying gun," the AH-64 Apache attack helicopter with its Hellfire missile, and the still more advanced projectiles with special sensors to find and destroy armor. No matter how many holes the Soviets plugged in their massive and expensive air defense network, the Americans seemed able to punch new ones with electronic countermeasures, nap-of-the-earth autopilots, and worst of all, the possibility of radarproof "stealth"

Boeing AH-64 Apache attack helicopter.

bombers that promised to reduce the whole investment to junk.

Meanwhile the Soviets' few forays into Great Leap Forward technology led nowhere. Their prospective moon rocket blew up in testing, their deep-diving titanium-hulled submarine was no more quiet than its steel-skinned cousins; even the Soviet MIRV, which did work, was purchased at a huge expenditure of effort.

The missing piece was electronics. In America, the stuff of arcade games, personal computers, and bankers' mainframes was turned back on military problems, providing the control and monitoring capabilities that enabled weapons to find targets by themselves and fly in regimes that defied physics. The USSR had almost none of this; so conditions in weapons acquisition were getting worse, not better. Looking back on the 1940s and 1950s when their war-ravaged economy had managed to generate jet fighters and bombers, ballistic missiles, and still more remarkably, nuclear weapons, the leadership must have found this a maddening development. After decades of sacrifice and hard work the Soviets were now truly losing the arms race, and losing it because of what that sacrifice and hard work had turned their empire into. Greek tragedy had no more excruciating punishment for its headstrong protagonists.

The period between 1964 and 1980 was a time of tragedy and frustration for the American military. The United States would not only lose its first war, but the nature of the Vietnam conflict made a mockery of our military power and seriously undermined domestic support for future defense spending. In the face of the Soviets' relentless arms buildup, the United States was increasingly perceived as a strategic Hamlet, if not quite the "pitiful, helpless giant" of Richard Nixon's rhetoric, then at least not up to an effective response. Instead, as the numbers of Soviet arms grew and the cost of American weapons multiplied, the whole nature of postwar hardware procurement came under increasing criticism.

Although the critics ultimately failed to exert any fundamental influence over the nature of U.S. arms, there was some evolution in design approach. Without sacrificing advanced technology, developers made a concerted effort to insure that superior performance had a significant impact on combat. Yet chronically escalating costs and shrinking numbers remained a fact of life. Meanwhile, at the strategic nuclear level, the United States continued developing and deploying weapons with little relevance to, and at times directly contradicting, its own military and even arms control strategies, but very much in consonance with raising the number of unaccountable variables and potential destructiveness of very large-scale war.

No development during the second phase of the great Cold War arms competition had more of a disruptive effect than multiple independently targeted vehicles. MIRV directly attacked the logic behind Mutually Assured Destruction (MAD), which mitigated against overkill and surprise attacks.[6] Although it was later claimed that MIRV was a response to Soviet efforts to create a ballistic missile defense (BMD), the evidence indicates that it was actually paced by our own Nike Zeus BMD, along with the possibilities revealed by American work during the early 1960s with penetration aids. It turned out to be a relatively short step from dispensing chaff and decoys to actually releasing multiple reentry vehicles, especially in a technological climate of reduced nuclear warhead size and increasing accuracy. Despite repeated statements from the highest levels of government that the United States did not desire a hard-target kill capability and simply wanted to maintain our deterrent, logic dictated that MIRV was primarily useful in eliminating multiple missile silos with a single launcher. This was destabilizing enough; but to make matters worse MIRV was plainly not considered in light of a potential Soviet equivalent. Nor did MIRV square with U.S. arms control strategy based on counting launchers or at least silos by satellite, since it was impossible to tell how many warheads lurked beneath each ICBM nose cone. In fact the only match between MIRV and traditional American policy was the consistent urge to deter through sheer destructiveness.

This was also evident in the reinvention of the cruise missile, the only major American weapons system deployed during the period because it was inherently cheap, promising, in the words of Pentagon Director of Research and Development John Foster, "more deterrent per dollar than any other of our schemes."[7] It was Department of Defense civilians who were behind the cruise missile; uniformed military showed little initial enthusiasm. Its heralded cheapness threatened the budgets of much more expensive service-sponsored strategic systems, and there was also the bitter memory of failed attempts in the 1950s to develop a workable cruise missile, epitomized by the crudely guided intercontinental Snark, one of which actually hit the wrong hemisphere in a test!

The electronics revolution changed this entirely, allowing the modern cruise missile equipped with miniaturized, super-accurate terrain contour matching (TERCOM) guidance to skim over the ground following its own internal "map" directly to the target. This diminutive guidance system, combined with the compactness of modern nuclear weapons, made it possible to employ a fuel-sipping 150-pound turbofan engine to complete a package whose small size was unmatched by any other strategic platform. Suddenly, not just the navy's ballistic missile submarines, but all of its submarines, even old battleships, became strategic platforms; not just the air force's bombers, but perhaps airliners; cruise missiles on land could be moved around in vehicles, or just hidden—qualities quickly reflected in contemplated numbers, which ballooned to almost 4,000 by late 1982. Metaphorically speaking there might be a cruise missile under every rock.

This was a possibility that, like MIRV, spelled chaos for U.S. arms control policy. Being so inconspicuous, numbers would be extremely hard to verify, while range and armament were nearly impossible to determine without literally taking the missile in question apart—negotiation show-stoppers at this point, since the secretive Soviets were firmly opposed to on-site inspection. As it happened, the cruise missile's future as a major pillar of the U.S. post–Cold War arsenal was almost exclusively nonnuclear. But at its inception this was barely contemplated. Instead, it was valued in part because it was unconstrained by the original Strategic Arms Limitation Treaty (SALT I) and could therefore be proliferated freely as a hedge against the Soviet nuclear buildup—"to improve," in the words of Carter

U.S. F-15 Eagle.

administration Defense Secretary Harold Brown, "the world's perceptions of the potency of [U.S.] forces."[8]

But if American strategy valued overkill over practically everything, the cruise missile's relatively low price tag was still a breath of fresh air in the budget-stifling climate of one weapon system cost overrun after another. Prices were escalating so fast that waggish but influential aerospace executive Norman Augustine was prompted tongue-in-cheek to project "in the year 2054, the entire defense budget will purchase just one tactical aircraft," which could then be shared by the services on alternating days.[9] Meanwhile, this cost curve had already translated into order-of-magnitude price increases for the next generation of strategic weapons. In 1980 each 17,000-ton *Ohio*-class ballistic missile submarine was projected at $1.5 billion as compared to the $110 million for its predecessor in the 1960s. Whereas 744 B-52s produced between 1954 and 1962 averaged under $8 million per copy, in 1975 it was estimated that the successor B-1 would run approximately $84 million (after being killed in 1977 and then revived by the Reagan administration, 100 B-1s were built for $260 million per plane). Worse still, the MX replacements for the $1 million Minuteman ended up costing close to $200 million apiece.

At least it could be argued that the diminishing buck still bought a prodigious (nuclear) bang at the strategic level. This was less evident with conventionally armed systems. Here, weapon for weapon, the cost curves

were nearly as steep and the numbers required plainly greater. At the top of
the feeding chain were surface warships, with a new aircraft carrier costing
several billion dollars, and the contemplated *Aegis* cruiser nearing the
$1 billion mark. If fighter aircraft ran a great deal less per unit, their
percentage increase was even more stratospheric. Over the twenty-seven
years that separated the World War II P-40 from the 1967 F-III swing-
wing attack plane, the real cost of procurement had multiplied eighty-fold.
And the incline showed no signs of abating during the 1970s, when the first
of the very advanced fourth-generation jet fighters, the F-14 and F-15,
came equipped with $28 and $13.5 million price tags. (By 1988 these figures
had grown to $50 and $38 million.)

Nor were the systems slated to reequip the army that much of a bargain
by comparison. The cost of a main battle tank had merely doubled from the
World War II vintage Sherman to the 1960s generation M-60; but it
would triple in the jump to the turbine-powered M-1. Still worse was the
projected $1 million price (the eventual figure was close to $2 million) of
the still maturing M-2/M-3 infantry fighting vehicle in comparison to the
$80,000 cost of the previous-generation M-113.

At the heart of the matter was the ever growing complexity and sophistication of the weapons themselves. But skyrocketing prices also reflected
inflation, which disguised the fact that by the mid-1970s real spending had
declined to approximately half of its peak during the Vietnam era. These
two trends combined to precipitously lower the quantities of new arms the
Pentagon was able to buy. In 1955 the air force had purchased 1,400 new
fighters; at the end of the 1970s yearly procurement stood at something
over 200. The problem extended throughout the arsenal. Virtually all military systems, even down to those costing less than $50,000, showed
substantial decreases in procurement rates and, eventually, inventory.

All of this was extremely upsetting to a growing segment of the American defense community, who believed numbers still mattered and that an
East-West showdown continued to be quite possible. Soon a cottage industry had sprung up in Washington and elsewhere dedicated to keeping careful account of relevant figures, and periodically issuing increasingly gloomy
assessments of the U.S.-Soviet military balance.

Concomitant with these appraisals was a critique of the basic soundness
of the American weapons acquisition philosophy by an amorphous group
known as the Military Reform Movement. Our penchant for superior technology, they argued, was a "vicious circle" in which diminishing numbers of
U.S. weapons were expected to defeat ever more numerous Soviet equiva-

lents. To illustrate the apparent aura of unreality surrounding this addiction to "force multipliers" defense analyst Frank Spinney fastened on one air force attrition model that called for an exchange ratio of 955 to 1 in favor of the new F-15. By contrast the reformers' own studies of military history indicated that in past wars, including some recent ones, superior weapons had made little difference in the outcome.

As American weapons drifted further from reality, the reformers argued, it was the Soviets who kept their heads. James Kehoe and K. S. Brower, after performing a broad and detailed examination of available Soviet weaponry, concluded that they were smaller, cheaper, easier to build, frequently more heavily armed, and entirely more practical under what were conceived as realistic wartime conditions. To drive their point home they cited an Israeli general who had used both brands: "American weapons are designed by engineers for other engineers, whereas Soviet weapons are developed for the combat soldier."[10] In a similar vein, a RAND Corporation study heaped praise on the MiG-25 as "unsurpassed in ease of maintenance, a masterpiece of standardization, and one of the most cost-effective combat investments in history"—all this about a plane essentially without a mission.[11]

Again and again, the reformers harked back to World War II, when masses of adequate but highly producible American weapons had overwhelmed the Axis powers. But what the reformers (and the Soviets) failed to accept was that this was not World War II; it was the nuclear age. Times had truly changed and so had the criteria for effective arms. As massive sustained warfare grew less probable, weapons tailored to such an environment became less relevant. Despite their cost and complexity, the performance of the new generation of American arms was extraordinary individually, the most obvious ground for comparison so long as very-large-scale war remained an abstraction.

Yet to another group of critics the far-fetched nature of our war plans was exactly the problem. The first president to really focus on the SIOP, the Single Integrated Operational Plan, was Jimmy Carter, the same man who had pressed hard for nuclear disarmament in his inaugural address. Behind this apparently fundamental change were two figures, National Security Advisor Zbigniew Brzezinski and his military aide, General William Odom. Brzezinski, a Polish émigré, was extremely interested in the details of the SIOP and during one briefing on criteria for destruction asked the startling question: "Where are the criteria for killing Russians?"—not Ukrainians or Moldavians, but Russians.[12] His logic was simple and brutal; you killed Russians first, particularly if they were communists or military. To do this, however, required an entirely more measured approach to waging nuclear war.

Since Robert McNamara's discovery of the "spasm" SIOP, the persistent search for a middle ground between capitulation and Armageddon had been thwarted by two stumbling blocks. First the American public interpreted graduated escalation as making nuclear conflict less unthinkable and, therefore, more likely. Second, when the services were encouraged to move away from pure "city busting" and to find militarily useful objectives, targets multiplied like rabbits, reportedly reaching 40,000 by 1980. And as the target list grew, so did the requirements for new warheads, making the prospect for a general exchange still more horrific.

Carter's advisors looked beyond these problems to the plans of the other side. "It's quite clear the Soviets . . . don't deploy their forces as if the world were going to end the day an exchange takes place," General Odom pointed out.[13] Rather, the nature of the USSR's arsenal suggested to the administration and increasing numbers of the cognoscenti that it was configured to fight and win a nuclear war. The simple logic of retribution argued that what they would do to America, America had a right to plan for them—at least when viewed through the U.S. strategic prism.

After conducting a series of tests and exercises, by late 1978 the Carter administration had settled on a course of action. New weapons were to be optimized for specifically military missions. Command, control, communications, and intelligence gathering (C3I) had to be made drastically more robust, wired to allow the leadership to simultaneously fight rationally and try to negotiate a way out. Finally, targeting policy had to be reworked to emphasize limited and sequential use of nuclear weapons.

Gradually, the pieces began to fall together. Nothing much happened until mid-1979, when Carter reluctantly approved the production and deployment of 200 mobile, ten-MIRV MX ICBMs, the first of the warfighting weapons, and halfheartedly accepted continued development of the Trident D-5, the first American submarine-launched ballistic missile with a point-target kill capability.

Six months later, in the face of increasingly ominous Soviet behavior that would culminate in the invasion of Afghanistan, the President's attitude had hardened considerably. In November Carter had signed a presidential directive setting forth clear objectives for U.S. C3I. Then on December 12 the NATO allies ratified American plans for European deployment of 464 Tomahawk cruise missiles and 108 Pershing IIs, all nuclear-armed and intended to counter the SS-20s. Next came retargeting.

Bad Missile

It was a chess player's weapon, brilliant in its conception but disastrous in the impression it left. The U.S. Pershing II was arguably the most dangerous weapon deployed during the Cold War. It wasn't an ICBM; it wasn't even a new system, but an upgrade of the 1960s-vintage Pershing mobile battlefield ballistic missile, with its range extended to just over 1,000 miles.

The devil, however, was in the details—particularly the guidance and warhead. Its single reentry vehicle was not only maneuverable, but guided itself to the target by making radar images of the terrain below and then comparing them with computer maps resulting in pinpoint accuracy. Equally ominous was its earth-penetrating W85 thermonuclear warhead with a dial-a-yield of from five to fifty kilotons (5,000 to 50,000 tons of TNT), which was plainly optimized to attack buried command and control bunkers. Finally, the Pershing II's fast accelerating solid-fueled propulsion, its inherently low trajectory, and its extremely short flight time promised almost no warning for the intended target. In short, it was the ideal instrument of preemption.

Yet what made the Pershing so dangerous was its brilliant deployment strategy. It did not simply counter the SS-20s, which were aimed at U.S. allies. It trumped them by targeting the USSR itself. Yet the impression it left was that of a strategic ice pick poised to strike the Soviet military brain, a message so stark that it provoked rather than deterred. Strategic communications had broken down in the most elemental way. The Soviet leadership, captured by their own Marxist and military delusions, refused to believe that American moves were simply equivalents of their own, and assumed instead that the United States was planning preemption. Pershing II drove the point home. Today, the last of the breed sits harmlessly on display in Washington at the National Air and Space Museum, next to its great Cold War rival, the SS-20. It is probably a better fate than both deserve.

In early 1980 Brzezinski charged Odom with developing a new doctrinal statement, the final draft of which Carter signed as Presidential Directive 59. The United States was officially serious about nuclear war fighting.

The entire scheme was based on the fear of a perceived Soviet capability to launch a devastating preemptive strike against the land-based Minutemen prior to the deployment of the reputedly less vulnerable MXs—the "window of vulnerability" projected to last through the late 1980s. Seen through less credulous eyes, however, the fabled window looked more like a kaleidoscope. The Soviets, like the Americans, were dealing with weapons that had never been used in earnest. ICBMs on both sides had been tested either west to east or east to west, rather than on the actual strike path over the extremely magnetic North Pole. Given the sensitivity of guidance packages and the necessary accuracies measured in yards over 5,000 miles, magnetic perturbations might well render incoming missiles useless against hard targets. Even presuming that hundreds, possibly thousands, of the USSR's ICBMs could be made to duplicate the performance of a relatively few test prototypes, the timing of a preemptive attack would have to insure that all targets were destroyed simultaneously, a task requiring incredible precision.

The window grew still more distorted when viewed from a general perspective. There was a tendency to treat the presumably invulnerable submarine-based retaliatory component as an afterthought, the assumption being that a preempted player would hesitate to launch these missiles, primarily useful as city busters, because they would inevitably call down a devastating counterblow. Yet the contemplated act of preemption involved a massive bombardment potentially indistinguishable from an all-out attack. Forbearance under such circumstances presumed almost superhuman rationality. The fragility of the entire scheme was exposed by the ease with which it could be circumvented; the preempted adversary had only to launch his threatened missiles on warning, leaving only empty silos as targets. Yet this concept also serves to illustrate the true nature of the strategic structure being built, its trip wires drawn tight and the entire framework suffused with a "use them or lose them" mentality. Even the prior edifice, with one side configured for preemption and the other committed to Mutually Assured Destruction, had the stabilizing virtue of each side knowing what to expect from the other. Its replacement was analogous to a Wild West shootout, with the side that drew first holding a major advantage. If Sam Colt's revolver was most at home in the O.K. Corral, then the Pershing II epitomized nuclear quick-draw.

Ronald Reagan arrived in the White House intent on reversing what he had characterized during his successful presidential campaign as a decided Soviet advantage in nearly every measure of military power. His approach could hardly have been more simple. He would sponsor a giant shopping spree for weapons. Yet so obvious were his initial moves that his critics and the Soviets persisted in reading deeper strategic motivation into his fervent buying. Reagan and his advisors were not by inclination strategic innovators. Rather the new administration can better be characterized as resorting to a more traditional American conception of deterrence.

Superior armaments were at the heart of the program, and, like those in the Kennedy administration, Reagan wanted more of them. Technologically, he took up basically what he found in the research and development hopper. The advice of the reformers was ignored; America would not cast aside its design philosophy and build simpler weapons, like those of the USSR. Lack of numbers would be addressed by spending more money, a great deal more. During the evening of January 30, 1981, the new secretary of defense, Caspar Weinberger, and his principal advisors met with Budget Director David Stockman and crafted a new five-year defense plan. When Stockman finished totaling the figures on his Hewlett-Packard calculator—he later noted the irony of its defense contractor origins—the bill came to $1.46 trillion, a figure with a message all its own.

Money may have talked, but what it bought was another matter. Between 1980 and 1985 yearly procurement roughly doubled, from under $50 billion to nearly $100 billion. Nonetheless, the quantities programmed into each of the budgets from 1983 to 1985—seven to thirty-four of the resurrected B-1 bombers, zero to forty MX missiles, thirty-nine to forty-eight F-15 fighters, 120 to 150 F-16 fighters, one to ten C-5B air transports, 720 to 855 M-1 tanks, and one *Ohio*-class ballistic missile submarine per year, among the top ten procurement programs—did not exactly add up to the plenitude of arms that Reagan and his contemporaries might have remembered from World War II's "arsenal of democracy."

Nevertheless, the impact of the buying spree altered perceptions dramatically. By the end of his first term it was widely held domestically that Ronald Reagan had managed to reverse the strategic tide, and that the United States was now abreast, or perhaps even ahead, of the Soviet Union. More to the point, this shift in psychology had a telling effect on the occupants of the Kremlin. Reagan and his advisors instinctively understood the

Soviets' respect for American economic strength and the repercussions of the decision to wield it. "We're already in an arms race," Reagan told anybody who would listen, "but only the Soviets are racing."[14] Tangibles were secondary; the key threat was bankruptcy.

If the realm of the symbolic had become the chief arena of the arms race, then the clash in the Bekaa Valley should not be underestimated. When the Israeli Defense Force moved north into Lebanon in June 1982 to confront the Palestine Liberation Army backed by Syrian units, it was reasonable to expect mixed results in terms of weapons performance. Prior to this, significant combat between forces using first-line American and Soviet equipment had occurred four times since World War II (Korea, Vietnam, the 1967 Arab-Israeli War, and the 1973 Yom Kippur War) and exchange ratios indicated a steady erosion of the U.S. qualitative edge. Now this was dramatically reversed. In the skies, Israeli pilots flying new American F-15s and F-16s and guided by U.S. E-2C Hawkeye sensor platforms inflicted an unprecedented rout of the Syrian air force, shooting down over eighty MiG-21s and 23s without suffering a single loss. While this was taking place, the Syrians' Soviet-equipped air defense network was shattered by a combination of guided bombs, antiradiation missiles, and remotely piloted vehicles. On the ground the story was the same: Israeli units chewed up Soviet armor, including the new and highly regarded T-72 tank, employing both their American-gunned Merkava tanks and a variety of sophisticated top-attack weapons. In all cases the Israelis were outnumbered but achieved such high attrition rates that it did not matter.

All told, the Bekaa campaign amounted to a stunning vindication of the American style of weapons acquisition that sent what former Israeli military intelligence chief Chaim Herzog described as "shock tremors through the Warsaw Pact over the poor performance of the Soviet equipment."[15] International orders for Soviet arms, particularly those involved in the Bekaa, declined precipitously. At the same time there was a reciprocal rush to "buy American," accompanied by an audible sigh of relief from United States—equipped allies.

The Bekaa might have been dismissed as a sideshow, a tiny war far removed from the massive, high-intensity campaign that Soviet equipment was designed to fight. Yet it was growing apparent that this clash of surrogates was typical of combat in an era of nuclear deterrence, the only kind of fighting likely to break out, and guaranteed to make American weapons look good. Like so much in the Reagan administration, the Bekaa demonstrated that looking good was tantamount to being good.

There were complications, however. If Ronald Reagan was able to intensify Soviet respect for American military potential, he also managed to thoroughly frighten the leaders in the Kremlin—though part of this was almost pro forma.

Upon entering office, the Reagan administration was confronted with the changes in American strategy embodied in Presidential Directive 59 and the new SIOP. They liked what they found. As visceral anti-communists, the eye-for-an-eye motivation behind the shift to war fighting struck the Reaganites as only logical and proper. They quickly ratified the new nuclear strategy in National Security Decisions 13 and 32 and hastened the purchase of the required tools. One hundred MX ICBMs were to be bought immediately and placed in fixed silos, rather than continuing the search for a viable mobile scheme; Trident D-5 development was accelerated; the hardening of the C3I net was continued; and the administration decided to go forward with procurement of the Tomahawks and Pershing IIs. Significantly, even administration hard-liners like Fred Iklé and Richard Perle had serious reservations over these intermediate nuclear force deployments and the strategic impression they might leave.

That impression was compounded by Ronald Reagan's tongue. From his first day as president, he subjected the Soviets to what amounted to an unprecedented hail of vilification, culminating on March 23, 1983, when he described the USSR as "the focus of evil in the modern world."[16] In reality the tone of Reagan's rhetoric differed little from the streams of anti-American propaganda that had been emanating from the USSR since the beginning of the Cold War. Therefore, just as in the case of the new strategy, it was logical to assume that the Kremlin leadership would have interpreted Reagan's phrases as a dose of their own medicine. They didn't. Instead, they responded with growing fury and unbridled suspicion.

In October 1982, after a series of disastrous harvests, the magnitude of the Soviet Union's economic predicament and the detrimental effects of the military effort were finally dawning on Leonid Brezhnev and the other members of the Kremlin's antique leadership. That month the ailing General Party Secretary gathered 500 generals in Moscow, and delivered his valedictory. Ostensibly, Brezhnev's aim was to congratulate them on their accomplishments and assure them of his commitment to match the concurrent U.S. buildup. But the implied message was that something had to be done about the economy, especially agriculture, and in the meantime the military's appetite would have to be curbed. A month later Brezhnev was dead, succeeded by Yuri Andropov, former head of the KGB. History revealed Andropov to be a deeply suspicious figure, a leader who as Soviet ambassador had crushed the Hungarian rebellion in 1956, and whose world-view was filtered through the thick lenses of Marxist-Leninist ideology.

Soviet policies took a dangerous turn when longtime Andropov hench-man Vladimir Kryuchkov, now head of the KGB and future leader of the coup against Gorbachev, told his senior espionage officers that an assessment had come from the Politburo that "Belligerent imperialist circles in the U.S.A. are getting ready for war and are preparing new weapons systems that could render a sudden attack feasible."[17] The leadership, therefore, had decided to mobilize its intelligence assets in a campaign of global vigilance, called Operation RYAN, the acronym standing for "nuclear missile attack" in Russian.

Behind RYAN was deep Soviet anxiety over the new American counter-force weapons, especially the Pershing II, which they had convinced them-selves was intended for a decapitating first strike against their national command and control centers buried around Moscow. As with RYAN in general, this judgment was irrational. Cold War historian Raymond Garthoff pointed out that while most Soviet references to Pershing II placed its range at 1,500 miles, sufficient to hit Moscow, U.S. testing patterns should have made it clear that the real figure was 400 miles short of the capital—though admittedly sufficient to blanket many targets in the western USSR.

Whatever its basis in fact, RYAN was not about to go away. Instead, it accelerated with regular reports being demanded every fortnight on any signs of war in Western capitals. In London, for example, the KGB was urged to monitor the patterns of work at 10 Downing Street and the Ministry of Defence, to watch for unusual movements of troops or aircraft, and to look out for signs of civil defense.

Meanwhile, U.S.-Soviet relations continued their downward spiral. By June 1983 Andropov was describing the situation as "marked by confrontation, unprecedented in the entire post war period."[18] Then on the first day of September an air defense interceptor deliberately shot down a Korean Airlines jumbo jet that had strayed into Soviet airspace, killing 269 people and provoking worldwide outrage. The incident was unconnected to RYAN; but the fact that it could occur and the Soviet leadership sought to justify it as an appropriate response to a "spy mission" illustrated in the starkest terms the paranoia prevalent among Soviet officialdom. On September 29 the by now very ill Andropov gave vent to the mood before the Supreme Soviet: "If anyone had any illusions about the possibility of an evolution for the better in the policy of the present American administration, recent events had dispelled them once and for all."[19] On November 7 Politburo member and fellow hard-liner Grigory Romanov called the international situation "white hot, thoroughly white hot."[20] This was no exaggeration from his perspective. The leadership was all but convinced that the Americans were about to launch an attack.

The pretext was supposed to be Able Archer, a NATO command post exercise to test nuclear release procedures, scheduled to run through November 11. As it proceeded American and British monitors were astonished to note a sharp increase in the volume and urgency of Eastern bloc communications, signs indicative of warnings sent of an imminent nuclear attack. Worse still, they observed a number of nuclear-capable planes being placed on standby at East German bases. The Warsaw Pact was on a strategic intelligence alert, the visible tip of an iceberg of fear.

Back in the Kremlin, fright was likely compounded by a leadership crisis. Andropov was growing sicker by the day. He was already suffering from diabetes and a heart condition, and by fall he was spending most of his time at a suburban Moscow clinic hooked up to a dialysis machine. In early November he was plainly dying, but still obsessed with RYAN. Probably, it devolved on his septuagenarian colleagues to carry out his orders. By November 9 they were apparently near panic. Moscow sent all KGB residencies in the unsuspecting capitals of the West an urgent call for any scrap of data that might pertain to an attack.

Disaster beckoned. Deterrence theorists' worst nightmares seemed about to materialize. For the first time since the Cuban Missile Crisis both sides were "eyeball to eyeball"—only now one had little idea what was going on, and the other was led by a dying man and acting crazy. The degree of the Soviets' strategic delusions should not be underestimated. Not one of the missiles that they so feared, the fast flying Pershing IIs, would be delivered to Europe before January of the new year. They had allowed fear of American technology, ideology, and Ronald Reagan to drive them to the brink.

Days passed and nothing happened. Able Archer wound down and still nothing happened. Gradually it must have dawned on the leadership that Armageddon was not upon them. The most dangerous nuclear crisis in the Cold War had passed. Yet those terrifying days in November were not easily forgotten, particularly it seems, by Mikhail Gorbachev, at the time a newly elevated member of the Politburo.

Reagan, though, continued to talk, only realizing the impact of his rhetoric when KGB insider Oleg Gordievsky defected in September 1985 and was debriefed by CIA director William Casey. Nearly two years had passed since Able Archer. Yet at least when the President met Gorbachev two months later in Geneva, each would understand the other in a way no American and Soviet leader had since Khrushchev and Kennedy. It is hard to believe that a mutual awareness of how close their countries had come to disaster did not pave the way to the intermediate nuclear force agreement and the elimination of the Pershing IIs and SS-20s. History has seldom witnessed a better deal.

RYAN marked a fundamental milestone (almost a tombstone) in the conduct of international affairs in the nuclear era. The crisis made it starkly apparent that the bridge of deterrence, while basically stable, could not bear every load. In an environment of sufficient hostility, perceptions are apt to become perilously distorted, and the rational actor could be transformed into a deluded heavy or could be simply oblivious to what was happening on the stage. Safety demanded that each side communicate its intentions openly and in the clearest possible terms. It also required that nuclear adversaries truly accept the absurdity of war at this level. Rational planning for nuclear war was an oxymoron. MAD worked precisely because it made the consequences so preposterously high. This was the end served by U.S. weapons acquisitions strategy, and why it ultimately succeeded in the Cold War environment.

By the beginning of 1984 the Soviets were coming to understand the magnitude of that success. They had avoided nuclear war, but fell further behind than ever in the arms race. In this sphere Ronald Reagan was far from finished with them; instead, he played upon their gloomy imagination with a dreamscape of futuristic military hardware—stealth aircraft and cruise missiles, a host of precision munitions, and above all his Strategic Defense Initiative (SDI).

Though prompted by his national security advisor, Robert McFarlane, Reagan's advocacy of a strategic defense against ballistic missiles was an unexpected gesture, and one he took up and ran with himself. "I call upon the scientific community in our country, those who gave us nuclear weapons," he beckoned in his famous "Star Wars" speech, "to turn their great talents now to the cause of mankind and world peace, to give us the means of rendering these nuclear weapons impotent and obsolete."[21] By all accounts Reagan was a passionate believer in SDI, but from a strategic perspective there could hardly have been a better bluff—"the greatest sting operation in history," McFarlane called the idea.[22]

The Soviets had no doubts about whose weapons he wanted "impotent and obsolete." Like nothing else Reagan did to them, SDI unmasked the depth of Soviet despair. When the leadership learned of SDI, reported former KGB general Oleg Kalugin, "They were convinced they would never be able to match the U.S. program for purely financial reasons." The USSR certainly had the scientific talent to effectively assess the possibilities of SDI. And in fact a number of the Soviet Union's most prestigious researchers, led by Yevgeni Velikhov and Roald Sagdeev, publicly maintained that such a defense could easily be defeated—either through saturation by increasing the number of offensive warheads, or through a variety of countermeasures. Yet these assurances always had a hollow ring. The very scale and intensity of the Soviet political and diplomatic campaign to stop SDI betrayed extreme anxiety over the program's possibilities.

Once again they were reacting to a mirage. The scheme—in Reagan's eyes an impenetrable shield floating above America—stood little chance of coming to fruition. Besides, a strategic missile defense was only threatening if deployed unilaterally, potentially allowing a preemptor to strike first and then activate his defenses to defeat retaliation. A missile defense on both sides would actually serve to deter a first strike, since neither party could be sure how well the other's defenses worked and thereby arrive at a firm estimate of initial attrition. Significantly, the side with the weaker technology would not necessarily be at a disadvantage so long as his defense remained minimally credible. Reagan even promised to share relevant technology. Yet the Soviets had neither the will nor the resources to respond. Barely discernible between their endless lines of anti-SDI invective could be seen waving, for the first time in the forty-year history of the great arms competition, the white flag of surrender.

"The Cold War is over," Ronald Reagan announced bluntly on the last day of his presidency—only to be promptly contradicted by his successor's vastly credentialed national security advisor. Reagan knew better. If anybody could claim credit for winning it, it was he. Of course, Lady Luck does have a weakness for movie stars.

The peaceful conclusion of the Cold War was a remarkable development. Never before had two such diametrically opposed powers waged such a prolonged economic and political struggle without it resulting in a major war. Since no other geopolitical factor had changed fundamentally, it is logical to assume that it was nuclear weapons, or at least the consequences of their use, that made the difference. Weapons development had always been an independent and even contradictory factor in the history of war, now it had apparently closed off the avenue to armed conflict at the highest level, at least so long as the potential combatants remained sane. The fact that we survived the Cold War is also due to at least some measure of good fortune. Nonetheless, by any realistic judgment, civilization's long-term prospects at its end appeared far better than at its beginning.

EPILOGUE

This train has reached the end of the line. At the last stop, the definitive end of the Cold War made it possible to draw some reasonably firm historical conclusions about the great U.S.-Soviet arms race. Beyond this, however, there is only rough and uncharted territory. War and weapons development plainly persist, but there is also a strong impression they have been knocked off kilter, that they are evolving rapidly but also with an edge of desperation.

Operationally, the 1991 Gulf War was supremely satisfying to the victors and among conventional war theorists; but its disappointing political results have raised serious suspicions that large combined arms campaigns are simply too blunt an instrument to achieve meaningful ends. This appears to be supported by the greater strategic success achieved by air operations in the Balkans and Afghanistan. The future of massed land forces is also challenged by the proliferation of guided munitions. Thus far, U.S. and allied ground forces have been fortunate to have been in sole possession of such arms in combat, thus avoiding the consequences of being on the receiving end. Ownership, however, is spreading, and it seems likely that when two similarly equipped forces clash, the results will be high, even paralyzing attrition. The alternative, much lighter, more mobile land forces operating in a nonlinear fashion, raises serious logistical questions, not to mention the need for protection against more conventional arms.

These contradictions appear symptomatic. The evolving military balance promises mostly paradox. In the developed world, power is increasingly spelled out in terms of social and economic factors, and war among the players in this fortunate sphere grows increasingly irrelevant and unlikely. Yet this is also the home of perhaps 80 percent of the planet's military power, used externally to deter war, stop it, or to reverse aggression—mostly among the poor and despotic.

In these circles, generally located in the vast and varied terrain of the developing world, military power retains far more perceived utility, but is in short supply. Hence organized violence gives the appearance of probing for weakness and dodging strength—especially the strength of advanced outsiders. Warfare here promises to move to complex terrain, particularly cities, where urban clutter will slow operations, thwart high-technology weapons, and above all inflict casualties. Weapons in places like Mogadishu and Grozny are not so much developed as employed with lethal ingenuity—rocket-powered grenades shot in barrages against helicopters, or mines set off by cell phones. These in turn are being countered in the West by ever more precision in targeting, by substituting robots for soldiers, and introducing that nearly oxymoronic instrument of coercion, the nonlethal weapon. Where this will leave us in another several decades is basically a matter of guessing.

Finally and most recently, we have been confronted by what amounts to state-of-the-art terrorism—strategic jujitsu so decoupled from any norms of human combat that it leaves us reeling. Weapons in this context take on a very peculiar cast, our multitrillion-dollar arsenals matched against al Qaida wielding exploding rubber boats and transforming our very means of transportation into flaming swords of mass destruction. Add to this the prospect of militant viruses and bacteria, heretofore considered too dangerous and uncontrollable to let loose. The undiluted urge to do us harm—to take us by surprise in our peaceful pursuits, to maim and kill not just our citizens but the symbols of our national life, and to pursue these objectives with an unwavering urge for self-destruction—leads to the frightening suspicion that we have been overtaken by something new, war dragged by zealotry into uncharted territory.

Yet in this very desperation there may be hope. If war has been marginalized to the point that it can only be waged enthusiastically by the truly evil and self-defeating, then it may well be a sign of its death throes. The evidence indicates that wealth and especially democracy have a fundamental effect in suppressing, if not eliminating, warfare. Both are spreading—not without setbacks—but their eventual triumph is hardly beyond imagining. Humans will remain individually violent, we will continue to be armed, but the future may still be one primarily of light not darkness.

Notes

CHAPTER 3: HISTORY POISED ON THE TIP OF A SPEAR

1. Cited in S. N. Kramer, *History Begins at Sumer* (Philadelphia: 1981), p. 123.

2. Homer, *Iliad*, trans. Richmond Lattimore (Chicago: 1951), 2:385–87.

3. Cited in Geoffrey Parker et al., *Warfare* (Cambridge, UK: 1995), p. 20.

CHAPTER 4: GHOST RIDERS

1. Herodotus, *Histories*, 4, 125–30.

2. Sun Tzu, *Art of War*, in R. D. Sawyer, *The Seven Military Classics of Ancient China* (Boulder: 1993), p. 161.

3. Herodotus, *Histories*, 4, 46–47, trans. Aubrey de Selincourt (Harmondsworth, UK: 1981), p. 286.

4. Plutarch, "Crassus," *Plutarch's Lives* (New York: 1992), Vol. 2, p. 738.

5. Georges Duby, *The Chivalrous Society* (London: 1977), p. 119.

6. Cited in Georges Duby, *William Marshal: The Flower of Chivalry* (London: 1986), p. 70.

7. Cited in Michael Prawdin, *The Mongol Empire* (London: 1940).

8. Cited in J. J. Saunders, *The History of the Mongol Conquests* (New York: 1971), p. 49.

CHAPTER 5: IMPERIAL TREADMILL

1. Bronze soldiered on and was given a big boost with the invention of firearms, especially cannon. The metal's resilience and ability to withstand sudden stress paid big dividends, so much so that as late as 1870, French gunners marched against the Prussians accompanied by artillery cast of the trusty metal. This marked the end of the line, with the fast-firing crucible-steel guns of Alfred Krupp blowing the Gallic bronzes into oblivion.

2. Cited in A. T. Olmstead, *The History of Assyria* (New York: 1923), pp. 648–49.

3. Livy 5.42.

4. Livy 31.34.4–6.

5. While puzzling signs of early contact with Eurasia do exist, most anthropologists and archaeologists continue to believe that the civilizations of the New World were a product of independent development.

6. Cited in G. W. Conrad and A. Demarest, *Religion and Empire: The Dynamism of Aztec and Inca Expansionism* (Cambridge, UK: 1984), p. 22.

7. Inga Clendinnen, *Aztecs: An Interpretation* (Cambridge, U.K.: Cambridge University Press, 1991), pp. 46, 237–38.

CHAPTER 7: TUBES OF FIRE

1. Egerton MS (British Museum, London) 2642, f. 150; cited in John U. Nef, *War and Human Progress* (New York: 1950).

2. Guns the size of bombards had not developed in China. This stemmed partly from China's unified power structure, which usually made it necessary to defend, not break into, cities; but also because the ancient *hang-t'u* (stamped earth) technique of building walls absorbed the shock of cannonballs far better than the thinner stone construction used in the West.

3. Blaise de Montluc, *Commentaries* (Paris: 1911), Book 1, 50.

4. Recorded in his notebook, the so-called *Codex Atlanticus.*

5. For 150 years this sequence explained why bursting projectiles were not extended to field pieces: their barrels were too long to fuse reliably. Experiments around 1740, however, demonstrated that the whole daredevil procedure was unnecessary. No matter what its position, escaping flames from the charge automatically lit the fuse. While this discovery was immediately applied to mortars, field pieces continued to fire primarily solid shot for over a century—a reflection, among other things, of the more predatory nature of siege warfare.

6. Niccolò Machiavelli, *The Prince,* trans. Luigi Ricci (New York: 1940), Chapter 4.

7. Hugo Grotius, *The Law of War and Peace,* Vol. 1, ed. William Wherwell (Cambridge, UK: 1853), p. lix.

CHAPTER 8: GUNS AWAY

1. Jeremy Black, *War and the World: Military Power and the Fate of Continents, 1450–2000* (New Haven: 1998), p. 58.

2. Its analogue, the all-big-gun, turbine-powered battleship of 1906, destined to set off the great capital-ship-building race leading up to World War I, bore the same name.

3. Garrett Mattingly, *The Armada* (Boston: 1959), p. 277.

4. Peter Kirsch, *The Galleon: The Great Ships of the Armada Era* (London: 1988), p. 28.

5. *The Diary of Christopher Columbus's First Voyage to America, 1492–1493,* abstracted by Fray Bartolomé de Las Casas, trans. Oliver Dunn and James E. Kelley, Jr. (Norman, Oklahoma: 1989), p. 67.

6. Aram Bakshian, Jr. "The Janissaries," *MHQ: The Quarterly Journal of Military History,* Vol. 4, No. 3 (Spring 1992), p. 38.

7. Black, *War and the World,* p. 46.

CHAPTER 9: GUN CONTROL

1. Cited in Geoffrey Parker, ed. *The Cambridge History of Warfare* (Cambridge, UK: 1995), p. 162.

2. Cited in David Patten, "Ferguson and His Rifle," *History Today* 28 (1978), p. 451.

3. Cited in John A. Lynn, "Vauban," *MHQ: The Quarterly Journal of Military History,* Vol. 1, No. 2 (1989).

4. Quoted in Bernard Brodie and Fawn Brodie, *From Crossbow to H-Bomb* (Bloomington: 1972), p. 94.

5. Cited in Geoffrey Parker, ed., *Warfare: The Triumph of the West* (Cambridge, UK: 1995), p. 127.

6. Cited in Ira Meistrich, "Trafalgar," *MHQ: The Quarterly Journal of Military History,*

Vol. 4, No. 1 (Autumn 1991), p. 91.

7. Napoleon Bonaparte, *Correspondence de Napoleon* (Paris: 1858), Vol. 32, p. 27.

8. Cited in Brodie and Brodie, *From Crossbow to H-Bomb,* p. 118.

9. Ibid., p. 127.

CHAPTER 10: DEATH MACHINES

1. Jefferson to John Jay, August 30, 1785, *Papers of Thomas Jefferson,* ed. Julian P. Boyd, Vol. 8, p. 455.

2. Jefferson to Monroe, January 1801, cited in Constance M. Green, *Eli Whitney and the Birth of American Technology* (Boston: 1956), p. 110.

3. Carrington Committee Report, January 6, 1827, cited in Merritt Roe Smith, *Harpers Ferry Armory and the New Technology* (Ithaca: 1977), p. 206.

4. Samuel Colt, "On the Application of Machinery to the Manufacture of Rotating Chambered-Breech Fire-Arms . . ." *Proceedings of the Institution of Civil Engineers* 11 (1851–52), p. 44.

5. Cited in William Hosley, *Colt: The Making of an American Legend* (Amherst: 1996), p. 45.

6. Grady McWhiney and Perry D. Jamieson, *Attack and Die: The Civil War, Military Tactics, and Southern Heritage* (Montgomery, Ala.: 1982), pp. 27–40.

7. General Hill cited in McWhiney and Jamieson, *Attack and Die,* p. 4.

8. Lyman cited in McWhiney and Jamieson, *Attack and Die,* p. 75.

9. Sherman cited in J. F. Marszalek, *Sherman: A Soldier's Passion for Order* (New York: 1993), p. 296.

10. Cited in Harold L. Peterson and Robert Elman, *The Great Guns* (New York: 1971), p. 140.

11. Letter, Gatling to Major General W. B. Franklin, April 26, 1868, cited in Paul Wahl and Donald R. Toppel, *The Gatling Gun* (New York: 1965), p. 36.

12. Cited in J. D. Scott, *Vickers: A History* (London: 1962), p. 25.

13. Nor did anything like the disaster of the Mitrailleuse befall the 75 in fighting along the Western Front. So long as the conflict remained fluid, the gun was devastating, at one point mowing down 2,000 members of the Lepel Brigade caught in a beet field in a matter of five minutes. At the Marne, the effects of the 75 likely saved Paris and perhaps France. But from that point, the gun became a victim of its own success, playing a major role in the German decision to dig in. Against trenches, the 75's mobility was largely irrelevant and its flat trajectory a distinct disadvantage. Plunging fire was needed here, and howitzers and other heavy pieces came to dominate. So in the end, Mademoiselle Soixante-quinze was left to become something of a wallflower.

CHAPTER 11: STEAMING THROUGH TROUBLED WATERS

1. William Hovgaard, *Modern History of Warships* (London: 1920), p. 15.

2. Richard Hough, *The Death of the Battleship* (New York: 1963), p. 6.

3. Cited in Dan van der Vat, *Stealth at Sea: The History of the Submarine* (Boston: 1995), p. 30.

4. Cited in Richard Knowles Morris, *John P. Holland* (Annapolis: 1966), p. 89.

5. Testimony of Admiral George Dewey taken by the House Committee on Naval Affairs, April 23, 1900, Dewey Papers, Library of Congress.

6. Cited in Sutherland Denlinger and Charles B. Gary, *War in the Pacific: A Study of Peoples and Battle Problems* (New York: 1936), p. 76.

7. Cited in Morris, *John P. Holland*, p. 7.

8. Wilson cited in Arthur J. Marden, *From Dreadnought to Scapa Flow*, 4 vols. (London: 1961), Vol. 1, p. 332.

9. Alfred T. Mahan, *From Sail to Steam* (New York: 1907), p. 197.

10. Alfred T. Mahan, *The Influence of Sea Power upon History: 1660–1783* (New York: 1890), p. 2.

11. "Report of the Reconciling Committee on Question 21," U.S. Naval War College, William Snowden Sims Papers, Library of Congress.

12. John A. Fisher, *Memories and Records*, Vol. 2 (New York: 1920), p. 197.

13. Arthur J. Marder, *Anatomy of British Sea Power* (New York: 1940), p. 422.

14. Cited in Richard A. Hough, *Admiral of the Fleet: The Life of John Fisher* (New York: 1969), p. 243.

15. Letter, Baron Friedrich von Holstein to Paul Hartzfeld, April 9, 1887, cited in Jonathan Steinberg, *Yesterday's Deterrent* (London: 1965), p. 117.

16. Alfred T. Mahan, "Reflections, Historic and Other, Suggested by the Battle of the Sea of Japan," *U.S. Naval Institute Proceedings*, June 1906, pp. 452, 461–62.

17. Winston S. Churchill, *The World Crisis*, Vol. 1 (New York: 1923), p. 33.

CHAPTER 12: FALSE PINNACLE

1. Hillaire Belloc, *The Modern Traveler*, cited in John Ellis, *The Social History of the Machine Gun* (New York: 1975), p. 94.

2. Cited in Jeremy Black, *War and the World: Military Power and the Fate of Continents, 1450–2000* (New Haven: 1998), p. 186.

3. Cited in E. P. Badesi, *From Adversaries to Comrades-in-Arms: West Africans and the French Military, 1885–1918* (Waltham, MA: 1979), p. 20.

4. *The Proceedings of the Hague Peace Conference* (London: 1920), pp. 286–87.

5. John Ellis, *The Social History of the Machine Gun* (New York: 1975), p. 70.

6. Cited in T. Ranger, *Revolt in Southern Rhodesia* (London: 1967), p. 121.

7. G. W. Stevens, *With Kitchener to Khartoum* (London: 1898), p. 300.

8. Cited in G. S. Hutchinson, *Machine Guns: Their History and Tactical Employment* (London: 1938), pp. 48–49.

9. Cited in Ellis, *The Social History of the Machine Gun*, pp. 98–99.

10. Cited in Ellis, *Social History of the Machine Gun*, p. 96.

11. Cited in Hubert C. Johnson, *Breakthrough!: Tactics, Technology, and the Search for Victory on the Western Front in World War I* (Novato, CA: 1994), p. 9.

12. Reinhard Scheer, *Germany's High Seas Fleet in the World War* (London: 1920), p. 11.

13. Ardant du Picq, *Battle Studies: Ancient and Modern Battles* (New York: 1921), pp. 101–2, 113, 181, 229.

14. Cited in William I. Hull, *The Two Hague Conferences and Their Contribution to International Peace* (Boston: 1908), p. 87.

15. "Declaration (IV, 2) Concerning Asphyxiating Gases," in Adam Robert and Richard Duelff, eds., *Documents on the Laws of War* (Oxford: 1982), p. 35.

16. I. S. Bloch, *The Future of War in Its Technical, Economic, and Political Relations*, abridged (Boston: 1902), pp. xvi, lxi–lxii.

17. A. Conan Doyle, *Danger* (London: 1913), p. 16.

18. Fitzgerald cited in I. F. Clarke, *Voices Prophesying War* (New York: 1993), p. 105.

19. H. G. Wells, *The World Set Free* (London: 1926), p. 116.

CHAPTER 13: ACCIDENTAL ARMAGEDDON

1. Cited in Henri Isselin, *The Battle of the Marne* (New York: 1966), p. 236.

2. Tappen to Reichsarchiv, July 16, 1930, cited in Ulrich Trumpener, "The Road to Ypres: The Beginnings of Gas Warfare in World War I," *Journal of Modern History* 47 (September 1975), p. 468.

3. Cited in J. F. C. Fuller, *The Conduct of War: 1789–1961* (New Brunswick, N.J.: 1961), p. 160.

4. Cited in Bruce I. Gudmundsson, "These Hideous Weapons," *MHQ: The Quarterly Journal of Military History,* Vol. 7, No. 1 (Autumn 1994), p. 73.

5. Cited in John Keegan, *The Face of Battle* (New York: 1976), p. 245.

6. John Jellicoe to the Admiralty, October 30, 1914.

7. Cited in Walter Gorlitz, *The Kaiser and His Court* (London: 1961), pp. 133–34; Alfred P. von Tirpitz, *My Memoirs,* Vol. 2, p. 424.

8. Cited in Geoffrey Bennett, *The Battle of Jutland* (Philadelphia: 1965), p. 84.

9. Reinhard Scheer, *Germany's High Seas Fleet in the World War* (London: 1920), p. 155.

10. Scheer cited in A. A. Hoeling, *The Great War at Sea,* (New York: 1965), pp. 150–51; Memorandum, Jellicoe to First Lord of the Admiralty, October 20, 1916.

11. Edward V. Rickenbacker, *Rickenbacker* (Englewood Cliffs: 1967), p. 106.

12. Cited in Ezra Bowen, *Knights of the Air* (Alexandria, VA: 1980), p. 19.

13. E. D. Swinton, "Notes on the Employment of Tanks," February 6, 1916.

14. Cited in Thomas Fleming, "Iron General," *MHQ: The Quarterly Journal of Military History* (Winter 1995), Vol 6, No. 4, p. 70.

15. Cited in ibid., p. 64.

CHAPTER 14: GRUDGE MATCH

1. Cited in Williamson Murray and Allan R. Millett, *Military Innovation in the Interwar Period* (Cambridge, U.K.: 1996), p. 272.

2. Winston Churchill, *The Second World War, Vol. 1: The Gathering Storm* (Boston: 1948), p. 156.

3. Cited in Ronald H. Bailey, *The Air War in Europe* (Alexandria, VA: 1979), p. 54.

4. Letter, Harris to Churchill, March 11, 1943.

5. Cited in Martin Caiden, *Black Thursday* (New York: 1981), p. 211.

6. Order cited in John Costello, *The Pacific War: 1941–1945* (New York: 1982), p. 105.

7. Cited in Len Giovannitti and Fred Freed, *The Decision to Drop the Bomb* (New York: 1965), pp. 194–95.

CHAPTER 15: COLD WAR—INFERNO OF ARMS

1. It now is known that the USSR had demobilized to a far greater extent than U.S. intelligence had originally estimated.

2. Cited in James Harford, *Korolev* (New York: 1997), p. 132.

3. Cited in Charles E. Bohlen, *Witness to History: 1929–1969* (New York: 1973), pp. 495–96. Now they set about making good on the promise.

4. Cited in W. F. Scott, "Illusion of Change," *Air Force Magazine,* March 1989, p. 71.

5. Interview, February 14, 1989.

6. So long as ballistic missiles carried but a single nuclear warhead, a preemptive assault against an adversary's ICBMs remained unattractive, since it would require at least one missile to kill its equivalent and in practice probably two or more. Unless radically outnumbered, both sides could thus count on weathering an initial attack and still having enough missiles to mount a devastating counterblow, presumably against enemy cities, which, because they were inherently soft targets, required proportionately fewer missiles. Hence, not only did striking first imply suicide, but accumulations beyond a certain point promised little payback.

7. Cited in Simon Rosenblum, *Misguided Missile: Canada, the Cruise, and Star Wars* (Toronto: 1985), p. 30.

8. Cited in *Department of Defense Annual Report, Fiscal Year 1979* (Washington, DC: 1978), p. 115.

9. Norman Augustine, *Augustine's Laws* (New York: 1982), p. 48.

10. Capt. J. W. Kehoe and K. S. Brower, "U.S. and Soviet Weapon System Design Practices," *International Defense Review* (6:1982), p. 706.

11. Arthur Alexander, "The Process of Soviet Weapons Acquisition," RAND paper presented to the European Study Commission, Paris 1977, p. 4.

12. Cited in Thomas Powers, "Planning the Strategy for World War III," *Atlantic Monthly* (November 1982), p. 86.

13. Cited in ibid., p. 94.

14. Cited in Ronnie Dugger, *On Reagan, the Man and His Presidency* (New York: 1983), p. 396.

15. Cited by Susan Poyas, *Reuters International News,* June 13, 1982.

16. *Public Papers of the Presidents of the United States: Ronald W. Reagan, 1983* (Washington, DC: 1984), p. 363.

17. Cited in Gordon Brook-Shepherd, *The Stormbirds: Soviet Postwar Defectors* (New York: 1989), p. 332.

18. Speech of General Secretary of the Communist Party Comrade Yu. V. Andropov, *Kommunist,* No. 9 (June 1983), p. 5.

19. Cited in Raymond Garthoff, *Detente and Confrontation: American-Soviet Relations from Nixon to Reagan* (Washington, DC: 1985), p. 1017.

20. Cited in Murray Marder, "Defector Told of Soviet Alert," *Washington Post,* August 8, 1986, p. A22.

21. Ronald Reagan, "Address to the Nation on Defense and National Security," *Historic Documents of 1983,* pp. 315–16.

22. Cited in Frances FitzGerald, *Way Out There in the Blue: Reagan, Star Wars and the End of the Cold War* (New York: 2000), p. 195.

Selected Bibliography

CHAPTER 1: A MOST ANCIENT TALISMAN

Anderson, Connie. "Predation and Primate Evolution." *Primates,* Vol. 27, No. 1 (1986).

Bettinger, R. L. *Archaeological and Evolutionary Theory* (New York: 1991).

Boyd, R., and P. J. Richerson, *Culture and the Evolutionary Process* (Chicago: 1985).

Cohen, M. N. *The Food Crisis in Prehistory* (New Haven: 1978).

Ehrenberg, Margaret. *Women in Prehistory* (London, 1989)

Ehrenreich, Barbara. *Blood Rites: Origins and History of the Passions of War* (New York: 1997).

Eibl-Eibesfeldt, I. *The Biology of Peace and War: Man, Animals, and Aggression* (New York: 1979).

———. *Human Ethology* (New York: 1989).

Falk, D. *Braindance* (New York: 1992).

Harris, M. *Our Kind: Who We Are, Where We Came From, and Where We Are Going* (New York: 1989).

Keeley, Lawrence H. *War Before Civilization: The Myth of the Peaceful Savage* (New York: 1996).

Liljegran, Ronnie. "Animals of Ice Age Europe." In Goren Burenhault, ed., *The First Humans: Human Origins and History to 10,000 B.C.* (San Francisco: 1993).

Lorenz, Konrad. *On Aggression.* Trans., Marjorie K. Wilson (New York: 1967).

McFarland, D., ed. *The Oxford Companion to Animal Behavior* (Oxford: 1981).

O'Connell, Robert L. *Of Arms and Men: A History of War, Weapons, and Aggression* (New York: 1989).

Ortega y Gasset, J. *Meditations on Hunting* (New York: 1985).

Riddle, John. *Contraception and Abortion from the Ancient World to the Renaissance* (Cambridge, MA: 1992).

Weissner, R. "Hxaro. A Regional System of Reciprocity for Reducing Risks Among the Kung." Ph.D. diss. University of Michigan, Ann Arbor, 1977.

Wenke, R. J. *Patterns in Prehistory: Humankind's First Three Million Years* (New York: 1990).

Wilson, E. *Sociobiology: The New Synthesis* (Cambridge, MA: 1975).

Zillman, D. *Hostility and Aggression* (Hillsdale, NJ: 1979).

CHAPTER 2: WAR'S ARRIVAL

Bar-Yosef, O. "The Walls of Jericho: An Alternate Interpretation." *Current Anthropology,* Vol. 27, No. 2 (1986).

Blumler, M., and R. Byrne. "The Ecological Genetics of Domestication and the Origins of Agriculture." *Current Anthropology,* Vol. 12, No. 1 (1991).

Budiansky, Stephen. *The Covenant of the Wild: Why Animals Chose Domestication* (New York: 1992).

Cribb, R. *Nomads in Archaeology* (Cambridge, UK: 1991).

Grunfeld, Foster. "The Unsung Sling." *MHQ: The Quarterly Journal of Military History,* Vol. 9, No. 1 (Autumn 1996).

Mellaart, J. *Çatal Hüyük: A Neolithic Town in Anatolia* (New York: 1967).

———. *The Neolithic of the Near East* (London: 1975).

O'Connell, Robert L. "The Mace." *MHQ: The Quarterly Journal of Military History,* Vol. 3, No. 3 (Spring 1991).

———. *Ride of the Second Horseman: The Birth and Death of War* (New York: 1995).

Rindos, D. *The Origins of Agriculture: An Evolutionary Perspective,* (Orlando: 1984).

Chapter 3: History Poised on the Tip of a Spear

Adams, R. M. "Developmental Stages in Ancient Mesopotamia." In S. Struever, ed., *Prehistoric Agriculture* (Garden City: 1971).

———. *The Evolution of Urban Society: Early Mesopotamia and Prehistoric Mexico* (Chicago: 1966).

Adcock, F. E. *The Greek and Macedonian Art of War* (Berkeley: 1957).

Andreski, Stanislav, *Military Organization and Society* (Berkeley: 1968).

Borza, Eugene N. "What Philip Wrought." *MHQ: The Quarterly Journal of Military History,* Vol. 5, No. 4 (Summer 1993).

Garlan, Yvon. *War in the Ancient World: A Social History* (London: 1975).

"Gilgamesh and Agga." Trans., S. N. Kramer. In J. B. Pritchard, *Ancient Near Eastern Texts: Relating to the Old Testament* (Princeton: 1950).

Hanson, Victor Davis. "Alexander the Killer." *MHQ: The Quarterly Journal of Military History,* Vol 10, No. 3 (Spring 1998).

———. "Genesis of Infantry" and "From Phalanx to Legion." In Geoffrey Parker, ed., *The Cambridge History of Warfare: The Triumph of the West* (Cambridge, UK: 1995).

———. "The Leuctra Mirage." *MHQ: The Quarterly Journal of Military History,* Vol 2, No. 2 (Winter 1990).

———. "Not Strategy, Not Tactics." *MHQ: The Quarterly Journal of Military History,* Vol. 1, No. 3 (Spring 1989).

———. *Western Way of War,* New York, 1989.

Homer. *Iliad.* Trans., Richmond Lattimore (Chicago: 1951).

The Mahabharata. Trans., J. A. B. van Buitenen (Chicago: 1978).

McNeill, W. H. *Plagues and Peoples* (Garden City: 1976).

Ober, Josiah. "The Evil Empire." *MHQ: The Quarterly Journal of Military History,* Vol. 10, No. 5 (Summer 1998).

O'Connell, Robert L. *Of Arms and Men: A History of War, Weapons, and Aggression* (New York: 1989).

Oman, Charles. *The Art of War in the Middle Ages.* Vol. 2, (Ithaca: 1953).

"Stele of Vultures." In Yigael Yadin, *The Art of Warfare in Biblical Lands: In Light of Archaeological Study.* Vol. 1 (New York: 1963).

Wilson, P. J. *The Domestication of the Human Species* (New Haven: 1988).

CHAPTER 4: GHOST RIDERS

Anthony, D. W. "The Bronze Age and Early Iron Age Peoples of Eastern Central Asia." In *Journal of Indo-European Studies Monograph* No. 26 (Philadelphia: 1998).

——. "The 'Kurgan Culture,' Indo-European Origins, and the Domestication of the Horse." *Current Anthropology* 265 (1986).

Anthony, D. W., and D. R. Brown. *Neolithic Horse Exploitation in the Eurasian Steppes: Diet, Ritual, and Riding.* To be published.

Contamine, Philippe. *War in the Middle Ages.* Trans., Michael Jones (Oxford: 1984).

Cribb, R. *Nomads in Archaeology* (Cambridge, UK: 1991).

Duby, Georges. *The Chivalrous Society* (London: 1977).

——. *William Marshal: The Flower of Chivalry* (London: 1986).

Herodotus. *The Histories,* trans. Aubrey de Sélincourt (New York: 1954).

Khazanov, A. *Nomads and the Outside World* (Cambridge, UK: 1984).

Legg, Stuart. *The Heartland* (New York: 1970).

Marshall, B. A. *Crassus: A Political Biography* (Amsterdam: 1976).

O'Connell, Robert L. "The First Warriors." *MHQ: The Quarterly Journal of Military History,* Vol. 8, No. 1 (Autumn 1995).

Okladnikov, A. P. "Inner Asia at the Dawn of History." In Denis Sinor, ed., *The Cambridge History of Early Inner Asia* (Cambridge, UK: 1990).

Plutarch, *The Lives of Noble Grecians and Romans.* 2 vols. Trans., John Dryden (New York: 1992.)

Saunders, J. J. *The History of the Mongol Conquests* (New York: 1971).

Shaughnessy, E. L. "Historical Perspectives on the Introduction of the Chariot into China." *Harvard Journal of Asiatic Studies,* Vol. 48, No. 1 (1988).

Sun Tzu. *Art of War.* In R. D. Sawyer, ed., *The Seven Military Classics of Ancient China* (Boulder: 1993).

Tarn, W. W. *Hellenistic Military and Naval Developments* (Cambridge, UK: 1930).

Verbruggen, J. F. *The Art of Warfare in Western Europe During the Middle Ages: From the Eighth Century to 1340* (Woodbridge, UK: 1997).

Warry, John, *Warfare in the Classical World* (London: 1980).

CHAPTER 5: IMPERIAL TREADMILL

Adams, R. M. *The Evolution of Urban Society: Early Mesopotamia and Pre-Hispanic Mexico* (Chicago: 1966).

Adcock, F. E. *The Roman Art of War Under the Republic* (Cambridge, MA: 1940).

Clendinnen, I. *Aztecs: An Interpretation* (Cambridge, UK: 1991).

Cockburn, T. A. "Infectious Disease in Ancient Populations." *Current Anthropology,* Vol. 12, No. 1 (1971).

Cogan, M., and I. Eph'al, eds. *Ah, Assyria: Studies in Assyrian History and Ancient Near Eastern Historiography Presented to Hayim Tadmor* (Jerusalem: 1999).

Díaz del Castillo, Bernal. *The Discovery and Conquest of Mexico* (New York: 1956).

Doyle, R. J. and N. Lee. "Microbes, Warfare, Religion, and Human Institutions." *Canadian Journal of Microbiology,* Vol. 32 (1986).

Flavius Vegetius Renatus. "The Military Institutions of the Romans." In T. R. Phillips, ed., *The Roots of Strategy* (Harrisburg: 1940).

Hassig, R. "Flower Wars." *MHQ: The Quarterly Journal of Military History,* Vol. 9, No. 1 (Autumn 1996).

——. *War and Society in Ancient Mesoamerica* (Berkeley: 1992).

Jankowska, N. B. "Some Problems of the Economy of the Assyrian Empire." In I. M. Diakonoff, ed., *Ancient Mesopotamia: Socio-Economic History* (Moscow: 1969).

Kemp, B. J. *Ancient Egypt: Anatomy of a Civilization* (London: 1989).

Kotker, N. "The Assyrians." *MHQ: The Quarterly Journal of Military History,* Vol. 3, No. 4 (Summer 1991).

Hall, A. R. "Military Technology." In C. Singer, E. J. Holmyard, and A. R. Hall, eds., *A History of Technology* (Oxford: 1954).

Herodotus. *The Histories,* trans. Aubrey de Sélincourt (New York: 1954).

Keightley, D. N., ed. *The Origins of Chinese Civilization* (Berkeley: 1983).

"The Legend of Sargon" and "Sargon of Agade." In J. B. Pritchard, ed., *Ancient Near Eastern Texts: Relating to the Old Testament* (Princeton: 1950).

Luttwak, E. *The Grand Strategy of the Roman Empire* (Baltimore: 1976).

Majino, G. *The Healing Hand: Man and Wound in the Ancient World* (Cambridge, MA: 1975).

McNeill, W. H. *Plagues and Peoples* (Garden City: 1976).

Meistrich, Ira. "The Mortuary Army of Ch'in Shih Huang Ti." *MHQ: The Quarterly Journal of Military History,* Vol. 6, No. 2 (Winter 1994).

O'Connell, Robert L. "The Insolent Chariot." *MHQ: The Quarterly Journal of Military History,* Vol. 2, No. 3 (Spring 1990).

Rindos, D. *The Origins of Agriculture: An Evolutionary Perspective* (Orlando: 1984).

Saggs, H. W. F. *The Might That Was Assyria* (London: 1984).

Sanders, W. T., J. R. Parsons, and R. S. Santley. *The Basin of Mexico: Ecological Processes in the Evolution of a Civilization* (New York: 1979).

Stewart, John. "The Elephant in War." *MHQ: The Quarterly Journal of Military History,* Vol. 3, No. 3 (Spring 1991).

Twitchett, D., and M. Loewe, eds. *The Cambridge History of China.* Vol. 1: *The Ch'in and Han Empires, 221 B.C.–A.D. 220* (Cambridge, UK: 1986).

Tyumenev, A. I. "The State Economy in Ancient Sumer." in I. M. Diakonoff, ed., *Ancient Mesopotamia: Socio-Economic History* (Moscow: 1969).

Warry, J. *Warfare in the Classical World* (London: 1980).

Yadin, Y. *The Art of Warfare in Biblical Lands: In Light of Archaeological Study.* 2 vols. (New York: 1963).

CHAPTER 6: AT SEA

Andreose, M., ed. *The Phoenicians* (New York: 1988).

Basch, L. "Phoenician Oared Ships." *Mariner's Mirror,* Vol. 55, No. 2 (1969).

Branigan, K., *The Foundations of Palatial Crete* (London: 1970).

Casson, Lionel, *The Ancient Mariners.* 2nd ed. (Princeton: 1991).

Hagg, R., and N. Marinatos, eds. *The Minoan Thalassocracy: Myth and Reality* (Göteborg: 1984).

Harden, D. *The Phoenicians* (New York: 1962).

Meijer, Fik. *A History of Seafaring in the Classical World* (London: 1986).

Mellersh, H. E. L. *The Destruction of Knossos: The Rise and Fall of Minoan Crete* (New York: 1970).

Morrison, J., and J. Coates. *The Athenian Trireme* (Cambridge, UK: 1986).

Roland, Alex. "Greek Fire." *MHQ: The Quarterly Journal of Military History,* Vol. 2, No. 3 (Spring, 1990).

Tarn, W. W. *Hellenistic Military and Naval Developments* (Cambridge, UK: 1930).

Tarn, W. W., and G. Griffith. *Hellenistic Civilization.* 3rd ed. (London: 1952).

Wallinga, H. *The Boarding Bridge of the Romans* (Groningen: 1956).

Warry, John. *Warfare in the Classical World* (London: 1980).

Whittaker, C. R. "Carthaginian Imperialism in the Fifth and Fourth Centuries." In P. D. A. Garnsey and C. R. Whittaker, eds., *Imperialism in the Ancient World: The Cambridge Research Seminar in Ancient History* (New York: 1978).

CHAPTER 7: TUBES OF FIRE

Adams, N., and S. Pepper. *Firearms and Fortifications: Military Architecture and Siege Warfare in Sixteenth Century Siena* (Chicago: 1986).

Black, Jeremy. *A Military Revolution? Military Change and European Society, 1550–1800* (Hampshire, UK: 1991).

Howard, Michael. *War in European History* (London: 1976).

Lieure, J. *Jacques Callot.* Vols. 7 and 8 (Paris: 1927).

MacCurdy, Edward. *The Notebooks of Leonardo da Vinci* (New York: 1937).

Machiavelli, Niccolò. *The Prince.* Trans., Luigi Ricci (New York: 1940).

McNeill, William H. *The Pursuit of Power* (Chicago: 1982).

Needham, Joseph. *Gunpowder as the Fourth Power, East and West* (Hong Kong: 1985).

———. *Science and Technology in Traditional China: A Comparative Perspective* (Cambridge, MA: 1981).

Nef, John U. *War and Human Progress* (New York: 1950).

Oman, Charles. *The Art of War in the Sixteenth Century* (New York: 1937).

Parker, Geoffrey. *The Military Revolution: Military Innovation and the Rise of the West, 1500–1800* (Cambridge, UK: 1988).

———. *The Thirty Years War* (London: 1984).

———. *Warfare: The Triumph of the West* (Cambridge, UK: 1995).

Roberts, Michael. *Gustavus Adolphus: A History of Sweden, 1611–1632.* Vol. 2 (London: 1958).

Schama, Simon. *The Embarrassment of Riches: An Interpretation of the Dutch Culture in the Golden Age* (Berkeley: 1988).

CHAPTER 8: GUNS AWAY

Arnold, Thomas. "Arms and Men: Floating Time Bombs." *MHQ: The Quarterly Journal of Military History,* Vol. 8, No. 4 (Summer 1996).

Ayalon, David. *Gunpowder and Firearms in the Mamluk Kingdom* (London: 1978).

Bakshian, Aram Jr. "The Janissaries." *MHQ: The Quarterly Journal of Military History,* Vol. 4, No. 3 (Spring 1992).

Black, Jeremy. *War and the World: Military Power and the Fate of Continents, 1450–2000* (New Haven: 1998).

Boots, J. L. "Korean Weapons and Armour." *Transactions of the Korean Branch of the Royal Asiatic Society,* Vol. 23, No. 2 (1934).

Brown, D. M. "The Impact of Firearms on Japanese Warfare, 1543–98." *Far Eastern Quarterly* 7 (1948).

Cipolla, Carlo M. *Guns and Sails in the Early Phase of European Expansion, 1400–1700* (London: 1965).

Conrad, G. W., and A. Demarest. *Religion and Empire: The Dynamism of Aztec and Inca Expansionism* (Cambridge, UK: 1984).

Dobyns, H. L., and P. L. Doughty. *Peru: A Cultural History* (New York: 1976).

Friel, Ian. *The Good Ship: Ships, Shipbuilding and Technology in England, 1200–1520* (London: 1995).

Guilmartin, John F. "The Galley in Combat." *MHQ: The Quarterly Journal of Military History,* Vol. 9, No. 2 (Winter 1997).

———. *Gunpowder and Galleys: Changing Technology and the Mediterranean Warfare at Sea in the Sixteenth Century* (Cambridge, UK: 1974).

Harrison, J. A. *Japan's Northern Frontier* (Gainesville: 1953).

Kirsch, Peter. *The Galleon: The Great Ships of the Armada Era* (London: 1990).

Las Casas, Bartolomé de. *The Diary of Christopher Columbus's First Voyage to America, 1492–1493.* Trans., Oliver Dunn and James E. Kelley, Jr. (Norman: 1989).

Lewis, Michael. *The Spanish Armada* (New York: 1968).

Mattingly, Garrett. *The Armada* (Boston: 1959).

McNeill, William H. *Venice: The Hinge of Europe* (Chicago: 1974).

Mura, J. V. "El 'Control Vertical' de un Máximo de Pisos Ecológicos en la Economía de las Sociedades Andinas." In J. V. Murra, ed., *Vistas de la Provincia de León de Huánuco* (Huánuco, Peru: 1972).

Parker, Geoffrey. *The Military Revolution: Military Innovation and the Rise of the West, 1500–1800* (Cambridge, UK: 1988).

———. "The Spanish Armada Revisited." *MHQ: The Quarterly Journal of Military History,* Vol. 10, No. 3 (Spring 1998).

———. "Why the Armada Failed." *MHQ: The Quarterly Journal of Military History,* Vol. 1, No. 1 (Autumn 1988).

———. *Warfare: The Triumph of the West* (Cambridge, UK: 1995).

Pierson, Peter. "Elizabeth's Pirate Admiral." *MHQ: The Quarterly Journal of Military History,* Vol. 8, No. 4 (Spring 1996).

———. "Lepanto." *MHQ: The Quarterly Journal of Military History,* Vol. 9, No. 2 (Winter 1997).

Unger, Richard W. *The Ship in the Medieval Economy, 600–1600* (Montreal: 1980).

Chapter 9: GUN CONTROL

Archibald, E. H. H. *The Wooden Fighting Ships in the Royal Navy: 892–1860* (New York: 1968).

Brodie, Bernard, and Fawn M. Brodie. *From Crossbow to H-Bomb* (Bloomington: 1972).

Chandler, David. *The Art of Warfare in the Age of Marlborough* (London: 1976).

———. *The Campaigns of Napoleon* (New York: 1966).

Childs, John. *Armies and Warfare in Europe: 1648–1789* (Manchester: 1982).

Clark, I. F. *Voices Prophesying War: 1763–1984* (London: 1966).

Clausewitz, Karl von. *On War* (Berlin: 1832).

Corbett, Julian S. *Some Principles of Maritime Strategy* (London: 1960).

Duffy, C. *Frederick the Great: A Military Life* (New York: 1985).

Dupuy, Trevor. *The Evolution of Weapons and Warfare* (Fairfax, VA: 1984).

Howard, Michael. *War and European History* (London: 1976).

Howarth, David. *Trafalgar: The Nelson Touch* (New York: 1969).

Kemp, P. K. *History of the Royal Navy* (New York: 1969).

Lynn, John A. "The Sun King's Star Wars." *MHQ: The Quarterly Journal of Military History*, Vol. 7, No. 4 (1995).

———. "Vauban." *MHQ: The Quarterly Journal of Military History*, Vol. 1, No. 2 (Winter 1989).

Lynn, John A., ed. *Tools of War: Instruments, Ideas and Institutions of Warfare, 1445–1871* (Urbana: 1990).

McNeill, William H. *Keeping Together in Time: Dance and Drill in Human History* (Cambridge, MA: 1995).

———. *The Pursuit of Power* (Chicago: 1982).

O'Connell, Robert L. "Brown Bess." *MHQ: The Quarterly Journal of Military History*, Vol. 1, No. 4 (Summer 1989).

———. "L'Armée Blanche." *MHQ: The Quarterly Journal of Military History*, Vol. 5, No. 1 (Autumn 1992).

Ropp, Theodore. *War in the Modern World* (Durham: 1959).

Rothenberg, Gunther E. *The Art of War in the Age of Napoleon* (London: 1977).

van Crevelt, Martin. *Technology and War: From 2000 B.C. to the Present* (New York: 1989).

CHAPTER 10: DEATH MACHINES

Brodie, Bernard, and Fawn M. Brodie. *From Crossbow to H-Bomb* (Bloomington: 1973).

Catton, Bruce. *This Hallowed Ground: The Story of the Union Side of the Civil War* (New York: 1956).

Crane, Stephen. *The Red Badge of Courage* (New York: 1963).

Dupuy, Trevor N. *The Evolution of Weapons and Warfare* (Fairfax, VA: 1984).

Grant, Ulysses S. *Personal Memoirs of U.S. Grant.* 2 vols. (New York: 1886).

Hallahan, William H. *Misfire* (New York: 1994).

Hanson, Victor Davis. *The Soul of Battle* (New York: 1999).

Harding, David, ed. *Weapons: An International Encyclopedia from 5000 B.C. to 2000 A.D.* (New York: 1990)

Hogg, Ian. *Artillery* (New York: 1972).

Hosley, William. *Colt: The Making of an American Legend* (Amherst: 1996).

Hounshell, David A. *From the American System to Mass Production, 1800–1932* (Baltimore: 1984).

James, Alfred P. "The Battle of the Crater." In T. Harry Williams, ed., *Military Analysis of the Civil War* (Millwood, NY: 1977).

Luvaas, Jay. *The Military Legacy of the Civil War* (Chicago: 1959)

McPherson, James M. *Battle Cry of Freedom: The Civil War Era* (New York: 1988).

McWhiney, Grady, and Perry D. Jamieson, *Attack and Die: The Civil War, Military Tactics, and Southern Heritage* (Montgomery, AL: 1982).

Mitchell, Reid. *Civil War Soldiers* (New York: 1988).

Mosby, John S. *The Memoirs of Colonel John S. Mosby* (Bloomington: 1959).

Peterson, Harold L., and Robert Elman. *The Great Guns* (New York: 1971).

Rogers, H. C. B. *A History of Artillery* (Secaucus: 1975).

Smith, Merritt Roe. *Harpers Ferry Armory and the New Technology* (Ithaca: 1977).

Wahl, Paul, and Donald R. Toppel. *The Gatling Gun* (New York: 1965).
Wilkinson, Frederick. *Guns and Rifles* (London: 1979).

CHAPTER 11: STEAMING THROUGH TROUBLED WATERS

Baxter, James P. *The Introduction of the Ironclad Warship* (New York: 1920).
Brown, David K. *Warrior to Dreadnought: Warship Development, 1860–1905* (London: 1997).
Churchill, Winston S. *The World Crisis.* Vol. 1 (New York: 1923).
Cuniberti, Vitorio. "An Ideal Battleship for the British Fleet." *Jane's Fighting Ships* (London: 1907).
Hough, Richard A. *Admiral of the Fleet: The Life of John Fisher* (New York: 1969).
———. *The Death of the Battleship* (New York: 1963).
Hovgaard, William. *Modern History of Warships* (London: 1920).
Mahan, Alfred T. *The Influence of Sea Power upon History: 1660–1783* (New York: 1890).
Marder, Arthur J. *From Dreadnought to Scapa Flow: The Royal Navy in the Fisher Era.* Vol. 1 (London: 1961).
Morison, Elting. *Admiral Sims and the Modern American Navy* (Cambridge, MA: 1942).
Morris, Richard Knowles. *John P. Holland* (Annapolis: 1966).
O'Connell, Robert L. *Sacred Vessels: The Cult of the Battleship and the Rise of the U.S. Navy* (New York: 1993).
Paine, Lincoln P. *Warships of the World to 1900* (Boston: 2000).
Puleston, W. D. *The Life of Admiral Mahan* (New Haven: 1939).
Ropp, Theodore. "Continental Doctrines of Sea Power." In Edward M. Earle, ed., *The Makers of Modern Strategy* (Princeton: 1943).
Rossler, Eberhard. *The U-boat: The Evolution and Technical History of German Submarines* (Annapolis: 1981).
Sandler, Stanley. *The Emergence of the Modern Capital Ship* (Newark, DE: 1979).
Steinberg, Jonathan. *Yesterday's Deterrent: Tirpitz and the Birth of the German Battle Fleet* (London: 1965).
Tuchman, Barbara. *The Guns of August* (New York: 1963).
van der Vat, Dan. *Stealth at Sea: The History of the Submarine* (Boston: 1995).

CHAPTER 12: FALSE PINNACLE

Angell, Norman. *The Great Illusion: A Study of the Relation of Military Power to National Advantage* (London: 1914).
Beal, Clifford. "Omdurman." *The Quarterly Journal of Military History,* Vol. 6, No. 1 (1993).
Black, Jeremy. *War and the World: Military Power and the Fate of Continents, 1450–2000* (New Haven: 1998).
Bloch, I. S. *The Future of War in Its Technical, Economic, and Political Relations.* Abridged (Boston: 1902).
Clarke, I. F. *Voices Prophesying War: 1963–1984* (London: 1966).
Doyle, A. Conan, *Danger* (London: 1913).
Ellis, John. *The Social History of the Machine Gun* (New York: 1975).
Johnson, Hubert C. *Breakthrough! Tactics, Technology and the Search for Victory on the Western Front in World War I* (Novato, CA: 1994).
Leed, Eric J. *No Man's Land: Combat and Identity in World War I* (Cambridge, UK: 1979).

Maxim, Hiram, P. *A Genius in the Family* (London: 1936).

Mueller, John. *Retreat from Doomsday: The Obsolescence of Major War* (New York: 1988).

Norman, E. H. *Japan's Emergence as a Modern State: Political and Economic Problems of the Meiji Period* (New York: 1940).

Spires, E. M. "The Use of the Dumdum Bullet in Colonial Warfare." *The Journal of Imperial and Commonwealth History* 9 (1975).

Trebilcock, C. "British Armaments and European Imperialism." *Economic History Review* 26 (1973).

Wahl, Paul, and Donald R. Toppel. *The Gatling Gun* (New York: 1965).

Wells, H. G. *The World Set Free* (London: 1926).

Winter, Jay, Geoffrey Parker, and Mary R. Habeck. *The Great War and the Twentieth Century* (New Haven: 2000).

CHAPTER 13: ACCIDENTAL ARMAGEDDON

Cross, Wilbur. *Challengers of the Deep: The Story of Submarines* (New York: 1959).

Fredette, Raymond. *The Sky on Fire: The First Battle of Britain, 1917–1918* (London: 1976).

Fussel, Paul. *The Great War in Modern Memory* (London: 1975).

Haber, L. F. *The Poison Cloud: Chemical Warfare in the First World War* (Oxford: 1986).

Hallion, Richard P. *The Rise of the Fighter Aircraft: 1914–1918* (Annapolis: 1984).

Herwig, Holger H. *The First World War: Germany and Austria-Hungary, 1914–1918* (London: 1997).

Hough, Richard. *Admiral of the Fleet: The Life of John Fisher* (New York: 1969).

———. *The Great War at Sea: 1914–1918* (New York: 1983).

Hynes, Samuel. *The Soldier's Tale: Bearing Witness to Modern War* (New York: 1997).

Johnson, Hubert C. *Breakthrough!: Tactics, Technology, and the Search for Victory on the Western Front in World War I* (Novato, CA: 1994).

Kilduff, Peter. *Germany's First Air Force: 1914–1918* (London: 1991).

Leed, Eric J. *No Man's Land: Combat and Identity in World War I* (Cambridge, UK: 1979).

Liddell Hart, B. H. *The Revolution in Warfare* (London: 1946).

Liddle, Peter, John Bourne, and Ian Whitehead, eds. *The Great World War, 1914–45*. Vol. 1. *Lightning Strikes Twice* (London: 2000).

Marder, Arthur J. *From Dreadnought to Scapa Flow: The Royal Navy in the Fisher Era.* Vol. 3 (London: 1961).

Nowarra, H. J., and Kimbrough S. Brown. *Von Richthofen and the Flying Circus* (Fallbrook, CA: 1964).

O'Connell, Robert L. *Sacred Vessels: The Cult of the Battleship and the Rise of the U.S. Navy* (New York: 1993)

Paschall, Rod. *The Defeat of Imperial Germany: 1917–1918* (Chapel Hill: 1989).

Rogers, H. C. B. *A History of Artillery* (Secaucus: 1975).

Rossler, Eberhard. *The U-boat: The Evolution and Technical History of German Submarines.* Trans., Harold Frenberg (Annapolis: 1981).

Schmitt, Bernadotte E., and Harold Vedeler. *The World in the Crucible* (New York: 1984).

Snyder, Jack. *The Ideology of the Offensive: Military Decision Making and the Disasters of 1914* (Ithaca: 1984).

Stokesbury, James L. *A Short History of World War I* (New York: 1981).

Tirpitz, Alfred von, *My Memoirs*. Vol. 2 (New York: 1919).

Tucker, Spencer C. *The Great War: 1914–1918* (London: 1998).

White, C. M. *The Gotha Summer* (London: 1986).

Winter, Jay, Geoffrey Parker, and Mary R. Habeck. *The Great War and the Twentieth Century* (New Haven: 2000).

CHAPTER 14: GRUDGE MATCH

Angelucci, Enzo. *Military Aircraft: 1914 to the Present* (New York: 1990).

Arpee, Edward. *From Frigates to Flattops: The Story of the Life and Achievements of Rear Admiral William Adger Moffett* (Lake Forest, IL: 1953).

Crow, Duncan. *Tanks of World War II* (New York: 1979).

Doughty, Robert A. "The Maginot Line." *MHQ: The Quarterly Journal of Military History,* Vol. 9, No. 2 (Winter 1997).

Erickson, John. *The Road to Stalingrad* (London: 1975).

Guderian, Heinz. *Panzer Leader* (New York: 1952).

Hanson, Victor Davis. *The Soul of Battle* (New York: 1999).

Honan, William H. "Bywater's Pacific War Prophecy." *MHQ: The Quarterly Journal of Military History,* Vol. 3, No. 3 (Spring 1991).

Horikoshi, Jiro. *Eagles of Mitsubishi: The Story of the Zero Fighter* (Seattle: 1980; trans. from 1970 Japanese edition).

Kahn, David. *Seizing the Enigma: The Race to Break the German U-boat Codes* (Boston: 1991).

Murray, Williamson, and Allan R. Millett. *A War to Be Won: Fighting the Second World War* (Cambridge, MA: 2000).

Murray, Williamson, and Allan R. Millett, eds. *Military Innovation in the Interwar Period* (Cambridge, U.K.: 1996).

O'Connell, Robert L. "The Cautionary Tale of the *Yamato*." *MHQ: The Quarterly Journal of Military History,* Vol. 3, No. 4 (Summer 1991).

——. "The Norden Bombsight." *MHQ: The Quarterly Journal of Military History,* Vol. 2, No. 4 (Summer 1990).

——. "T-34: Hammer of the Proletariat." *MHQ: The Quarterly Journal of Military History,* Vol. 2, No. 2 (Winter 1990).

Overy, Richard. *Why the Allies Won* (New York: 1995).

Polmar, Norman. "Torpedoes That Think." *MHQ: The Quarterly Journal of Military History,* Vol. 9, No. 4 (Summer 1997).

Rhodes, Richard. *The Making of the Atomic Bomb* (New York: 1986).

Roskill, Stephen. *Naval Policy Between the Wars*. 2 vols. (London: 1976).

Smith, Malcolm. *British Air Strategy between the Wars* (Oxford: 1984).

Spector, Ronald. *Eagle Against the Sun: The American War with Japan* (New York: 1985).

Speer, Albert. *Inside the Third Reich* (New York: 1970).

Stockholm International Peace Research Institute. *The Problem of Chemical and Biological Warfare* (Stockholm: 1971).

Strahan, Jerry E. *Andrew Jackson Higgins and the Boats That Won World War II* (Baton Rouge: 1994).

Toland, John. *The Rising Sun: The Decline and Fall of the Japanese Empire, 1936–1945* (New York: 1970).

Turnbull, Archibald D., and Clifford Lord. *History of United States Naval Aviation* (New Haven: 1940).

Van Aller, Chris. "Democracy's Weapon: The De Havilland Mosquito." *Defense Analysis,* Vol. 15, No. 2 (1999).

Weinberg, Gerhard L. *A World at Arms: A Global History of World War II* (Cambridge: 1994).

Whaley, Barton. *Covert German Rearmament, 1919–1939: Deception and Misperception* (Frederick, MD: 1984).

CHAPTER 15: COLD WAR—INFERNO OF ARMS

Aman, Ronald, and Julian Cooper. *The Technological Level of Soviet Industry* (New Haven: 1977).

Arkin, William, and Peter Pringle. *SIOP: The Secret U.S. Plan for Nuclear War* (New York: 1983).

Augustine, Norman. *Augustine's Laws* (New York: 1982).

Ball, Desmond. *Politics and Force Levels: The Strategic Missile Program of the Kennedy Administration* (Berkeley: 1980).

Betts, Richard K. *Cruise Missiles: Technology, Strategy, Politics* (Washington, DC: 1981).

Brook-Shepherd, Gordon. *The Stormbirds: Soviet Postwar Defectors* (New York: 1989).

Brzezinski, Zbigniew. *Power and Principle: Memoirs of the National Security Advisor, 1977–1981* (New York: 1983).

Collins, John M. *American and Soviet Military Trends Since the Cuban Missile Crisis* (Washington, DC: 1978).

Dugger, Ronnie. *On Reagan, the Man and His Presidency* (New York: 1983).

Evangelista, Matthew. "How Technology Fuels the Arms Race." *Technology Review,* Vol. 91, No. 5 (July 1988).

———. *Innovation and the Arms Race: How the United States and the Soviet Union Develop New Military Technology* (Ithaca: 1988).

FitzGerald, Frances. *Way Out There in the Blue: Reagan, Star Wars and the End of the Cold War* (New York: 2000).

Furlong, R. D. M. "Israel Lashes Out." *International Defence Review* 8 (1982).

Gallagher, Matthew P., and Karl F. Spielmann. *Soviet Decision-Making for Defense: A Critique of U.S. Perspectives on the Arms Race* (New York: 1972).

Gansler, Jacques S. *Affording Defense* (Cambridge, MA: 1986).

Garthoff, Raymond. *Detente and Confrontation: American-Soviet Relations from Nixon to Reagan* (Washington, DC: 1985).

Harford, James. *Korolev* (New York: 1997).

Hart, Gary. *America Can Win: The Case for Military Reform* (Bethesda: 1986).

Herken, G. *The Winning Weapon: The Atomic Bomb and the Cold War, 1945–1950* (New York: 1980).

Holloway, David. *The Soviet Union and the Arms Race* (New Haven: 1983).

Jenkins, D. H. C. "T-34 to T-80." *International Defense Review* (December 1981).

Just, Ward. "Soldiers." *Atlantic Monthly* (November 1970).

Khrushchev, Nikita S. *Khrushchev Remembers: The Last Testament.* Trans., Strobe Talbott (Boston: 1974).

Morris, Edmund. *Dutch: A Memoir of Ronald Reagan* (New York: 1999).

Oberg, J. E. *Red Star in Orbit* (New York: 1981).

Powaski, Ronald. *March to Armageddon: The United States and the Nuclear Arms Race, 1939 to the Present* (New York: 1987).

Powers, Thomas. "Planning the Strategy for World War III." *Atlantic Monthly* (November 1982).

Rosenberg, David Allen. "The Origins of Overkill: Nuclear Weapons and American Strategy, 1945–1960." *International Security,* Vol. 7, No. 4 (Spring 1983).

————. "U.S. Nuclear Stockpile, 1945–1950." *Bulletin of Atomic Scientists* 38 (May 1982).

Sapolsky, Harvey M. *The Polaris System Development: Bureaucratic and Programmatic Success in Government* (Cambridge, MA: 1972).

Snyder, Glenn H. "The New Look." In Warner R. Schilling, Paul Y. Hammond, and Glenn H. Snyder, eds., *Strategy, Politics, and Defense Budgets* (New York: 1962).

Spinney, F. *Defense Facts of Life: The Plans/Reality Mismatch* (Boulder: 1985).

Sprey, Pierre. "Mach 2: Reality or Myth? The Progression of Maximum Speeds in Fighter Aircraft." *International Defense Review* 8 (1980).

Stockman, David A. *The Triumph of Politics: Why the Reagan Revolution Failed* (New York: 1986).

Velikhov, Yevgeni, Roald Sagdeev, and Andre Kokoshin. *Weaponry in Space: The Dilemma of Security* (Moscow: 1986).

Index

Page numbers in *italics* refer to illustrations.